MULTIPHYSICS SIMULATIONS IN AUTOMOTIVE AND AEROSPACE APPLICATIONS

MULTIPHYSICS SIMULATIONS IN AUTOMOTIVE AND AEROSPACE APPLICATIONS

Edited by

Mojtaba Moatamedi
Thurai Rahulan
Hassan Khawaja

ELSEVIER

ACADEMIC PRESS
An imprint of Elsevier
elsevier.com/books-and-journals

Academic Press is an imprint of Elsevier
125 London Wall, London EC2Y 5AS, United Kingdom
525 B Street, Suite 1650, San Diego, CA 92101, United States
50 Hampshire Street, 5th Floor, Cambridge, MA 02139, United States
The Boulevard, Langford Lane, Kidlington, Oxford OX5 1GB, United Kingdom

Notices

Knowledge and best practice in this field are constantly changing. As new research and experience broaden our understanding, changes in research methods, professional practices, or medical treatment may become necessary.

Practitioners and researchers must always rely on their own experience and knowledge in evaluating and using any information, methods, compounds, or experiments described herein. In using such information or methods they should be mindful of their own safety and the safety of others, including parties for whom they have a professional responsibility.

To the fullest extent of the law, neither the Publisher nor the authors, contributors, or editors, assume any liability for any injury and/or damage to persons or property as a matter of products liability, negligence or otherwise, or from any use or operation of any methods, products, instructions, or ideas contained in the material herein.

British Library Cataloguing-in-Publication Data
A catalogue record for this book is available from the British Library

Library of Congress Cataloging-in-Publication Data
A catalog record for this book is available from the Library of Congress

ISBN: 978-0-12-817899-7

For Information on all Academic Press publications visit our website at https://www.elsevier.com/books-and-journals

Publisher: Matthew Deans
Acquisitions Editor: Carrie Bolger
Editorial Project Manager: Charlotte Rowley
Production Project Manager: Sojan P. Pazhayattil
Cover Designer: Alan Studholme

Typeset by Aptara, New Delhi, India

Contents

Contributors

Essam A. Al-Bahkali
Department of Mechanical Engineering, King Saud University, Riyadh, Saudi Arabia

Zahra Andleeb
Ghulam Ishaq Khan Institute of Engineering Sciences and Technology, Topi, Swabi, Pakistan

Basem Alzahabi
Al Ghurair University, Dubai, United Arab Emirates

Gernot Boiger
Institute of Computational Physics, Zurich University of Applied Sciences, Winterthur, Switzerland

Daniel Brunner
Institute of Computational Physics, Zurich University of Applied Sciences, Winterthur, Switzerland; UiT - The Arctic University of Norway, Tromsø, Norway

Hisham Elkenani
Department of Mechanical Engineering, King Saud University, Riyadh, Saudi Arabia

Fouad Erchiqui
Université du Québec en Abitibi-Témiscamingue, École de génie, 445, boul. de l'Université, Rouyn-Noranda, (Québec) J9X 5E4, Canada

Arash Soleiman Fallah
Department of Mechanical, Electronic, and Chemical Engineering, OsloMet, Oslo, Norway

Ghulam Hussain
Ghulam Ishaq Khan Institute of Engineering Sciences and Technology, Topi, Swabi, Pakistan

Dg Hafizah Kassim
Centre for Advanced Research on Energy, Universiti Teknikal Malaysia Melaka, Durian Tunggal, Melaka, Malaysia

Hassan Abbas Khawaja
UiT - The Arctic University of Norway, Tromsø, Norway; Al Ghurair University, Dubai, United Arab Emirates

Andreea Koreanschi
Aeronautical and Mechanical Engineering, University of Salford, Salford, United Kingdom

Sondre Ludvigsen
UiT - The Arctic University of Norway, Tromsø, Norway

Sohail Malik
Ghulam Ishaq Khan Institute of Engineering Sciences and Technology, Topi, Swabi, Pakistan

Navid Mehreganian
Department of Civil and Environmental Engineering, Imperial College London, London, United Kingdom

Mojtaba Moatamedi
Oslo Metropolitan University, Oslo, Norway; Al Ghurair University, Dubai, United Arab Emirates

Thaynan A. Oliveira
Unité de Mécanique de Lille, Université de Lille, Villeneuve-d'Ascq, Lille, France

Azma Putra
Centre for Advanced Research on Energy, Universiti Teknikal Malaysia Melaka, Durian Tunggal, Melaka, Malaysia

Mhamed Souli
Unité de Mécanique de Lille, Université de Lille, Villeneuve-d'Ascq, Lille, France

Oliviu Şugar-Gabor
Aeronautical and Mechanical Engineering, University of Salford, Salford, United Kingdom

Preface

The industrial revolution led to the age of steam which enabled the introduction of the first generation of railway vehicles and ships powered by external combustion engines. About 120 years ago, the advent of the reciprocating internal combustion engine, with higher power-to-weight factors, enabled the pioneering of two new types of vehicles, which quickly captured the public imagination as they facilitated private transport and longer journeys. The motor car enabled the citizens to travel to desired destinations at their convenience, and later to do so in an enclosed environment protected from the elements. The aeroplane was able to save time by traveling at speeds much higher than were possible using land-based vehicles and did not need extensive pathways to be prepared on land except for the construction of runways and aprons.

The people of the world quickly fell in love with both these forms of transport as they provided new found freedom and dizzying speeds, enabling travel over greater distances, with improved comfort. From humble beginnings, the automotive and aerospace industries rapidly expanded and grew in complexity during the first few decades. The development of the gas turbine during the second world war saw its use being gradually introduced to power aeroplanes as they became larger, faster, and more sophisticated, but proved to be impractical for use in road vehicles. The railway industry was the first to pioneer the use of the onboard internal combustion engine and electric motor combination mainly to improve the power transmission characteristics to the wheels. Development of the overhead gantry to provide mains electricity then negated the need for heat engines thus greatly simplifying the distribution system and maintenance requirements.

Increasing concern over environmental issues caused by exhaust emissions forced a rethink of alternate propulsion units for cars and aeroplanes on a large scale around the turn of the 20th century and this led to the current developments in rechargeable battery technologies and advanced solid-state power control systems. Because of the dramatic increase in the integration of technologies during the design stages of modern machines, a number of new numerical methods were developed to simultaneously address the physics of many systems where the intricacies of coupling equations describing the behavior of each subsystem were tackled, especially those exhibiting highly nonlinear characteristics. Dramatic increases in the speed of arithmetic processing in modern computing systems enabled the implementation of these complex algorithms to obtain key results whose validity could be verified against test data obtained from experiments conducted under controlled conditions.

These new ideas of combining several mathematical models into one realm and generating meaningful solutions led to the new discipline of multiphysics which has now become firmly established. The International Society of Multiphysics (www.multiphyiscs.org) has organized annual conferences in different parts of the world since 2006 to enable like-minded mathematicians and scientists to present their findings and exchange ideas in this new arena of unification.

This book which focuses on the multidisciplinary aspects is part of the Elsevier series on Multiphysics: Advances and Applications. The idea of collecting together the latest scientific findings relating to automobile and aerospace technologies in one publication suitable for aspiring engineers, young and old alike, was mooted a while ago by the editors and this textbook which is divided into two sections is the result. All scientists researching in multidisciplinary sciences will find the concepts described in this publication very useful.

The first section comprises five chapters dealing with studies focused on automotive technologies starting with occupant safety. Chapter 1 addresses the physics of airbag inflation in cars using a novel formulation which negates the need to generate a mesh when performing fluid-structure interaction calculations. Chapter 2 is focused on an equally important aspect of car safety concerning the road grip characteristics of conventional pneumatic tyres when compared to the new generation airless tyres. Chapter 3 addresses the acoustic energy absorption characteristics of multilayer materials used in the design of quieter environment in cars. Chapter 4 deals with the assessment of the health status of engine lubricant using a new approach based on the torsional resonance caused by fluid-structure interaction. Chapter 5 is presented in two parts, in which sound energy absorption capabilities are analyzed with respect to changes in the reverberating characteristics of interior acoustic environments.

The second section also comprises five chapters dealing with aerospace technologies starting with the behavior of structures subjected to varying moments and forces. Chapter 6 explores failure mechanisms of structures subjected to nonsteady forces with applications to aircraft seat support tubes and undercarriage. Chapter 7 considers nonlinear extensions to steady-state and unsteady classical theories, derived from both inviscid analytical methods, and two-dimensional numerical approaches, which possess good stability and convergence properties. Chapter 8 is somewhat unusual and equally interesting in that it addresses the design of a multipreg-laminate based carbon fiber reinforced polymer frame for use by rotorcraft for vertical lifting of external loads. Chapter 9 focuses on the forming process of bio composite-based components used in remotely controlled air vehicles which require advanced manufacturing methods. Chapter 10 examines the highly nonlinear effects of fluid-structure shock wave interaction by assessing the effects of an explosion inside an aircraft fuselage.

The editors would very much like to thank all the authors and co-authors of chapters in this book, in addition to many others who have contributed so much toward the study of the relatively new field of multiphysics. Their most valuable contributions in the development of combining mathematical models describing a bewildering variety of physical disciplines, and programming them into computer codes readily usable by scientists the world over, are much appreciated.

M. Moatamedi
T. Rahulan
H. Khawaja

CHAPTER ONE

Investigation of particle method and FEM coupling method for automotive airbag

Essam A. Al-Bahkali [a], Hisham Elkenani [a], Mhamed Souli [b], Fouad Erchiqui [c], Mojtaba Moatammedi [d,e]
[a] Mechanical Engineering Department, King Saud University, Riyadh, Saudi Arabia
[b] Universite de Lille, LML UMR CNRS, Lille, France
[c] Université du Québec en Abitibi-Témiscamingue, Rouyn-Noranda, QC, Canada
[d] Oslo Metropolitan University, Oslo, Norway
[e] Al Ghurair University, Dubai, United Arab Emirates

1.1 Introduction

Theoretical and experimental analysis of fluid structure interaction (FSI) for airbag and membrane deployment have been considered by several researchers over the past decades, using uniform pressure methods. The uniform pressure formulation does not consider the gas simulation, and no computational fluid dynamics (CFD) modeling is involved. Experiments have shown that the resulting flow is quite complex, involving several physical phenomena as damping effects. Numerical simulation using appropriate hydrodynamic equations for the gas helps to describe these phenomena, and also minimize the number of tests required that are very costly. Once simulations are validated by test results, it can be used as design tool for the improvement of the system structure involved. Initially finite element methods (FEMs) Lagrangian were used to simulate these problems, unfortunately classical Lagrangian methods cannot resolve large mesh distortion, runs are stopped before reaching termination time, due to negative Jacobian in highly distorted element. Arbitrary Lagrangian Eulerian (ALE) multimaterial description of motion developed in Aquelet et al. [1] can be used as an alternative for the simulation of high explosive phenomena. The ALE formulations have been developed to overcome the difficulties due to large mesh distortion. For some applications, including underwater explosions and their impact on the surrounding structure, engineers have

Multiphysics Simulations in Automotive and Aerospace Applications.
DOI: 10.1016/B978-0-12-817899-7.00002-2

switched from ALE to smooth particle hydrodynamic (SPH) method to reduce (CPU) time and save memory allocation.

It is well known, see Ozdemir et al. [15], that the classical FEM Lagrangian method is not suitable for most of the FSI problems due to high mesh distortion in the fluid domain. In many applications, the ALE formulation has been the only alternative to solve FSI for engineering problems. For the last decade, SPH method has been usefully used for engineering problems to simulate high velocity impact problems, high explosive detonation in soil, underwater explosion phenomena, and bird strike in aerospace industry. SPH is a mesh free Lagrangian description of motion that can provide many advantages in fluid mechanics and for modeling large deformation in solid mechanics. Unlike ALE method, and because of the absence of the mesh, SPH method suffers from a lack of consistency than can lead to poor accuracy. In the literature there are different formulations of meshless methods, the meshless local Petrov–Galerkin method for solving the bending problem of a thin, see Han and Atluri [8] and Long and Atluri [11] and also meshless finite volume method developed for solving elastostatic, see Atluri et al. [2]. In meshless local Petrov–Galerkin method mixed approach, both the strains as well as displacements are interpolated, at randomly distributed points in the domain, through local meshless interpolation schemes such as the moving least squares or radial basis functions.

The aim of this chapter devoted to ALE and SPH formulations for FSI problems, the mathematical and numerical implementation of the ALE and SPH formulations are described. From different simulation, it has been observed that for the SPH method to provide similar results as ALE formulation, the SPH meshing, or SPH spacing particles need to be finer than the ALE mesh. To validate the statement, we perform a simulation of a membrane deployment generated by pressurized gas. For this problem, the particle spacing of SPH method needs to be at least two times finer than ALE mesh. A contact algorithm is performed at the fluid structure interface for both SPH and ALE formulations. Then, the governing equations of the ALE formulation are described. The advection algorithms used to solve mass, momentum, and energy conservation in the multimaterial formulation are discussed. After that, the SPH formulation is described. Unlike ALE formulation which is based on the Galerkin approach, SPH is a collocation method. The last section is devoted to numerical simulation of a membrane deployment under high pressurized gas, using both ALE and SPH methods. The numerical results will be compared to experimental data.

1.2 ALE formulation

Fluid problems, in which interfaces between different materials (gas and ambient air) are present, are more easily modeled by using a Lagrangian mesh. However, if an analysis for complex tank geometry is required, the distortion of the Lagrangian mesh makes such a method difficult to use many remeshing steps are necessary for the calculation to continue. Another method to use is the Eulerian formulation. This change from a Lagrangian to an Eulerian formulation, however, introduces two problems. The first problem is the interface tracking Hallquist [7] and the second problem is the advection phase or advection of fluid material across element boundaries.

To solve these problems, an explicit FEM for the Lagrangian phase and a finite volume method (flux method) for the advection phase are used. We can refer to several explicit codes such as Pronto, Dyna3D, and LS-DYNA; see Hallquist [7], and Von Neumann and Richtmeyer [17] for a full description of the explicit FEM. The advection phase has been developed for extending the range of applications that cannot be used with the Lagrangian formulation. Current applications include sloshing involving a "free surface," and high velocity impact problems where the target is modeled as a fluid material, thus providing a more realistic representation of the impact event by capturing large deformations.

An ALE formulation contains both pure Lagrangian and pure Eulerian formulations. The pure Lagrangian description is the approach that: the mesh moves with the material, making it easy to track interfaces and to apply boundary conditions. Using an Eulerian description, the mesh remains fixed while the material passes through it. Interfaces and boundary conditions are difficult to track using this approach; however, mesh distortion is not a problem because the mesh never changes. In solid mechanics, a pure Eulerian formulation is not useful because it can handle only a single material in an element, while an ALE formulation is assumed to be capable of handling more than one material in an element.

In the ALE description, an arbitrary referential coordinate is introduced in addition to the Lagrangian and Eulerian coordinates. The material derivative with respect to the reference coordinate can be described in Eq. (1.1). Thus, substituting the relationship between the material time derivative and the reference configuration time derivative derives the ALE equations.

$$\frac{\partial f(X_i,t)}{\partial t} = \frac{\partial f(x_i,t)}{\partial t} + w_i \frac{\partial f(x_i,t)}{\partial x_i} \tag{1.1}$$

where X_i is the Lagrangian coordinate, x_i is the Eulerian coordinate, w_i is the relative velocity. Let denote by v the velocity of the material and by u the velocity of the mesh. In order to simplify the equations, we introduce the relative velocity $w = v - u$. Thus, the governing equations for the ALE formulation are given by the following conservation Eqs. (1.2)–(1.4):

(i) Mass equation.

$$\frac{\partial \rho}{\partial t} = -\rho \frac{\partial v_i}{\partial x_i} - w_i \frac{\partial \rho}{\partial x_i} \tag{1.2}$$

(ii) Momentum equation.

The strong form of the problem governing Newtonian fluid flow in a fixed domain consists of the governing equations and suitable initial and boundary conditions. The equations governing the fluid problem are the ALE description of the Navier–Stokes equations:

$$\rho \frac{\partial v_i}{\partial t} = \sigma_{ij,j} + \rho b_i - \rho w_i \frac{\partial v_i}{\partial x_j} \tag{1.3}$$

Boundary and initial conditions need to be imposed for the problem to be well posed.

(iii) Energy equation.

$$\rho \frac{\partial E}{\partial t} = \sigma_{ij} v_{i,j} + \rho b_i v_i - \rho w_j \frac{\partial E}{\partial x_j} \tag{1.4}$$

where ρ is the density, σ_{ij} is the total Cauchy stress, b_i is the body force, and E is the internal energy per unit mass.

Note that the Eulerian equations commonly used in fluid mechanics by the CFD community are derived by assuming that the velocity of the reference configuration is zero and that the relative velocity between the material and the reference configuration is therefore the material velocity. The term in the relative velocity in Eqs. (1.2)–(1.4) is usually referred to as the advective term, and accounts for the transport of the material past the mesh. It is the additional term in the equations that makes solving the ALE equations much more difficult numerically than the Lagrangian equations, where the relative velocity is zero.

In the second phase, the advection phase, transport of mass, internal energy, and momentum across cell boundaries are computed; this may be thought of as remapping the displaced mesh at the Lagrangian phase back to its original or arbitrary position.

From a discretization point of view of Eqs. (1.2), (1.3), and (1.4), one-point integration is used for efficiency and to eliminate locking. The zero energy modes are controlled with an hourglass viscosity Benson, see Libersky et al. [9]. A shock viscosity, with linear and quadratic terms, is used to resolve the shock wave, see Lucy [14]; a pressure term is added to the pressure in the energy Eq. (1.4). The resolution is advanced in time with the central difference method, which provides a second-order accuracy in time using an explicit method in time. For each node, the velocity and displacement are updated using second-order time integration as follows:

$$u_i^{n+1/2} = u_i^{n-1/2} + \Delta t.M^{-1}.\left(F_i^{\text{ext}} + F_i^{\text{int}}\right) \tag{1.5}$$

where F_i^{ext} is the nodal external force, F_i^{ext} is the nodal internal force, and M_i is the nodal mass.

The multimaterial formulation is attractive for solving a broad range of nonlinear problems in fluid and solid mechanics, because it allows arbitrary large deformations and enables free surfaces to evolve. The Lagrangian phase of the volume of fluid method is easily implemented in an explicit ALE FEM. Before advection, special treatment for the partially voided element is needed. For an element that is partially filled, the volume fraction satisfies $V_f \leq 1$, and the total stress by σ is weighed by volume fraction $\sigma_f = \sigma \cdot V_f$.

In the second phase, the transport of mass, momentum, and internal energy across the element boundaries is computed. This phase may be considered as a "remapping" phase. The displaced mesh from the Lagrangian phase is remapped into the initial mesh for the Eulerian formulation, or an arbitrary undistorted mesh for the ALE formulation.

In this advection phase, we solve a hyperbolic problem, or a transport problem, where the variables are density, momentum, and internal energy per unit volume, where the Donor Cell algorithm, a first-order advection method and the Van Leer algorithm, a second-order advection method, see Von Neumann and Richtmeyer [17], are used. As an example, Eq. (1.6) is used for mass conservation. Furthermore, Fig. 1.1 describes the two phases for a one step explicit calculation.

$$\frac{\partial \rho}{\partial t} + \nabla.(\rho u) = 0 \tag{1.6}$$

1.3 SPH formulation

The SPH method developed originally for solving astrophysics problem has been extended to solid mechanics by Libersky et al. [9] to model

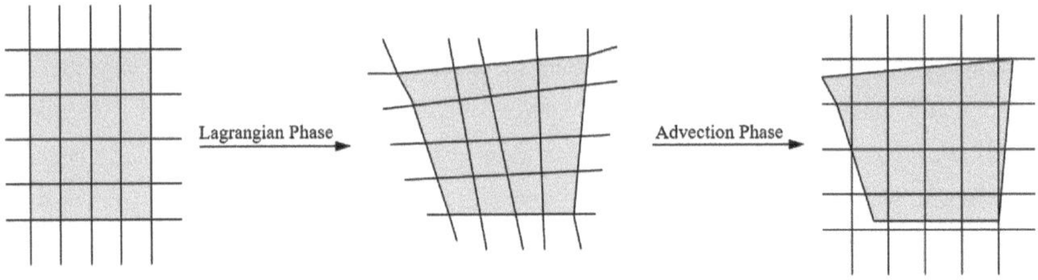

Figure 1.1 Lagrangian and advection phases in multimaterial ALE formulation.

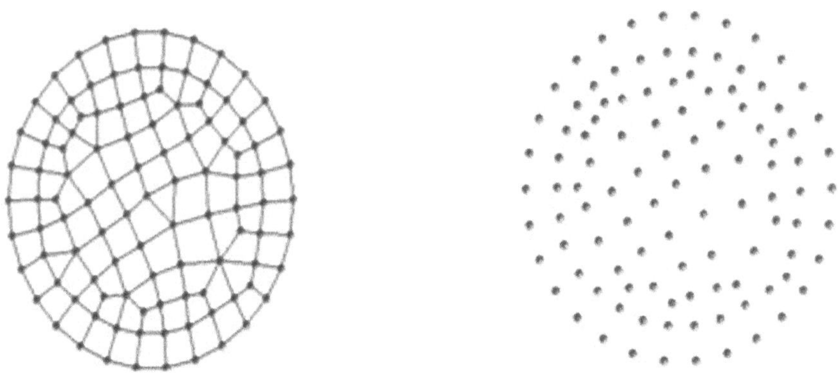

Figure 1.2 FEM model, mesh, and nodes (left) and SPH model, particles (right).

problems involving large deformation including high velocity impact. SPH method provides many advantages in modeling severe deformation as compared to classical FEM formulation which suffers from high mesh distortion, Fig. 1.2. The method was first introduced by Lucy [14] and Gingold and Monaghan [6] for gas dynamic problems and for problems where the main concern is a set of discrete physical particles than the continuum media. The method was extended to solve high velocity impact in solid mechanics, CFD applications governed by Navier–Stokes equations and FSI problems.

It is well known from previous papers that SPH method suffers from lack of consistency, which can lead to poor accuracy of motion approximation. Unlike finite element, interpolation in SPH method cannot reproduce constant and linear functions.

A detailed overview of the SPH method is developed by Liu and Liu [10], where the two steps for representing of function u, an integral interpolation and a kernel approximation are given by:

$$u(x_i) = \int^{u} (y).\delta(x_i - y)dy \tag{1.7}$$

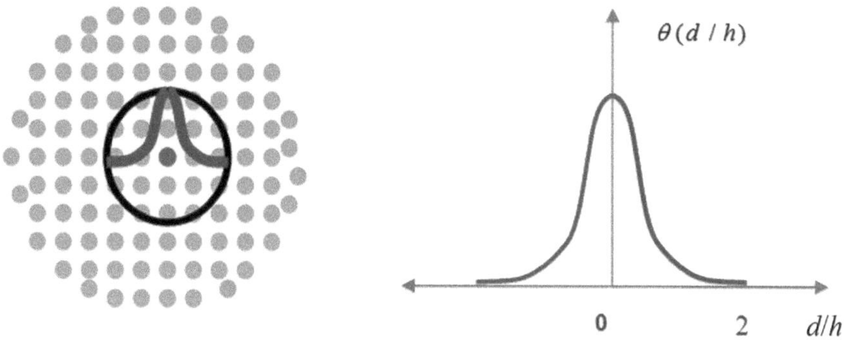

Figure 1.3 Kernel function and its support domain for a 2D function.

where the Dirac function satisfies:

$$\begin{aligned} \delta(x_i - y) &= 1, \ \text{if } x_i = y \\ \delta(x_i - y) &= 0, \ \text{if } x_i \neq y \end{aligned} \tag{1.8}$$

The approximation of the integral function Eq. (1.7) is based on the kernel approximation W, which approximates the Dirac function based on the smoothing length h.

$$W(d, h) = \frac{1}{h^\alpha}.\theta\left(\frac{d}{h}\right) \tag{1.9}$$

that represents support domain of the kernel function, see Fig. 1.3.

So that Eq. (1.7) becomes,

$$\langle u(x_i) \rangle = \int u(y).W(||x - y||, h)dy \tag{1.10}$$

Taking in consideration the support domain of the kernel function, the SPH approximation of a particle x_i is obtained discretizing the integral into a sum over the particles that are within the kernel support domain as it is shown in Fig. 1.3.

$$u_{\text{sph}}(x_i) = \sum_{j \in D_i} \omega_j u_j W\left(||x_i - x_j||, h\right) \tag{1.11}$$

where the weight $\omega_j = m_j/\rho_j$ is the volume of the particle.

Integrating by part Eq. (1.10) and considering the properties of the SPH interpolation and that

$$\nabla(u) = u.\nabla(1) - 1.\nabla(u) \tag{1.12}$$

The SPH approximation for the gradient operator of a function is given by,

$$\nabla u_{\text{sph}}(x_i) = \sum_{j \in D_i} \omega_j.\left(u_i - u_j\right).\nabla W\left(||x_i - x_j||, h\right) \tag{1.13}$$

Considering that

$$\frac{\nabla(P)}{\rho} = \frac{P}{\rho^2}\nabla(\rho) + \nabla\left(\frac{P}{\rho}\right) \tag{1.14}$$

Applying the SPH interpolation on Navier–Stokes equations, one can derive a symmetric SPH formulation for Navier–Stokes equations such that the principle of action and reaction is respected, and that the accuracy is improved. Finally, we have the following discretized set of equations:

(i) Mass equation.

$$\frac{D\rho_i}{Dt} = \rho_i \sum_{j \in D_i} \omega_j.\left(v_i^\beta - v_j^\beta\right).\frac{\partial W\left(||x_i - x_j||, h\right)}{\partial x_i^\beta} \tag{1.15}$$

(ii) Momentum equation.

$$\frac{Dv_i^\alpha}{Dt} = \sum\nolimits_{j \in D_i} m_j.\left(\frac{\sigma_i^{\alpha\beta}}{\rho_i^2} + \frac{\sigma_j^{\alpha\beta}}{\rho_j^2}\right).\frac{\partial W\left(\left\|x_i - x_j\right\|, h\right)}{\partial x_i^\beta} + f_{\text{ext}} \tag{1.16}$$

(iii) Energy equation.

$$\frac{De_i}{Dt} = \frac{1}{2}\sum_{j \in D_i} m_j.\left(\frac{P_i}{\rho_i^2} + \frac{P_j}{\rho_j^2}\right).\left(v_i^\beta - v_j^\beta\right)\frac{\partial W\left(||x_i - x_j||, h\right)}{\partial x_i^\beta} + \frac{\mu_i}{\rho_i}\varepsilon_i^{\alpha\beta}\varepsilon_i^{\alpha\beta} \tag{1.17}$$

1.4 ALE penalty coupling algorithm

Penalty coupling behaves like a spring system and penalty forces are calculated proportionally to the penetration depth and spring stiffness. The head of the spring is attached to the structure or slave node and the tail of the spring is attached to the master node within a fluid element that is intercepted by the structure, as illustrated in Fig. 1.4.

Similarly, to penalty contact algorithm, the coupling force proportional to the penetration depth is applied to both the fluid and the structure nodes.

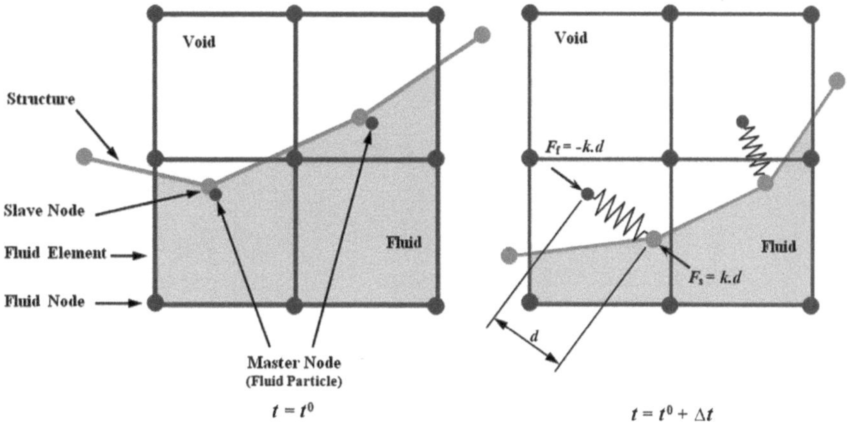

Figure 1.4 Description of penalty coupling algorithm.

Therefore, penalty force that is applied to both the fluid and structure nodes in opposite directions is described by:

$$F = k.d \tag{1.18}$$

where k represents the spring stiffness and d the penetration. The force F in Fig. 1.4 is applied to both master and slave nodes in opposite directions to satisfy force equilibrium at the interface coupling, and thus the coupling is consistent with the fluid–structure interface conditions namely the action–reaction principle.

The main difficulty in the coupling problem comes from the evaluation of the stiffness coefficient k in Eq. (1.18). The stiffness value is problem dependent, a good value for the stiffness should reduce energy interface in order to satisfy total energy conservation, and prevent fluid leakage through the structure. The value of the stiffness k is still a research topic for explicit contact-impact algorithms in structural mechanics. In this chapter, the stiffness value is similar to the one used in Lagrangian explicit contact algorithms, described in Benson [3]. The value of k is given in term of the bulk modulus K of the fluid element in the coupling containing the slave structure node, the volume V of the fluid element that contains the master fluid node, and the average area A of the structure element connected to the structure node.

$$k = p_f.\frac{K.A^2}{V} \tag{1.19}$$

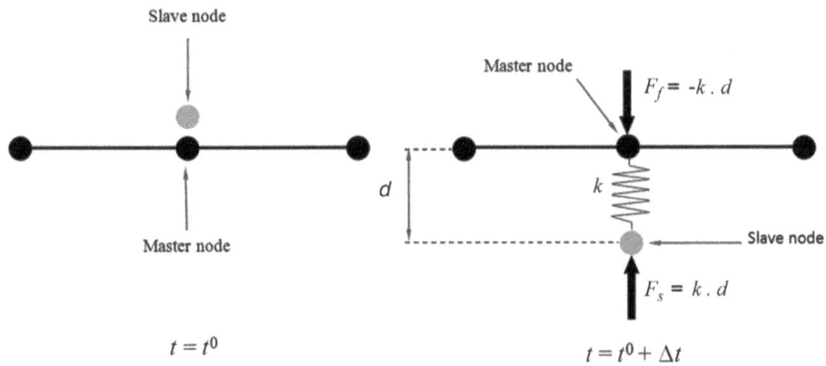

Figure 1.5 Description of penalty contact algorithm between slave particle and master structure.

1.5 SPH contact algorithm

Several contact methods have been published in the literature between different structure material parts. Classical implicit and explicit coupling are described in detail in Longatte et al. [12] and Longatte et al. [13], where hydrodynamic forces from the fluid solver are passed to the structure solver for stress and displacement computation. Here, a coupling method based on contact algorithm is used. Since the coupling method described is based on the penalty method for contact algorithms, the contact approach is a good introduction to this method. In contact algorithms, a contact force is computed proportional to the penetration vector, the amount the constraint is violated. In contact algorithms, one surface is designated as a slave surface, and the second as a master surface. The nodes lying on both surfaces are also called slave and master nodes, respectively. In an explicit FEM method, contact algorithms compute interface forces due to impact of the slave node on the master node, these forces are applied to the slave and master nodes in contact in order to prevent a node from passing through contact interface. The penalty method imposes a resisting force to the slave node, proportional to its penetration through the master segment, as shown in Fig. 1.5 describing the contact process. This force is applied to both the slave node and the nodes of the master segment in opposite directions to satisfy equilibrium.

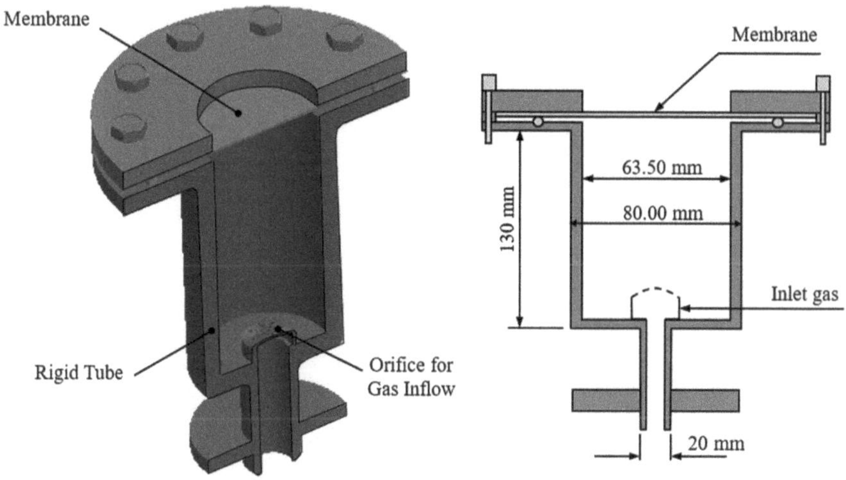

Figure 1.6 Test rig used in the experiment.

1.6 Description of the experimental setup

Erchiqui et al. [4] experimentally investigated the inflation behavior in a thermoplastic membrane under the combined effects of applied stress and temperature using bubble inflation tests. The results were used to identify the material constants embedded in the constitutive models of the polymeric materials in this work. The polymeric material, acrylonitrile-butadiene-styrene (ABS), was tested under biaxial deformation using the bubble inflation technique. The initial ABS sheet thickness was 1.57 mm. The experimental setup is shown in Fig. 1.6.

The circular membranes were mounted between two metal disks containing a circular aperture and subsequently clamped onto a support. The exposed circular area of diameter $D = 63.5$ mm was heated in an infrared heating chamber to the softening point. When the temperature became uniform over the flat sheet, the circular area was inflated using compressed air under a controlled flow rate. The applied inflation pressure, which was uniform over the membrane, was measured with a pressure sensor during the testing. The height at the hemispheric pole (or the center of the membrane) and the time were recorded simultaneously using a video camera and a data acquisition system. For most inflation tests, testing ends when the bubble bursts. Fig. 1.7 shows the inflation profile of a circular plane membrane.

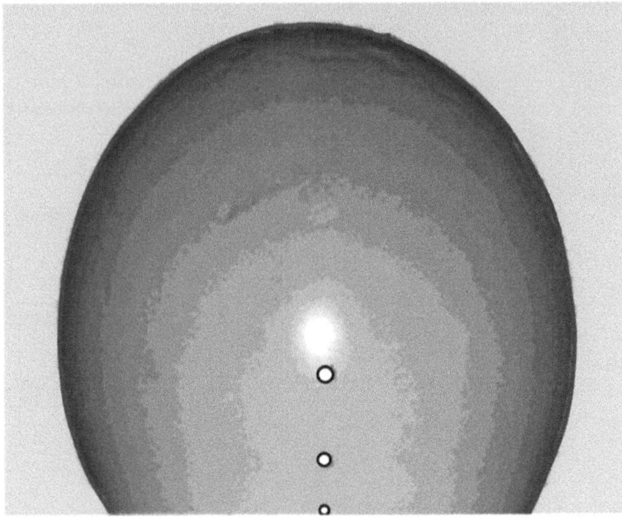

Figure 1.7 Inflation of a flat circular ABS membrane.

Table 1.1 Air properties.

Material behaviors	Value, SI units
Density, ρ	1.2047 kg/m^3
Viscosity, μ	1.8205×10^{-5} N/m^2 S
EOS	$C_0 = 343.210$ m/s (Us – Up)

1.7 Numerical simulation using SPH method

The model setup including all simulation approaches used as well as the boundary conditions applied to the model are described in detail below.

1.7.1 Meshing and material modeling

The model, as shown in Fig. 1.8, consists of four parts, a membrane, fluid domain, sidewall, and a rigid plate acts as a piston. The fluid domain in this case has the property of air. Table 1.1 denotes the details of the property values.

The membrane is modeled using ABS. The material constants for Moony–Rivlin model are $C_{10} = 105{,}000$ Pa and $C_{01} = 10{,}500$ Pa, see Souli and Zolesio [16], corresponding to $\alpha = 0.1$ ($\alpha = C_{01}/C_{10}$). The theoretical material constants are obtained by fitting pressure deformation experimental data with a membrane diameter 63.5 mm and a sheet thickness of 1.57 mm. Fitting curves for various values for α of isotropic hyperelastic model at

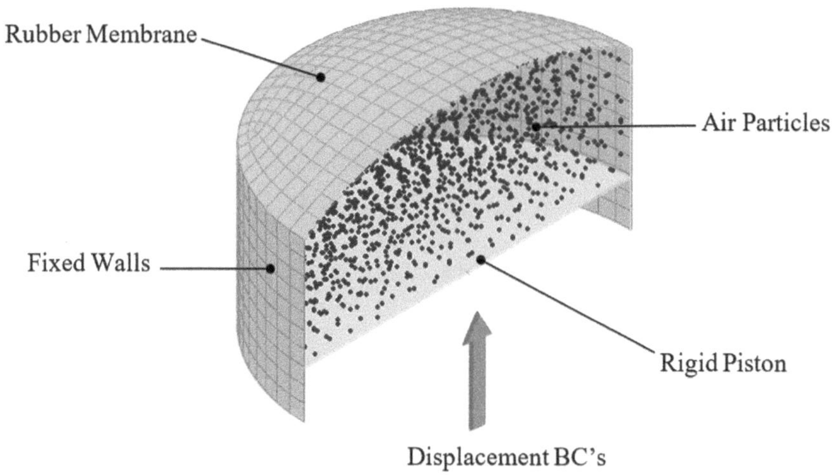

Figure 1.8 Model used for membrane inflation simulation.

Table 1.2 Parts mesh detailed.

Part name	Element type	No of elements
Membrane	Linear quadrilateral shell element	1104
Air	Linear hexahedral solid element (converted into 1-node element)	34,960

143°C shows that the value of $\alpha = 0.1$ is the best fit for the experimental data.

Fig. 1.9 compares the predicted pressure versus the bubble height data at the pole of the bubble using the Mooney–Rivlin model for the experimental ABS measurements. The measurement error is less than 5% during the pressure rise. However, the numerical model for the ABS provides a better fit to the experimental data, as shown in Fig. 1.9. It should be noted that at the maximum pressure, the bubble height is of the order of the initial membrane radius. Beyond this critical point, the height increases rapidly while the pressure falls. This unstable behavior continues until the bubble bursts.

The membrane is meshed with shell elements. For the air, to ensure uniform distribution for the air particles using SPH method (see Table 1.2), it initially meshed with solid elements, and then conversion functionality is used to convert the solid element into SPH particles, Fig. 1.10.

We can also specify the time when the conversion of all the elements in the affected element set is to take place. In our case, the conversion time

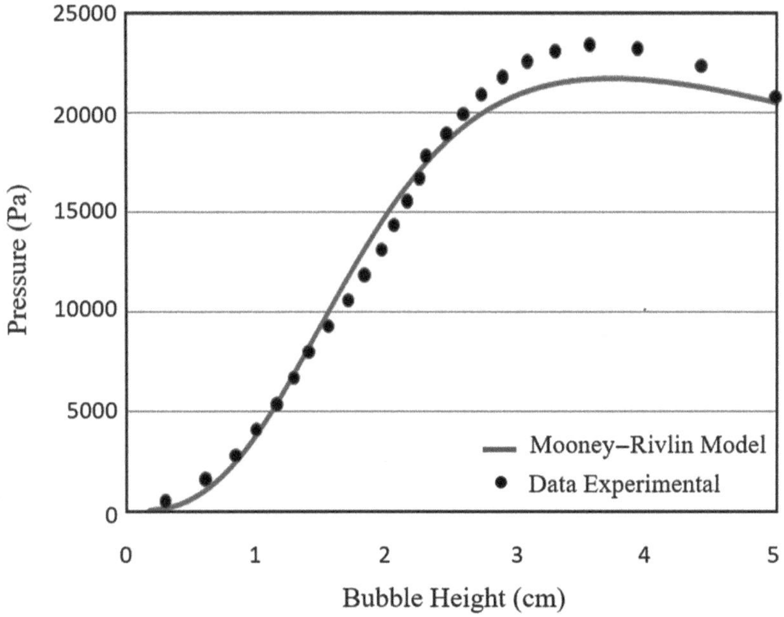

Figure 1.9 Experimental and predicted pressure versus bubble height for ABS at 143°C.

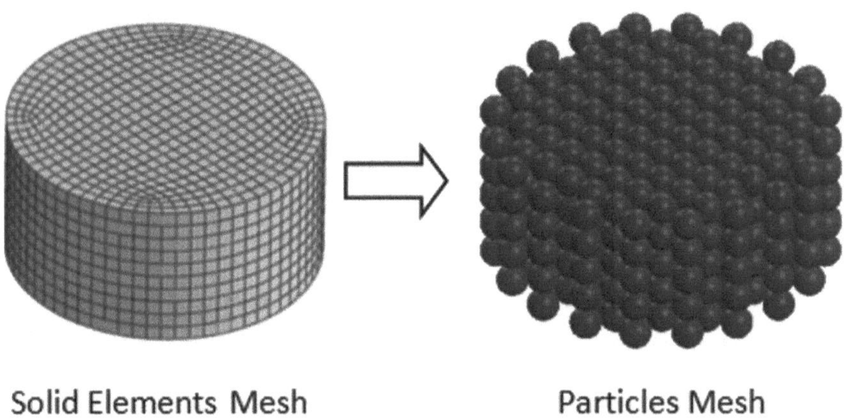

Figure 1.10 Element conversion from FEM mesh to SPH particles.

is specified as zero, then the conversion takes place at the beginning of the analysis. The rigid plate which acts as a piston is modeled as a rigid body as there is no need to introduce it in the solution. It just acts as inflator's piston. So, no material assigned to it.

1.7.2 Boundary conditions and initial conditions

For the membrane, the boundary conditions of its circumferential are fixed in all degree of freedom as well as the side wall. The rigid plate is fixed in all its degree of freedom expect the sliding direction.

A displacement boundary condition is applied to the rigid plate in the sliding direction for a distance equal to 27 mm, which is sufficient to deliver the volume flow rate necessary to inflate the membrane to the desired height.

When more than one part involved in a simulation, interaction properties need to be defined. In this case, the interaction property has been defined as frictionless tangential behavior. This property has been applied to all surfaces on the model.

1.8 Numerical results and observations

A dynamic explicit step with a total duration time equal to 0.6 s is trigged to simulate the inflation process. Fig. 1.11 shows the initial air particles distribution after the element conversion process at the beginning of the simulation. Fig. 1.12 shows the membrane inflation at 0.3 s and at the end of simulation ($t = 0.6$ s), respectively.

The equation of state used to define the air material behavior is Us–Up with the parameter shown in Table 1.1. As shown in Fig. 1.13, the numerical results obtained using higher number of elements are closer to the experimental data than the numerical results obtained with lower number of elements.

In the case of SPH method, the pressure of the fluid is represented by the equivalent pressure on the particles. Fig. 1.14 shows the equivalent pressure curves for a sample of particles represent the air fluid. Good correlation has been obtained between experimental and numerical displacement of the center of the structure as shown in Fig. 1.15. Consequently, the trend of all curves is relatively coincident with experimental pressure data.

1.9 Conclusions

The chapter presents the ALE and SPH methods as well as their limitations for specific problems lime airbag inflation and membrane deployment. Mine explosion, underwater explosion, and bird impact on structures are commonly solved using ALE formulation, in defense industry; some of these problems are solved using SPH method. For the last decade, SPH

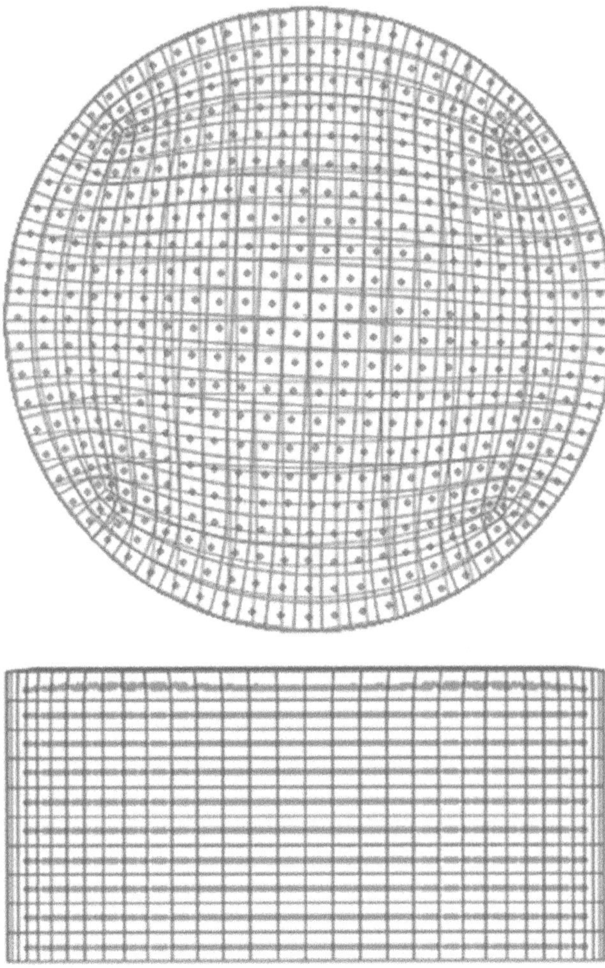

Figure 1.11 Initial air particles distribution at $t = 0$ s.

methods are gaining in accuracy numerical stability, and the use of SPH method is becoming more common in industry for solving fluid structure coupling problems. Analysis shows ALE formulation is more CPU time consuming and requires more memory allocation than that of the SPH method. In our study, first, we describe both ALE and SPH methods, and we compare numerical results between the two methods using similar mesh size, each ALE element is replaced by an SPH particle at the element center. Using a simple FSI problem, it has been observed that using same mesh size for methods, numerical results, displacement, and gas pressure using SPH

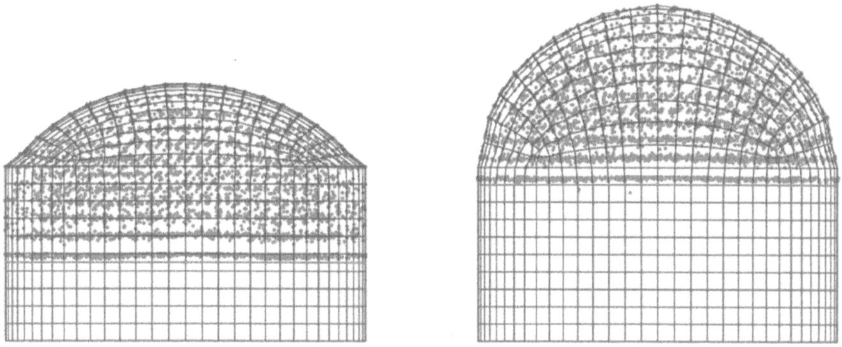

Figure 1.12 Fluid material and structure displacement at $t = 0.3$ s, and $t = 0.6$ s.

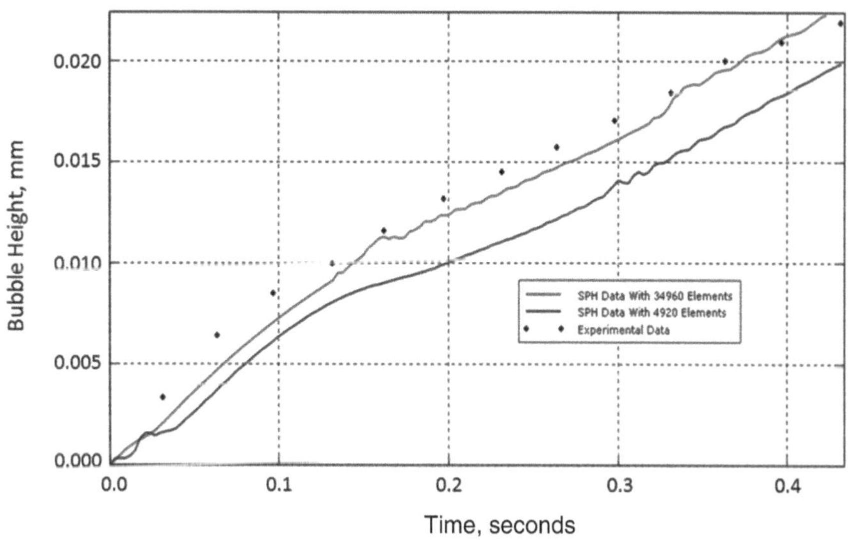

Figure 1.13 Bubble height time evolution with different mesh resolution comparing with experimental data.

method, provide good correlation with experimental data. For this particular application, without refining the SPH particles, using same number and elements for SPH and ALE, results from SPH method are in good correlation with those from ALE simulation; in terms of displacement, velocity, and Von Mises stress on the structure. Since the ultimate objective is the design of structure resisting to load blast, numerical simulations from ALE and SPH methods can be included in shape design optimization with shape optimal

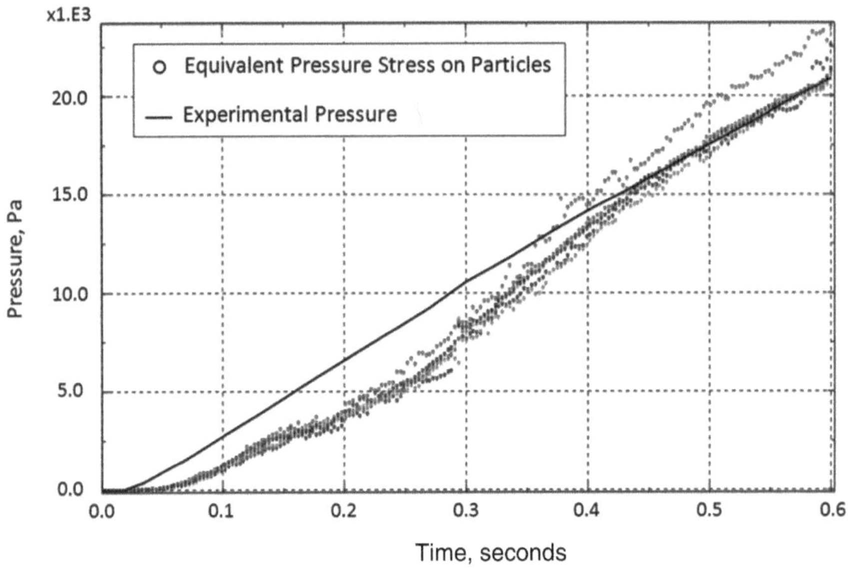

Figure 1.14 Equivalent pressure stress for a sample of elements represent the air.

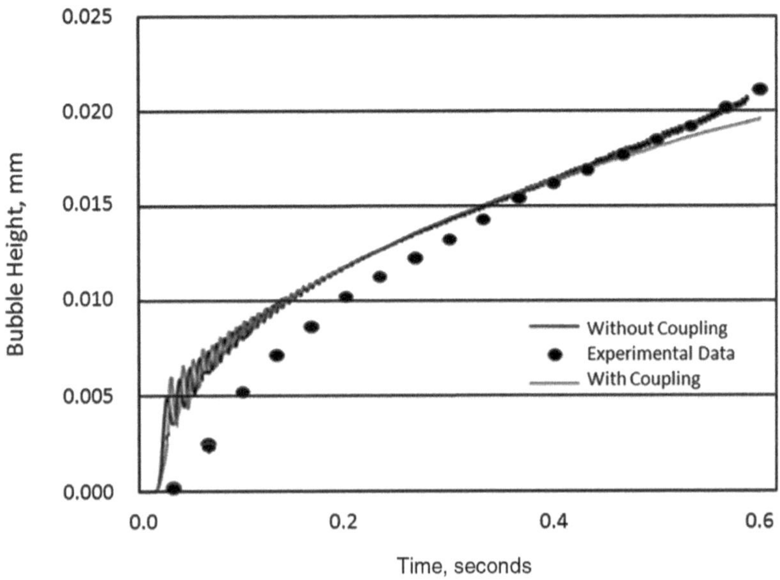

Figure 1.15 Bubble height time evolution for $C_{10} = 115{,}590$ Pa obtained with ALE method.

design techniques, see Souli and Zolesio [16], and material optimization, see Erchiqui et al. [5]. Once simulations are validated by test results, they can be used as design tool for the improvement of the system structure being involved. This method can be intensively used in automotive industry such as airbag system, sloshing in vehicle fuel tanks and truck fluid tanks.

References

1. N. Aquelet, M. Souli, L. Olovson, Euler Lagrange coupling with damping effects: application to slamming problems, Comput. Methods Appl. Mech. Eng. 195 (2005) 110–132.
2. S.N. Atluri, Z.D. Han, A.M. Rajendran, A new implementation of the meshless finite volume method, through the MLPG 'mixed approach', CMES: Comput. Model. Eng. Sci. 6 (6) (2004) 491–513.
3. D.J. Benson, Computational methods in Lagrangian and Eulerian hydrocodes, Comput. Methods Appl. Mech. Eng. 99 (1992) 235–394.
4. F. Erchiqui, A. Derdouri, A. Gakwaya, E. Verron, Free inflation of flat thermoplastic membrane: numerical analysis and experimental validation, Entropie 235/236 (2001) 118–125.
5. F. Erchiqui, M. Souli, R. Ben Yedder, Nonisothermal finite-element analysis of thermoforming of polyethylene terephthalate sheet: incomplete effect of the forming stage, Polym. Eng. Sci. 47 (12) (2007) 2129–2144.
6. R.A. Gingold, J.J. Monaghan, Smoothed particle hydrodynamics: theory and applications to non-spherical stars, Month. Note. R. Astron. Soc. 181 (1977) 375–389.
7. J.O. Hallquist, LS-DYNA Theory Manual, Livermore Software Technology Corporation, 7374 Las Positas Road, Livermore, CA 94551, 1998.
8. Z.D. Han, S.N. Atluri, On the meshless local Petrov-Galerkin MLPG-Eshelby method in computational, finite deformation solid mechanics – part II, CMES: Comput. Model. Eng. Sci. 97 (3) (2014) 199–237.
9. L.D. Libersky, A.G. Petschek, T.C. Carney, J.R. Hipp, F.A. Allahdadi, High strain Lagrangian hydrodynamics: a three-dimensional SPH code for dynamic material response, J. Comput. Phys. 109 (1993) 67–75.
10. M.B. Liu, G.R. Liu, Smoothed particle hydrodynamics (sPH): an overview and recent developments, Arch Comput. Methods Eng. 17 (2010) 25–76.
11. S. Long, S.N. Atluri, A meshless local Petrov-Galerkin method for solving the bending problem of a thin plate, CMES: Comput. Model. Eng. Sci. 3 (1) (2002) 53–63.
12. E. Longatte, Z. Bendjeddou, M. Souli, Application of arbitrary Lagrange Euler formulations to flow-induced vibration problems, J. Pressure Vessel Technol. 125 (4) (2003) 411–417.
13. E. Longatte, V. Verreman, M. Souli, Time marching for simulation of fluid-structure interaction problems, J. Fluids Struct. 25 (1) (2009) 95–111.
14. L.B. Lucy, A numerical approach to the testing of fission hypothesis, Astronom. J. 82 (1977) 1013–1024.
15. Z. Ozdemir, M. Souli, Y.M. Fahjan, Application of nonlinear fluid-structure interaction methods to seismic analysis of anchored and unanchored tanks, Eng. Struct. 32 (2) (2010) 409–423.
16. M. Souli, J.P. Zolesio, Shape derivative of discretized problems, Comput. Methods Appl. Mech. Eng. 108 (1993) 187–199.
17. J. Von Neumann, R.D. Richtmeyer, A method for the numerical calculation of hydrodynamical shocks, J. Appl. Phys. 21 (1950) 232.

CHAPTER TWO

Multiphysics analysis of contact pressure profile of airless tires as compared to conventional tires

Sondre Ludvigsen[a], Zahra Andleeb[b], Hassan Abbas Khawaja[a,d], Mojtaba Moatamedi[c,d], Basem Alzahabi[d]
[a] UiT-The Arctic University of Norway, Tromsø, Norway
[b] Ghulam Ishaq Khan Institute of Engineering Sciences and Technology, Topi, Swabi, Pakistan
[c] Oslo Metropolitan University, Oslo, Norway
[d] Al Ghurair University, Dubai, United Arab Emirates

2.1 Introduction

The cold climatic conditions of the Arctic severely affect many areas of life. Accumulation of ice during winters is a common phenomenon that hinders the normal operations of various mechanical structures. Researchers are continuously proposing new ways to cope with it [2,3,35,36,37][39–43]. Cold conditions also make driving significantly harder and more unpredictable. The first winter tires used for road cars were applied during the mid-1930. The difference to these tires compared to the summer tires was the enlargement of the grooves. Deeper lateral groves gave better traction as it allowed the tread blocks to dig deeper into the snow-covered roads. In 1961, the first tires with metal studs were introduced [6]. The studs help the tire gaining a grip on hard-packed snow and ice as the studs' claws into the ice and increases friction. The focus during the last 50 years has been on improving traction by optimizing rubber compositions and treading. This was done by changing parameters like elasticity, hardness, adding of studs as well as modifying tread pattern design. In recent years, the government and manufacturers have been looking toward stud-free winter tires due to environmental benefits. The implementation of studs is not in practice nowadays due to the ever-increasing focus on air pollution and environmental hazards caused by road dust produced by studded tires [15,17].

2.1.1 Common definition of grip

The general perception of having a good grip is when two objects with forces parallel to their contact area, do not move relative to each other. In

DOI: 10.1016/B978-0-12-817899-7.00006-X

other words, the grip is the force that gives the object the ability to stay in contact with surfaces without slipping, referred to as frictional force [22]. Grip can be broadly differentiated into three categories: *no grip, sliding*, and *sticking*. No grip means no resistance against the movement. Sliding friction is having a sliding motion but with friction working against the direction of travel. This friction force will limit the velocity by transferring the kinetic energy into potential energy, in form of heating generation on the contact surface and the object. Sticking grip is where there is enough frictional force to prevent any movement between the two objects. This means, from a stationary position the frictional force is greater than the force trying to move the object. The surface friction coefficient is the surface property stating the gripping ability against each other. The friction coefficient is higher if the contact surface is rough and dry, compared to a lower friction coefficient if it is smooth and covered with a lubricant such as a grease or another slippery fluid [4]. In addition to the above, the frictional force is directly proportional to the normal force or applied pressure.

There are two factors that change the level of grip or friction: surface coefficient of friction and normal force, as illustrated in Eq. (2.1).

$$F_{\text{friction}} = \mu \cdot N \quad (2.1)$$

where F_{friction} is the frictional force, μ is the surface friction coefficient, and N is the normal force as shown in Eq. (2.2),

$$N = \int_A P \, dA \quad (2.2)$$

where P is the pressure and A is the surface area.

2.1.2 Tire components

The components of the tire listed below can be seen graphically in Fig. 2.1 [24].

1. *Inner liner*: The first material from the inside is a synthetic rubber providing an airtight layer. It is to prevent any leakage of air, which would lead to a pressure loss inside the tire.
2. *Carcass*: The carcass is made with strong textile fiber cords implemented in a rubber housing. Its objective is to maintain the tire's shape under internal pressure. It also ensures that the tire would not bulge out when inflating it.
3. *Beads*: The beads are steel wires included in the part of the tire sidewall in contact with the rim. The purpose of the wires is to ensure an airtight

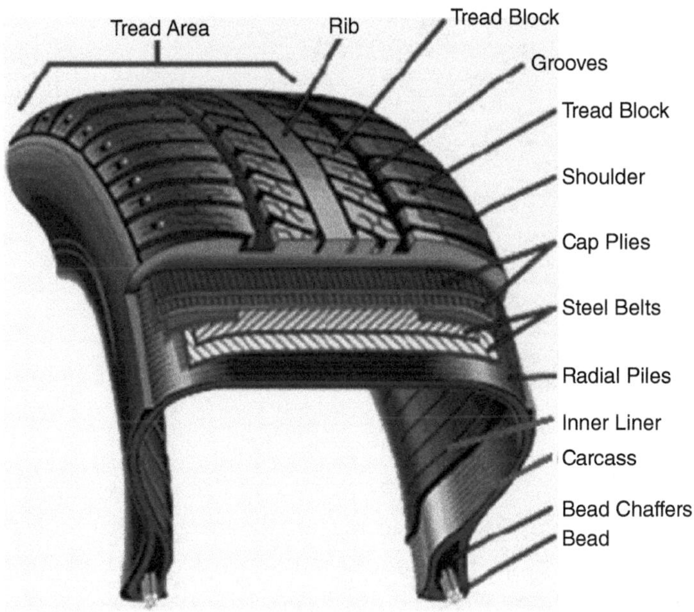

Figure 2.1 Tire's anatomy [24].

contact between the tire and the rim to avoid any air leakage. A set of wires included in the tire can be, in some cases, withstand 10 times the weight of the car [1].

4. *Sidewalls*: The sidewall is where the logo of the manufacturer and the details about the tire and its production is printed. The details are preferable; dimensions of the tire, speed rating, preferred rolling direction, and the month and year of its production.
5. *Steel belts*: Steel belts are bounded into the rubber, providing strength. It makes the tire's ability to handle the strain from turning and preventing the tire from expanding from the centrifugal force caused by fast rotation.
6. *Cap plies*: The cap plies are rubber layers with integrated nylon that stretches around the circumference of the tire, located between the steel belts and the treading. Both layers add resistance against expansion and reduce heating induced by friction.
7. *Tread*: The tread is the part of the tire in contact with the road and is the visible part from outside. It is usually monitored to determine the quality and characteristics of tire performance. The objective of the tread is to provide grip against the road surface while providing a low level of abrasion and heat generation.

Figure 2.2 Comparison of summer tire versus winter tire [11].

8. Grooves are the cuts in the tread, making the tread patterns. The treading is responsible for road noise mitigation, water diversion, and to provide a large contact area with a correct frictional coefficient. Grooves are explicitly designed to provide sufficient grip in given temperatures and conditions.

2.1.3 Tread differences

There are some key differences to a winter tire compared to a summer tire other than the possible addition of studs, that is, rubber composition, presence of pores over the thickness of tire and having sipes in the tread blocks [11,24]. The most noticeable being the shape and pattern of the treading. As Fig. 2.2 shows, the summer tire has large tread blocks divided by wide grooves. These groves are mainly oriented in the longitudinal direction (rolling direction) for maximum water displacement at higher speeds. The tread blocks are also smooth for the maximum contact area.

As seen in Fig. 2.2, the winter tire's treads have more, but narrower grooves. The grooves are a little deeper than on winter tires and oriented in different directions, including lateral movement (perpendicular to the rolling direction). It makes for an efficient deflection of water and slushy snow, providing the tread a better contact with the road without hydroplaning. It is more crucial in Arctic conditions due to more exposure to water. Due to the tire pressure, the snow melts due to the regelation phenomenon as the freezing point of water is lowered by applying pressure, hence creating a water film between the tires and road surface. This water needs to be moved

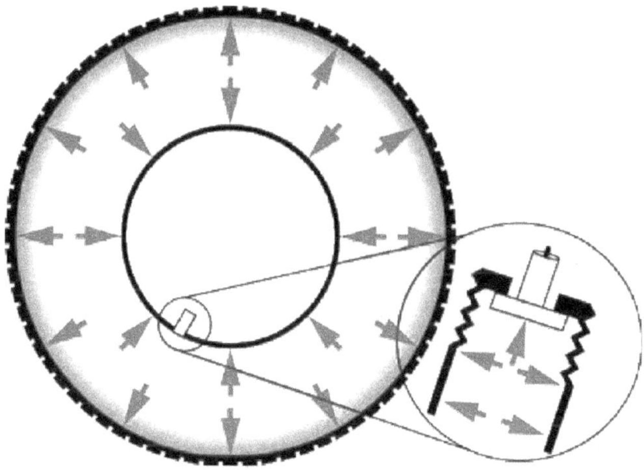

Figure 2.3 Illustration of uniformly distributed tire pressure [10].

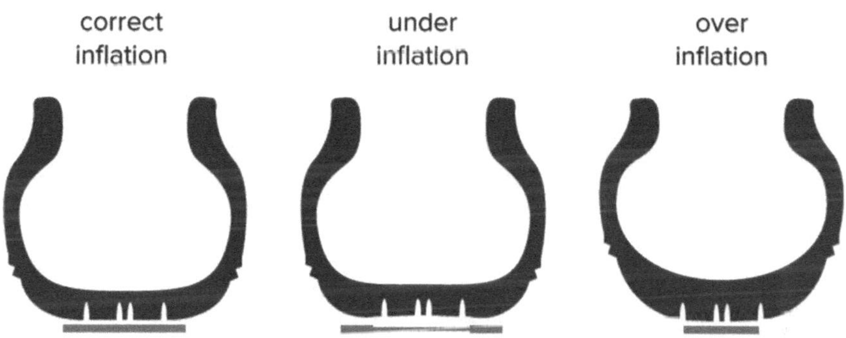

Figure 2.4 Illustration of tire pressure changing the contact area [14].

from underneath the tread surface to prevent hydroplaning [16,28] and is effectively done by the unique tread patterns of winter tires.

2.1.4 Physics of pressure profile

The pressure inside a car tire can be illustrated in Fig. 2.3. As shown pressure on every part of the surface is equal and acting perpendicular to the area. So even with a deformation of the tire due to it being pushed on the ground, the pressure is still equal throughout its inner surface.

This behavior is as predicted for isotropic elastic–plastic solids at finite strain and arbitrary pressure is likely to give an equal uniform pressure profile over the contact area with the ground [30]. Some variations can occur due to the inflation pressure values, as seen in Fig. 2.4 [34]. A drawback of using

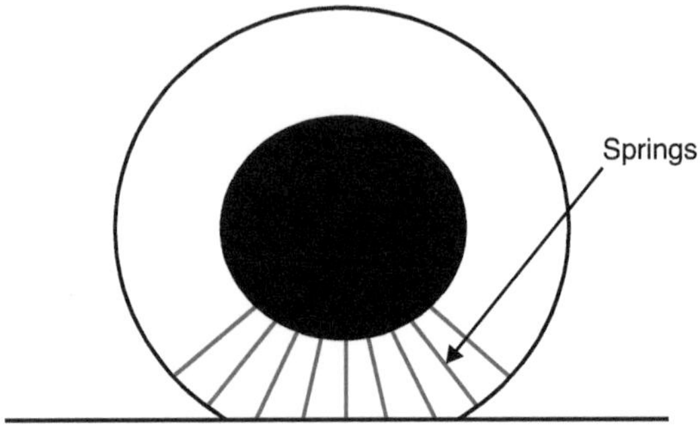

Figure 2.5 Illustration of the springs in a nonpneumatic tire being compressed against the road.

pressurized air inside the tires, especially in Arctic conditions, is that the air pressure is dependent on the temperature of the atmosphere [5,33]. This temperature effect can be neglected by changing to a nonpneumatic tire with suitable spring materials.

2.1.5 Nonpneumatic contact profile theory

The pressure profile from a nonpneumatic tire might not be the same as for pneumatic tire as we have springs transferring the weight of the car to the ground instead of air [18]. Fig. 2.5 shows an exaggerated picture of a nonpneumatic tire being pushed down to the ground, where the red lines are symbolizing the tire springs closest to the contact area. As Hooke's law [23] explains, the force on a spring with linear behavior (constant spring constant) is determined by the displacement. It means that the springs that are being compressed the most will have the highest force. The force from these springs passes through the rubber and down to the ground, influencing the contact pressure between the tread and the road [13]. The spring being compressed the most, in a stationary situation, is always the one in the middle. This gave a higher-pressure concentration in the middle of the contact pressure profile.

2.2 Methodology

This work focuses on ensuring that nonpneumatic tires improve mechanical grip as compared to pneumatic tires. As conventional pneumatic tires contain pressurized air, the pressure is always uniformly distributed throughout the inside of the tire, giving it a more evenly distributed contact pressure profile as shown in Fig. 2.6. The study was performed to change the pressure profile by concentrating the pressure from the tire down to

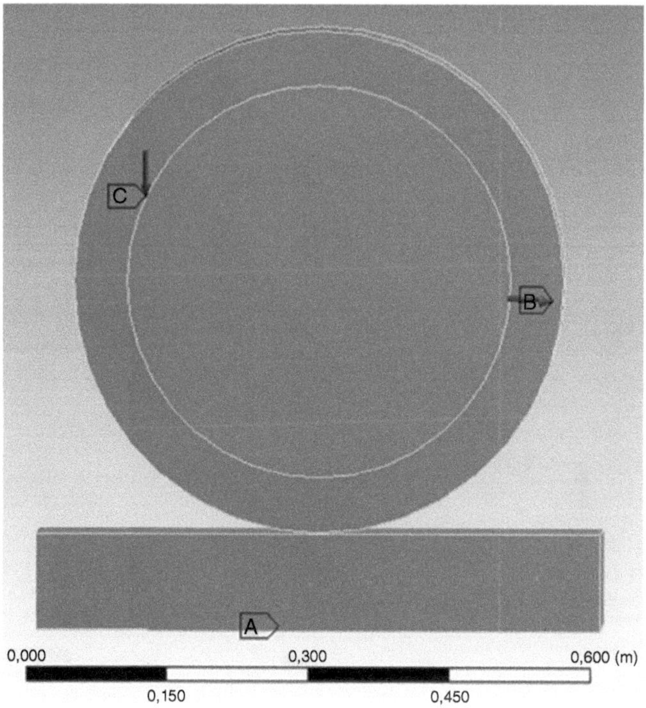

Figure 2.6 A schematic model displaying the fixed constraint "A" at the bottom of the ground block, uniform inflation pressure "B," and applied internal pressure on pneumatic tire "C."

the contact area; it was done by replacing the pressurized air with springs. This work involves a finite element analysis [25,26], computer-aided design (CAD) models were developed in Inventor®, FEM analysis model was developed in ANSYS® Workbench [12]. Pneumatic and nonpneumatic tires' results were compared. Similar studies have been performed in ANSYS® Workbench to analyze the materials that show the most promising results for aquaculture purposes, to simulate micro vibrations on a nontrivial torsionally oscillating structure and to optimize elstomeric microfluidic valve dimensions [8,19,27].

2.2.1 Materials

All the materials for this study were taken from the ANSYS® Workbench library. The material properties of concrete, hard, and soft rubbers and custom material for springs are presented in Table 2.1.

Table 2.1 Material used for modeling in ANSYS® Workbench.

Material	Volume applied to	Young's module (MPa)	Poisson ratio
Concrete	Ground	30,000	0.18
Hard rubber	Pneumatic tread	400	0.47
Soft rubber	Pneumatic sidewalls	10	0.47
Custom material	Springs for nonpneumatic model	2600	0.4

Figure 2.7 Tread pattern.

2.2.2 Finite element analysis modeling

2.2.2.1 Treading

The idea behind the tread pattern is to make a high surface area while having grooves oriented in both longitudinal and lateral direction. The tread pattern was made in Inventor® CAD software as shown in Fig. 2.7, the right half of the blue highlighted sketch was mirrored to create the left half of the tread but with a longitudinal offset of 10.5 mm. With a complete drawing covering the tire's width, an extrusion of 8 mm was made into the tire's surface to separate the tread blocks with grooves. This extrusion was then applied a circular pattern around the tire's circumference with a number high enough to make the pattern interfere with itself again, making up a tread pattern that is the same all around. When having the right pattern, the fillet option was used to trim the edges of the tire, making realistic tire shoulders. These were made by an arched line with a radius of 8 mm. Depth was chosen to have a smooth transition between the tread and the sidewalls.

Table 2.2 Car tire numbers explained.

205	Width in millimeters
55	Aspect ratio (section height/width)
R	Construction radial
16	Rim diameter in inches

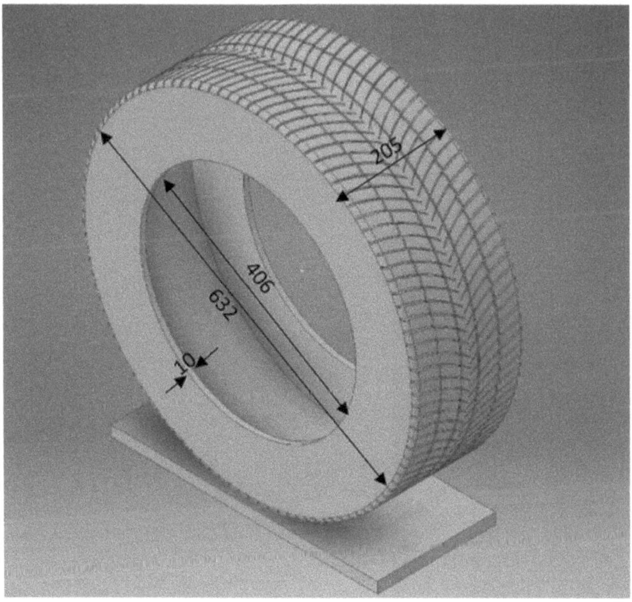

Figure 2.8 Pneumatic tire CAD model.

2.2.2.2 *Pneumatic tire*

The pneumatic tire model was made like a 205/55/R16 radial tire as shown in Table 2.2. As discussed in tire anatomy, there are many components embedded in the rubber of the tire. In the CAD model, only those were included that has an impact on the characteristic of a tire when applied with vertical force. The features that were developed were the sidewalls, steel belts, cap plies, carcass, and treads. The finished tire model is shown in Fig. 2.8, and features dimensions are provided in Table 2.3.

2.2.3 *Nonpneumatic tire*

The CAD model of the nonpneumatic tire was also made in Inventor®, as shown in Fig. 2.9. The dimensions of 205/55/R16 tire model are followed as shown in Table 2.2. Treads from the pneumatic model and features inside the rubber's inner circumference were redesigned. It included the removal

Table 2.3 Dimensions used for pneumatic tire sketch.

Part of tire	Size (mm)
The outer diameter of the tire	632
Tire width	205
Groove depth winter tire	8
Total rubber thickness	15
The radius of the rounded tire shoulder	8
Sidewall thickness	10
Sidewall height	105
Rim diameter	406

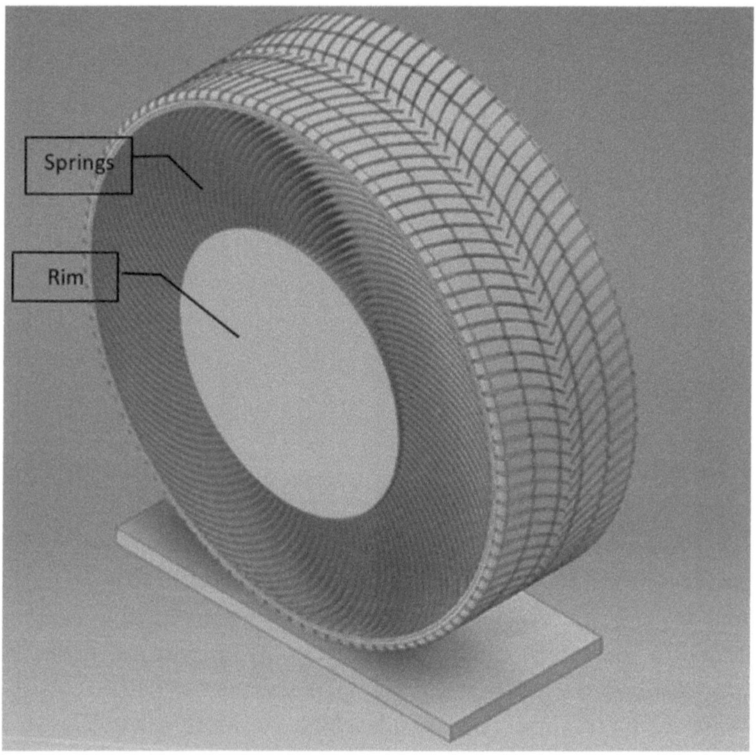

Figure 2.9 Nonpneumatic tire CAD model.

of the sidewalls, carcass, steel belt, and cap plies from the pneumatic tire. Rim was made, represented by a solid cylinder in the center. Between the rim and the rubber, curved plates were added to act as springs, these curves had an angle of 136° and a thickness of 2 mm, making a total number of 120 springs. Dimensions of the features are shown in Table 2.4.

Table 2.4 Dimensions of nonpneumatic sketch.

Part of tire	Size (mm)
Total diameter	632
Tire width	205
Rim diameter	330
Ring thickness connecting springs and rubber	2
Height of spring area	134
Groove depth	8
Total rubber thickness	15

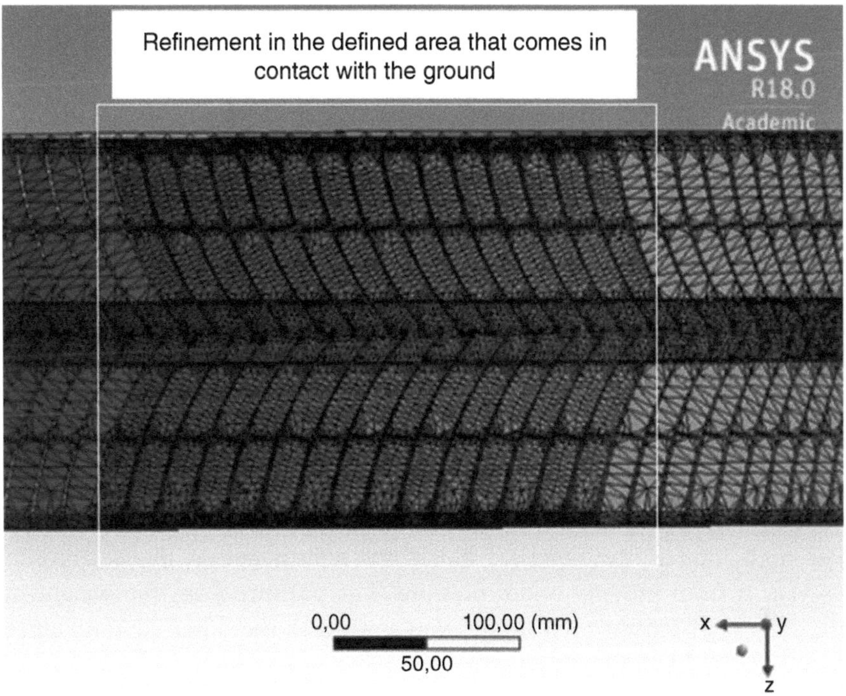

Figure 2.10 Meshing of the tread on pneumatic and nonpneumatic model.

2.2.3 Meshing

After importing the CAD model into ANSYS® and defining the materials, meshing is done. The refinement was set to 1, which means that each element is divided one time over both in the *X*-axis and *Y*-axis. By doing this, the number of elements was increased four times in the defined area. Fig. 2.10 shows how the tread was meshed, as the tread had the same mesh for both models. The structure of the elements gave tetrahedral elements on the tire and quadrilaterals for the ground block.

2.2.3.1 Mesh sensitivity analysis

The mesh sensitivity analysis was performed on the nonpneumatic model. Two size parameters were changed when performing this analysis; the relevance center and the span angle center, at the levels; coarse, medium, and fine. The mesh element type was set to automatic, mainly generating tetrahedral elements. The mesh sensitivity analysis graphs are presented in Fig. 2.11. The mesh sensitivity analysis was performed on equivalent stress (Fig. 2.11(A)), force reaction (Fig. 2.11(B)), structural error (Fig. 2.11(C)), and strain energy (Fig. 2.11(D)).

This analysis shows a significant difference in result values over the different mesh densities. The relevance center is the most significant factor, compared to the span angle center, as it determines the sizing of the elements. When increasing the relevance center one level (e.g., coarse to medium), the number of elements and nodes almost doubles, while increasing the span angle center gives an increase in element number of around 10%. It is clear that the coarse meshes fluctuate at high levels, while medium and fine mesh converge better. These mesh settings were applied when solving the simulations in ANSYS® Workbench.

2.2.4 Boundary condition and body contacts

For both models, the concrete block was constrained at the bottom, to prevent it from moving under pressure. The pressure from the weight of the car, calculated for each model, was applied to the outer circumference area of each rim, acting downward.

2.2.4.1 Pneumatic model

The contact between the tread and the sidewall were set to be "bonded contact." It was done to ensure that there was no material or pneumatic pressure leakage through the model under applied pressure and strain. The contact between the tread and the ground was set to frictional contact. A fixed constraint was applied to the top of the sidewall to prevent the pneumatic tire from moving as shown in Fig. 2.12. As the tire was fixed, the external force had to come from the ground.

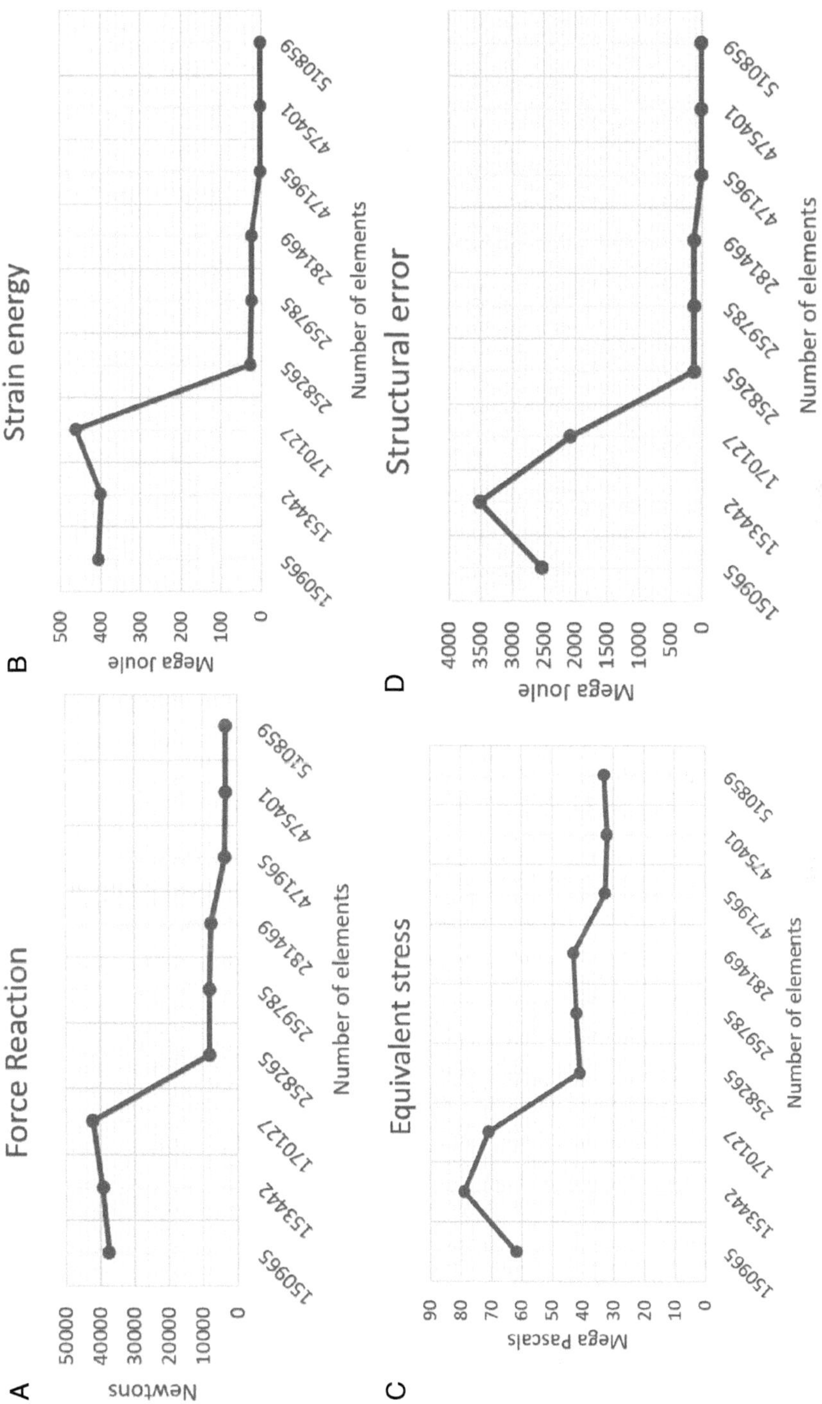

Figure 2.11 Mesh sensitivity analysis graphs.

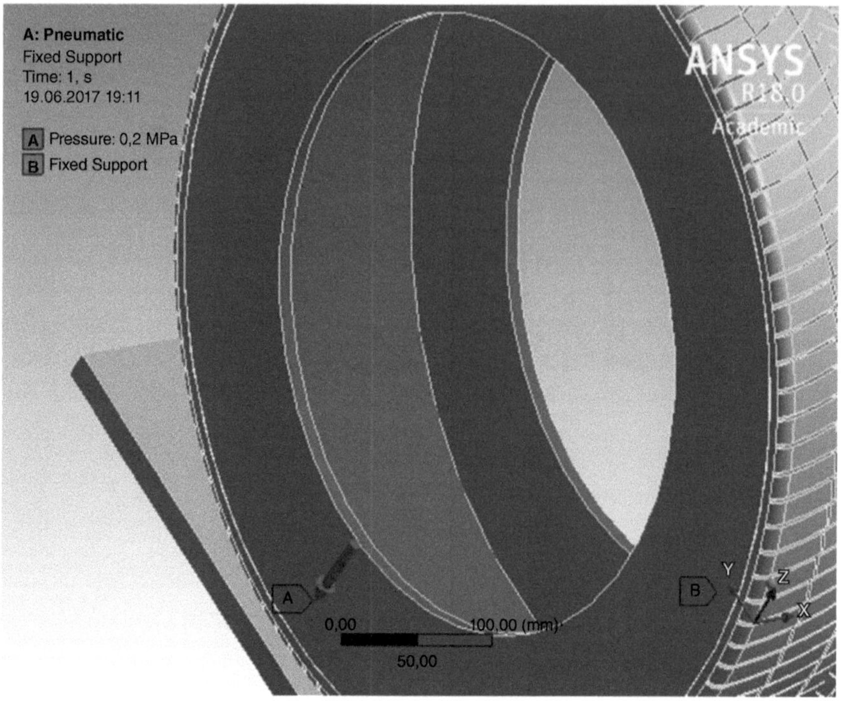

Figure 2.12 Fixed support and internal pressure applied to the pneumatic model.

2.2.4.2 Force and displacement

For the tire to maintain its structure, a pneumatic pressure had to be applied. A pressure of 0.2 MPa, or 2 bars, was applied to the inner part of the rubber acting outward (as seen in red in Fig. 2.12). A 2-bar pressure was used as this pressure value is representative for a 205/55R16 tire with a car weighing around 1200 kg. With both fixed support and a force replacing air pressure, the ground plate was moved toward the tire, aiming for a reaction force equal to one-fourth of a car's weight. The reaction force is calculated by ANSYS® Workbench from the displaced body. A vehicle with balanced weight distribution on each wheel consists of reaction force per wheel = 2974.6 N, where the mass on each wheel $= = \frac{2974.6\,N}{9.81\,m/s^2} = 303.4\ kg.$ So total mass of the vehicle = 1213.6 kg.

2.2.4.3 Nonpneumatic model

The nonpneumatic model was structurally more complex than the pneumatic model because of the presence of more number of bodies, that is, springs. There were a total of 120 springs in this model as shown in Fig. 2.13.

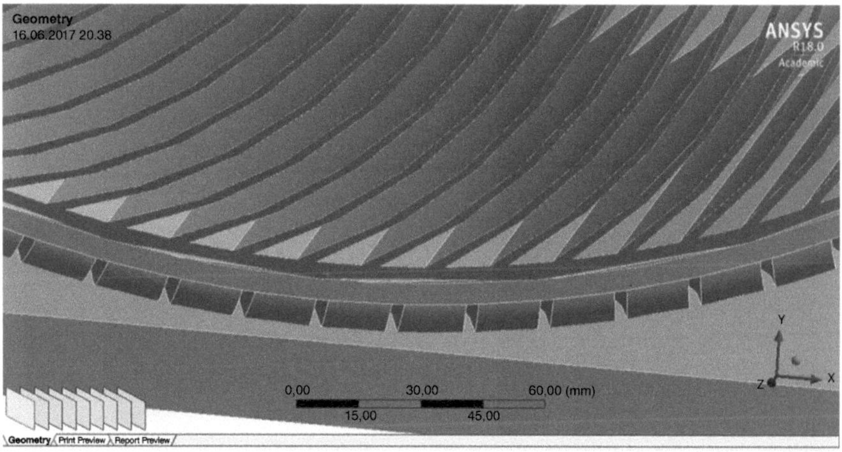

Figure 2.13 Springs and connected ring merged to one body.

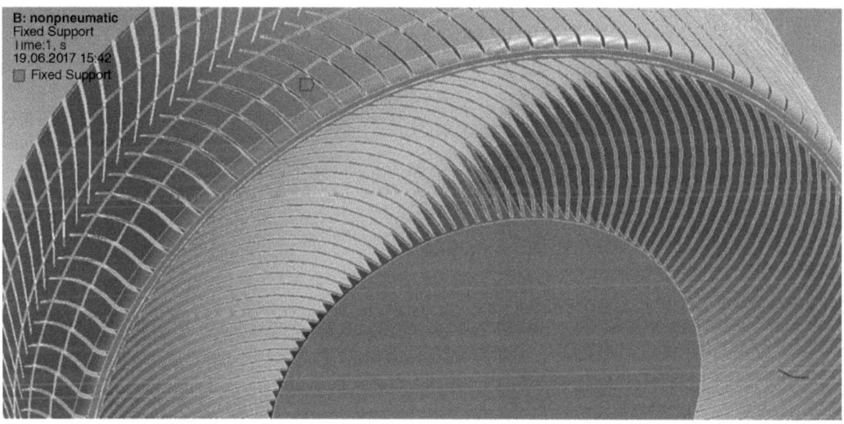

Figure 2.14 Fixed support, highlighted blue, added to the nonpneumatic model.

Before importing the CAD model to ANSYS® workbench, the bodies had to be defined in order to specify the material later. The rim, springs, and a ring around the outer circumference of the springs were combined into one body. It left the model with three bodies: the springs, rubber tread, and the ground block.

Similar to the pneumatic model, the tire was constrained. The rim surface was applied a "fixed support," highlighted in blue in Fig. 2.14, to ensure correct behavior and zero movements under strain from the ground plate. The springs and the rubber were "bonded contacts" to prevent relative movement between the two. The last contact between the tread and the ground was again set to be a "frictional contact."

2.2.4.4 Force and displacement

The force in this model is defined as a displacement of the ground. The level of displacement was set to match the total deformation of the pneumatic model. As the pneumatic model had deformation caused by both internal force and external displacement, the total deformation was 7.9 mm. By having the displacement determined, the aim was to get a reaction force close to the pneumatic model of 2979.3 N. As all inputs were the same between the models, the Young's modulus (value of stiffness) for the springs was the only parameter separating the two models. After optimizing the Young's modulus, the reaction force was found to be 2976.6 N. It was only 3.3 N different from the pneumatic model, 2976.6 N divided by the gravity of 9.81 m/s^2 gives a weight of 303.7 kg. Multiplied by the four tires gives a total car mass of 1214.8 kg. It is only 0.01% different from the pneumatic model; this confirms the validity of the numerical model.

2.3 Results and discussion

2.3.1 Pneumatic model

2.3.1.1 Deformation

The deformation is affected by two external factors; the pressure applied from the inside, simulating inflation pressure, and the strain from the plate pushing the tire surface radially inward as shown in Fig. 2.15. The rubber characteristics worked adequately, ensuring that the tire radius at the contact patch was less than the free radius around the rest of the unloaded circumference. The total deformation under the additional stress from the ground plate's displacement looks realistic. The highest deformations happen close to the contact area with a maximum deformation of 7.9 mm located at about 20 cm to both sides of the contact, shown in red in Fig. 2.15. The smooth transitions between the colors indicate that the solution has converged fully and has a satisfying mesh density.

2.3.1.2 Pressure profile

The pressure profile is given in Fig. 2.16. Pressure distribution through the contact area between the tire and the ground can be seen. The tire is experiencing low-pressure values in the blue zone, increasing toward the red. The average pressure over the middle area is not very high at pressures below

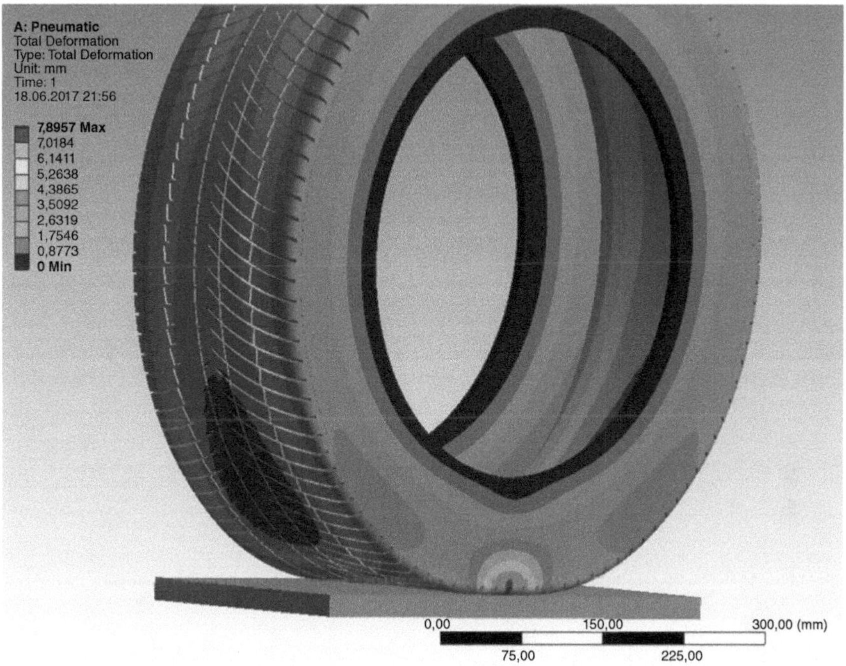

Figure 2.15 Deformation of pneumatic model.

1.2 MPa. Two areas experience a pressure above 2 MPa, and that is on both sides of the tire, just below where the sidewall ends. Low tire pressure can give better traction on the loose surface such as snow and gravel due to the increase in the contact area and the tire's ability to deform over the surface unevenness. However, it is not recommended to have a too low pressure for everyday driving as the wear on edges increases and heat generations happens quicker with low tire pressure, leading to a risk of the tire cracking [7,29].

2.3.1.3 Equivalent stress (Von-Mises stress)

The equivalent stresses are displayed in Figs. 2.17 and 2.18. This can be interpreted that some of the high stresses are located inside the tire over the contact area, with a maximum value of 12 MPa. The other regions experiencing high pressures are in the grooves of the treading [9], which have been offset to the contact area. This is the area where the max stress of 31 MPa occurs. Another thing to notice is that the stresses are generally higher in the center of the width than on the tire shoulders. The stresses on the pneumatic model are a result of the internal and external forces.

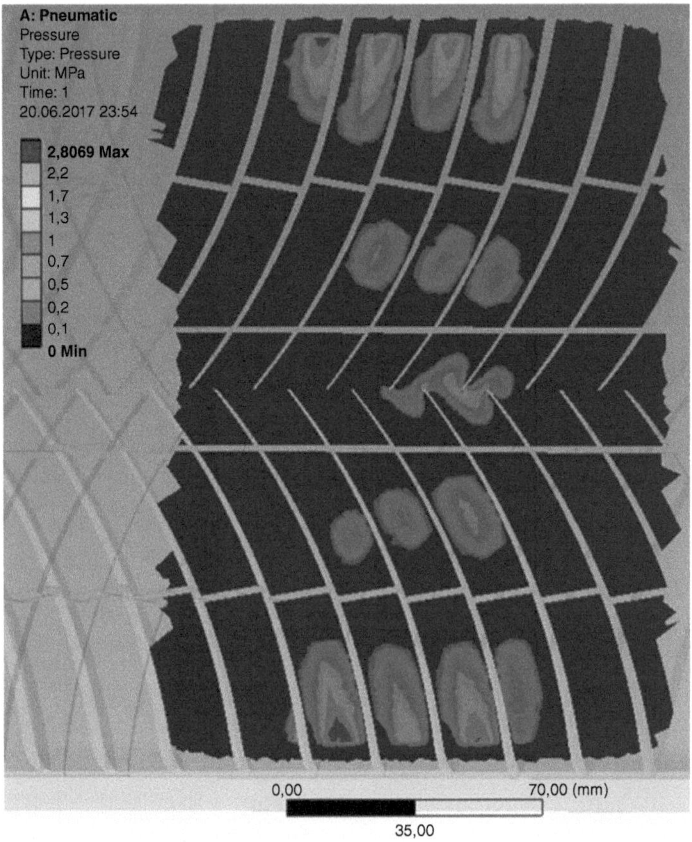

Figure 2.16 Pressure profile for pneumatic model.

If the model runs without including the ground plate, the stress would be distributed evenly at given width over the tire's circumference because of the pneumatic pressure. Khawaja et al. performed similar work to discuss 2-D approximation technique for solving stress analyses in FEM [20] and the study of CRFP shell structures under dynamic loading in shock tube setup [21], his work proved materials, that is, CFRP behave linearly under dynamic loading [38].

2.3.2 Nonpneumatic model

2.3.2.1 Deformation

The deformation for the nonpneumatic model is shown in Fig. 2.19. The plate's displacement was set to 7.9 mm to match the total deformation

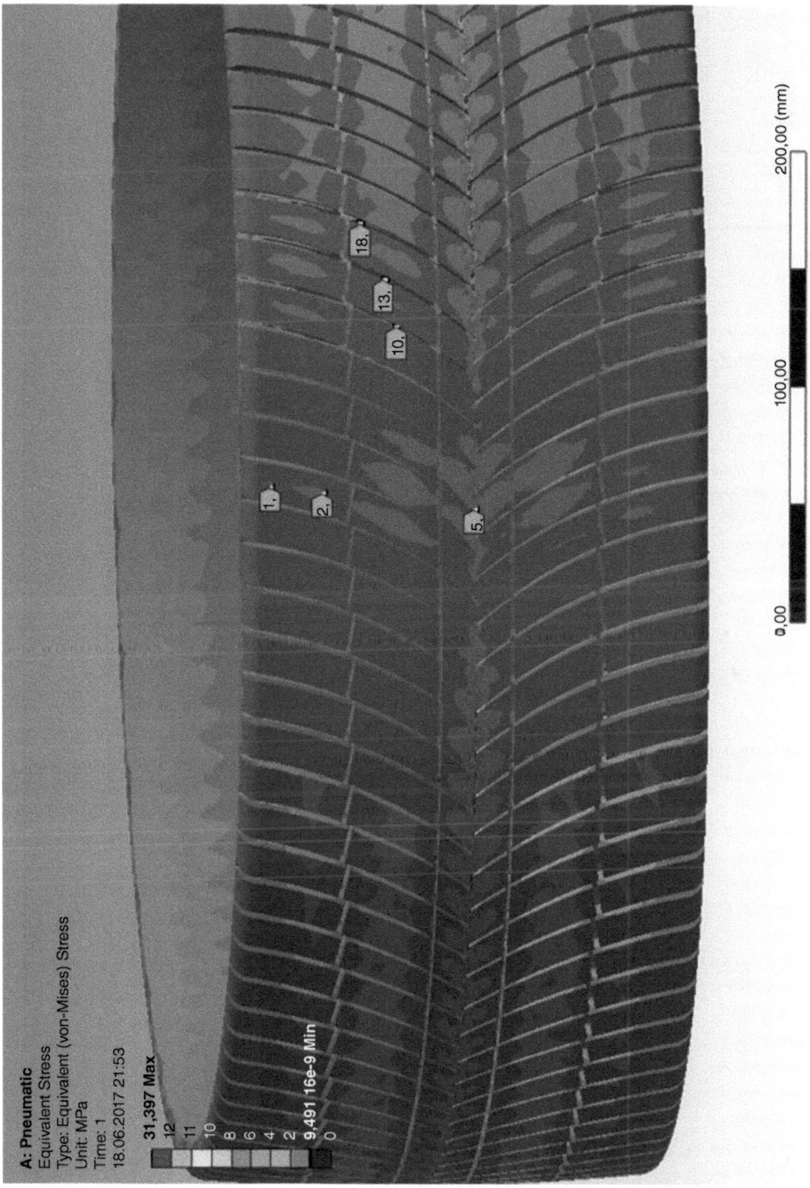

Figure 2.17 Equivalent stress on the outside of the pneumatic tire.

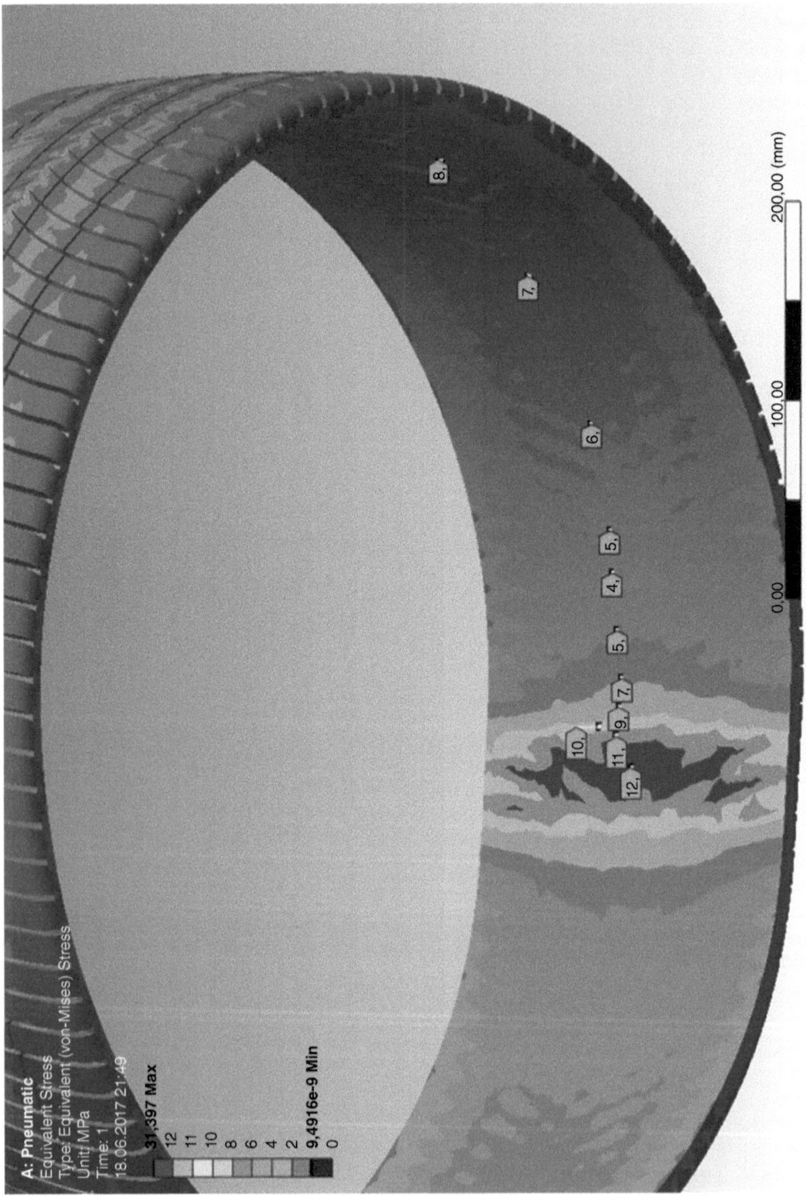

Figure 2.18 Equivalent stress on the inside of the pneumatic model.

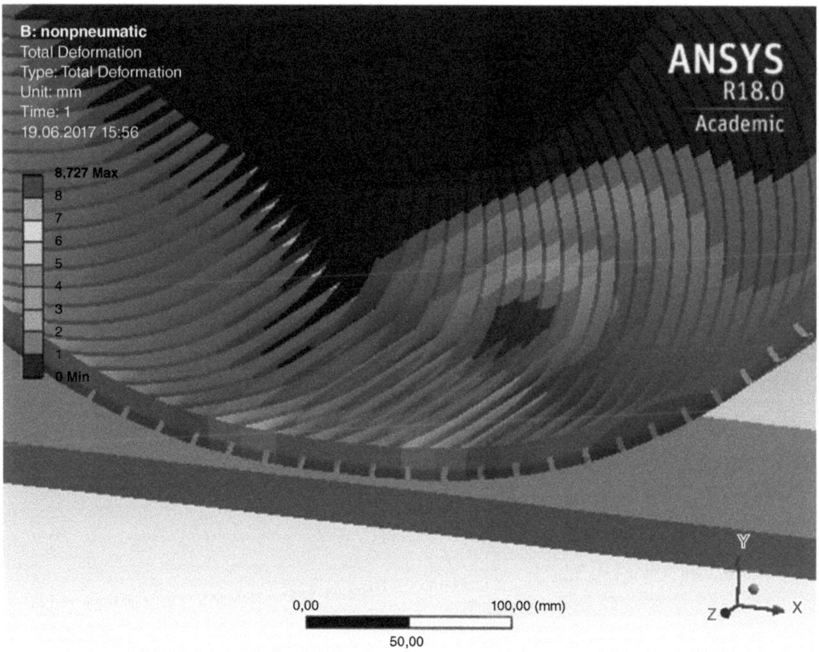

Figure 2.19 Deformation of nonpneumatic model.

of the nonpneumatic model. The highest deformation happens at the springs that are connected over the contact area. Even with deformation of 8.7 mm, the springs still have proper clearance between them. The deformations of the model were as expected when the design was made. The springs behaved predictably, just as visualized when sketching the model. The most considerable deformation happened in the middle of the spring connected to the rubber above the contact area. This deformation was more significant than the displacement of the ground plate. It indicates the modeling accuracy of the nonpneumatic tire model.

2.3.2.2 Pressure profile

Fig. 2.20 shows the pressure profile for the nonpneumatic model. For the most part, the pressure is uniformly distributed over the width of the tire and the contact area. The pressures are above 1.6 MPa on every tread block over that width, which results in pressure concentration through the center of the contact area. The maximum pressure of 4.94 MPa occurs in the center of the profile.

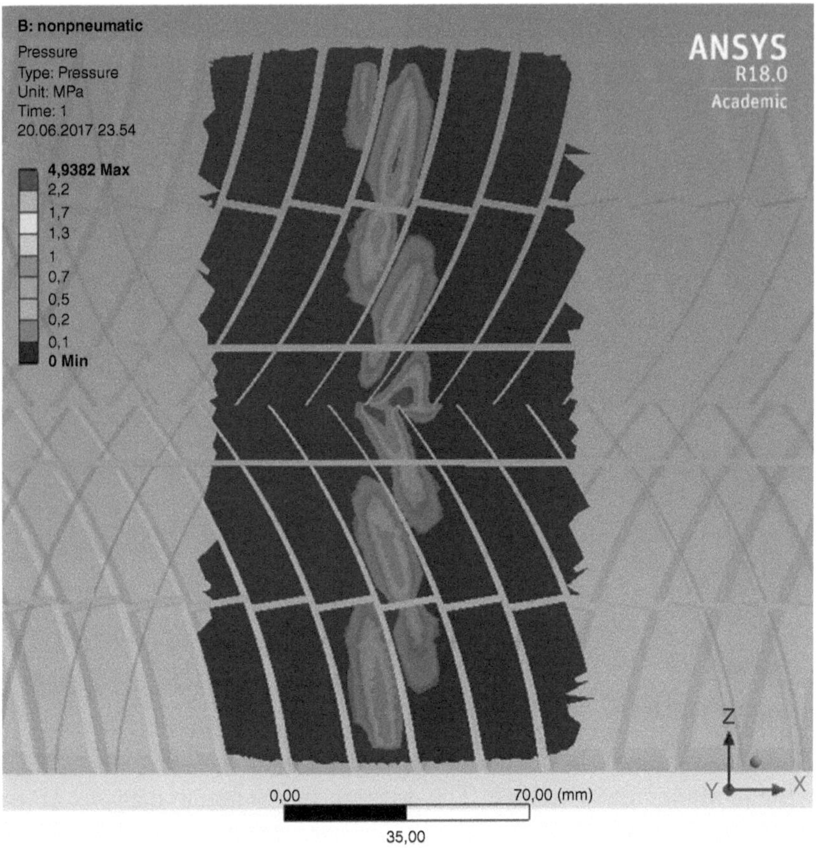

Figure 2.20 Pressure profile for the nonpneumatic model showing high-pressure concentration at the contact area.

2.3.2.3 Equivalent stress

The equivalent stress is displayed in Fig. 2.21. The labels show the stresses are higher at points where the springs are attached to the rim and the rubber compared to the center of the individual spring. The maximum stress appears inside the rubber in the center of the contact area. This is the same point that had both the maximum deformation and maximum pressure. Stange et al. performed similar work to visualize stresses in steel samples using IR Thermography [31,32].

2.4.1 Comparison of pressure profiles

When comparing the two pressure profiles side-by-side, as shown in Fig. 2.22, we can note that the pneumatic model's contact area is larger than the nonpneumatic model. The difference is visible as the pneumatic profile

Figure 2.21 Equivalent stress for nonpneumatic model.

has eight blocks in contact with the road than the nonpneumatic model's six blocks when counting in the longitudinal direction (left to right).

By analyzing the pressure profile curves as shown in Fig. 2.23(A) and (B), for pressure values across the mid-section of the tires, there is a clear trend. The pressure of the nonpneumatic model on the left is much more concentrated toward the center and has higher pressure values. The pneumatic model's pressure concentrations cover the longitudinal length to a more considerable extent than the nonpneumatic model, with approximately four tread blocks compared to about 2. This is a positive result as the aim of this work is to prove that the nonpneumatic tire has a more concentrated pressure profile with higher pressure values.

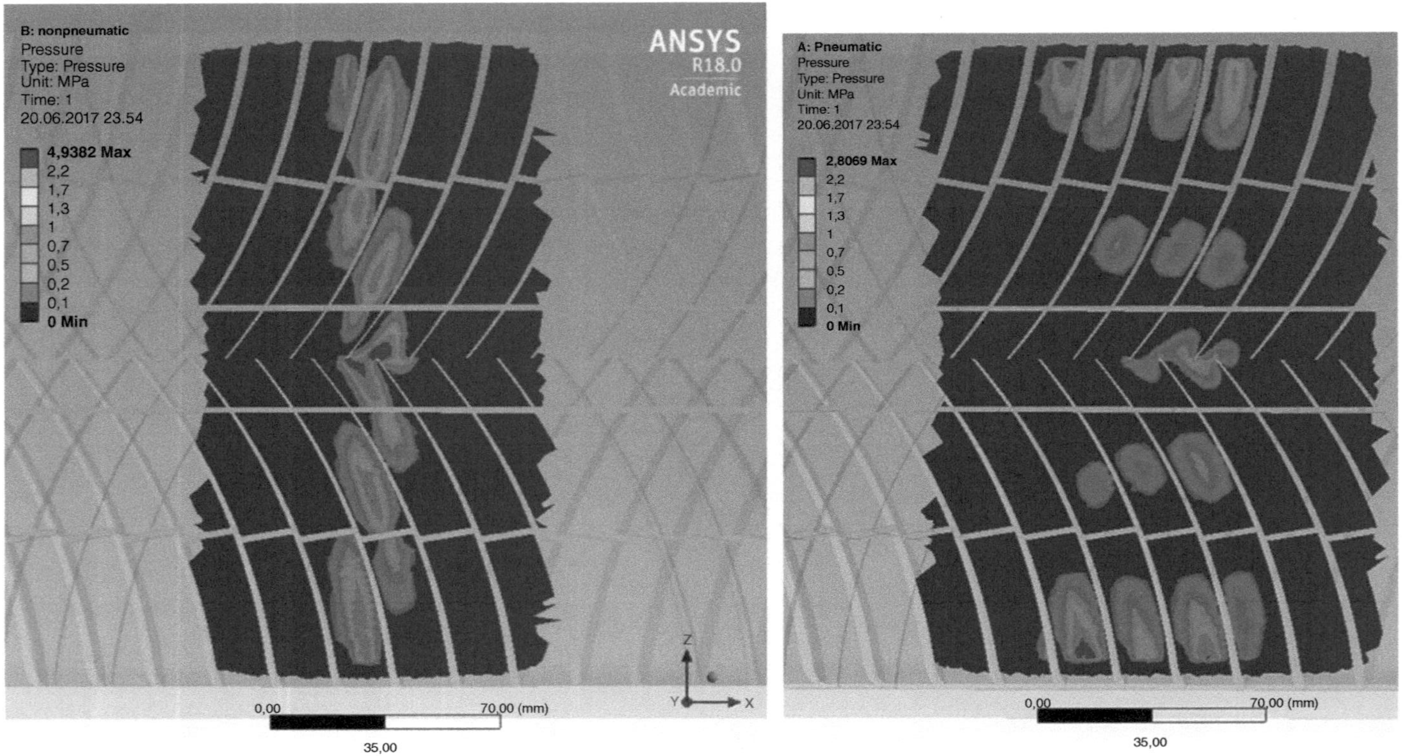

Figure 2.22 Side-by-side comparison of the pressure profiles: (A) nonpneumatic and (B) pneumatic.

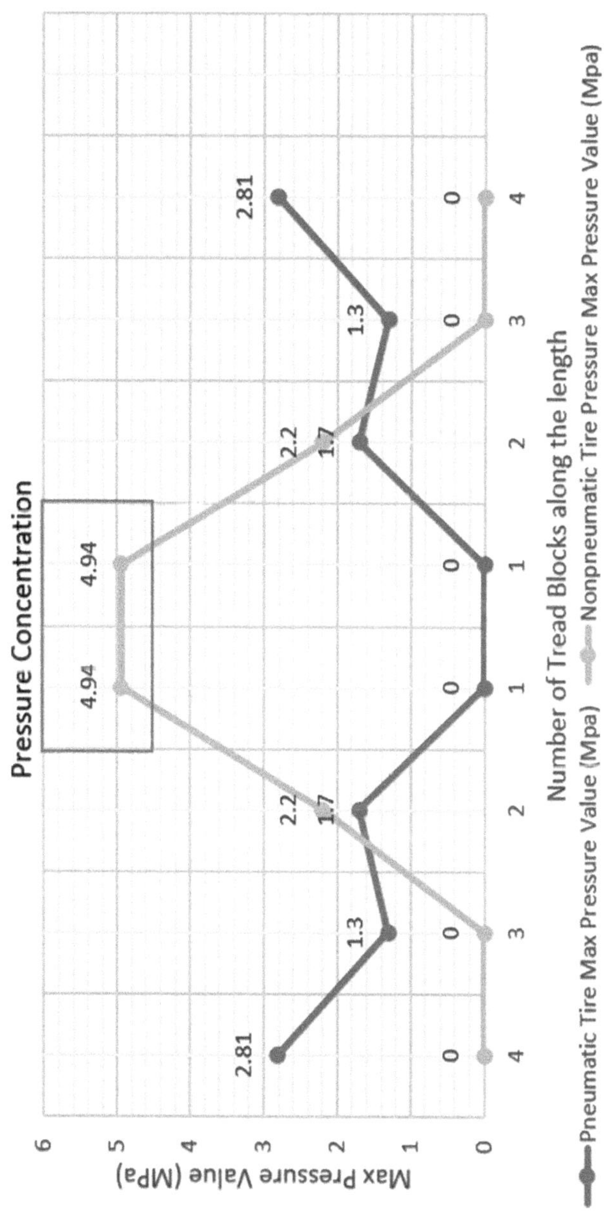

Figure 2.23 Pneumatic and nonpneumatic pressure profile curves plotted together.

2.5 Conclusions

This study confirmed that finite element analysis is a valid approach to analyze the pressure profile of two kinds of tires, and to draw a comparison based on their performance in cold climatic conditions. However for the pneumatic model, the longitudinal contact area is larger than the nonpneumatic model but the pressure on the nonpneumatic tire is more evenly distributed over the tire width, with both higher average and maximum pressure values of 4.94 MPa as compared to 2.81 MPa for pneumatic tire. This concentration of pressure is believed to increase the grip of a tire as the frictional force is a function of pressure. The increase in the grip of a nonpneumatic tire significantly increases the friction and reduces the chances of sliding over snow and ice in cold weather conditions. Also as nonpneumatic tires are based on material springs instead of air, the tire characteristics are not affected by temperature fluctuations as is the case with the air pressure inside a pneumatic tire. This research work proved that nonpneumatic tires are a better choice for use in winters. Carefully designed nonpneumatic tires will be useful in avoiding the detrimental impact loads inflicted on the roads by either studded or chained tires.

References

1. S. Ag, G. Ferrer, The Economics of Tire Remanufacturing, INSEAD, Fontainebleau, France, 1993.
2. T. Ahmad, et al., Study of wind chill factor using infrared imaging, Int. J. Multiphys. 10 (3) (2016) 325–341, doi:10.21152/1750-9548.10.3.325.
3. T. Ahmad, H. Khawaja, Review of low-temperature crack (LTC) developments in asphalt pavements, Int. J. Multiphys. 12 (2) (2018) 169–187, doi:10.21152/1750-9548.12.2.169.
4. A. Albers, et al., Numerical investigations in mixed friction systems, Int. J. Multiphys. 4 (1) (2016) 83–94, doi:10.1260/1750-9548.4.1.83.
5. Z. Andleeb, et al., Multiphysics analysis of CFRP Charpy tests by varying temperatures, Int. J. Multiphys. 14 (2) (2020) 143–160, doi:10.21152/1750-9548.14.2.143.
6. M.J. Angerinos, et al., A Synthesis on Studded Tires, Washington State Department of Transportation, Olympia, Washington, 1999 Available at https://www.wsdot.wa.gov/research/reports/fullreports/471.1.pdf. (Accessed date: 08 May 2021).
7. L. Ann Myers, et al., Measurement of contact stresses for different truck tire types to evaluate their influence on near-surface cracking and rutting, Transp. Res. Rec. 1655 (1) (1999) 175–184, doi:10.3141/1655-23.
8. D. Brunner, et al., CFD modelling of pressure and shear rate in torsionally vibrating structures using ANSYS CFX and COMSOL Multiphysics, Int. J. Multiphys. 12 (4) (2018) 349–358, doi:10.21152/1750-9548.12.4.349.
9. B. Clapson, G. Lake, Truck tire groove cracking theory and practice, Rubber Chem. Technol. 44 (5) (1971) 1186–1202, doi:10.5254/1.3544800.

10. J.D. Cullen, et al., In-field trials of a tyre pressure monitoring system based on segmented capacitance rings, Measurement 32 (3) (2002) 181–192, doi:10.1016/S0263-2241(02)00010-6.
11. J.A. Davisson, Design and Application of Commercial Type Tires 690001, SAE Trans. V78 (A) (1969) 1–32, doi:10.4271/690001.
12. ANSYS Inc. Engineering Simulation & 3D Design Software | Ansys (no date). Available at: https://www.ansys.com/. (Accessed 8 July 2020).
13. T.H. Fay, S.D. Graham, Coupled spring equations, Int. J. Math. Educ. Sci. Technol. 34 (1) (2003) 65–79, doi:10.1080/0020739021000029258.
14. B.T. Fijalkowski, B.T. Fijalkowski, Basics of Automotive Vehicle Braking, Automotive Mechatronics: Operational and Practical Issues., Springer, the Netherlands, 47, 2011, pp. 435–451, doi:10.1007/978-94-007-0409-1_28.
15. N. Fukuzaki, T. Yanaka, Y. Urushiyama, 'Effects of studded tires on roadside airborne dust pollution in Niigata, Japan', Atmos. Environ. (1967) 20 (2) (1986) 377–386, doi:10.1016/0004-6981(86)90041-7.
16. B.M. Gallaway, et al., Pavement and Geometric Design Criteria for Minimizing Hydroplaning. A Technical Summary., National Technical Reports Library, National Technical Reports Library, United States, 1979.
17. B. Jancsek-Turóczi, et al., Sampling and characterization of resuspended and respirable road dust, J. Aerosol Sci. 65 (2013) 69–76, doi:10.1016/j.jaerosci.2013.07.006.
18. J. Ju, D.M. Kim, K. Kim, Flexible cellular solid spokes of a non-pneumatic tire, Compos. Struct. 94 (8) (2012) 2285–2295, doi:10.1016/j.compstruct.2011.12.022.
19. Khawaja, et al., Optimization of elstomeric micro-fluidic valve dimensions using non-linear finite element methods, Int. J. Multiphys. 3 (2) (2009) 187–200, doi:10.1260/175095409788837847.
20. H. Khawaja, Application of a 2-D approximation technique for solving stress analyses problem in FEM, Int. J. Multiphys., 9, 2015, pp. 317–324, doi:10.1260/1750-9548.9.4.317.
21. H.A. Khawaja, et al., Study of CRFP shell structures under dynamic loading in shock tube setup, J Struct. 2014 (2014) 1–6, doi:10.1155/2014/487809.
22. H.A. Khawaja, K. Parvez, Validation of normal and frictional contact models of spherical bodies by FEM analysis, Int. J. Multiphys. 4 (2) (2010) 175–185, doi:10.1260/1750-9548.4.2.175.
23. C.K. Lee, et al., Use of Hooke's law for stabilizing future smart grid the electric spring concept, in: 2013 IEEE Energy Conversion Congress and Exposition, ECCE 2013, 2013, pp. 5253–5257, doi:10.1109/ECCE.2013.6647412.
24. S. Ludvigsen , Improving Mechanical Grip on Winter Tires. Finite Element Analysis on Pressure Profile of Airless Tire Compared to Conventional Tire Using ANSYS Workbench, UiT – the Arctic University of Norway, Norway, 2017.
25. M. Moatamedi, H. Khawaja, Finite Element Analysis, CRC Press, Boca Raton-United States, 2018.
26. U. Mughal, H. Khawaja, M. Moatamedi, Finite element analysis of human femur bone, Int. J. Multiphys. 9 (2) (2015) 101–108, doi:10.1260/1750-9548.9.2.101.
27. O. Myrli, H. Khawaja, Fluid-structure interaction (FSI) modelling of aquaculture net cage, Int. J. Multiphys. 13 (1) (2019) 97–111, doi:10.21152/1750-9548.13.1.97.
28. G.P. Ong, T.F. Fwa, Wet-pavement hydroplaning risk and skid resistance: modeling, J. Transp. Eng. 133 (10) (2007) 590–598 (2007)133:10(590), doi:10.1061/(ASCE)0733-947X.
29. P. Owende, et al., Minimizing distress on flexible pavements using variable tire pressure, J. Transp. Eng. 127 (3) (2001) 254–262 (2001)127:3(254), doi:10.1061/(ASCE)0733-947X.
30. B. Raniecki, H.V. Nguyen, 'Isotropic elastic-plastic solids at finite strain and arbitrary pressure', Arch. Mech. (Poland) J. Volume: 36 (5–6), (1984) pp. 687–704. Available at: https://www.osti.gov/etdeweb/biblio/5726644 (Accessed 8 August 2020).

31. E. Stange, et al., Multiphysics study of tensile testing using infrared thermography, Int. J. Multiphys. 13 (2) (2019) 191–202, doi:10.21152/1750-9548.13.2.191.
32. E. Stange, Z. Andleeb, H.A. Khawaja, Qualitative visualization of the development of stresses through infrared thermography, Vestnik MGTU 22 (4) (2019) 503–507, doi:10.21443/1560-9278-2019-22-4-503-507.
33. C.H. Strand, Z. Andleeb, and H.A. Khawaja, 'Multiphysics impact analysis of carbon fiber reinforced polymer (CFRP) shell', Explos. Shock Waves High Strain Rate Phenom., 13, (2019), pp. 115–120. doi:10.21741/9781644900338-20.
34. T. Umeno, et al., Observer based estimation of parameter variations and its application to tyre pressure diagnosis, Control Eng. Pract. 9 (6) (2001) 639–645, doi:10.1016/S0967-0661(01)00037-5.
35. M. Virk, et al., Multiphysics based numerical study of atmospheric ice accretion on a full scale horizontal axis wind turbine blade, Int. J. Multiphys. 10 (3) (2016) 237–245, doi:10.21152/1750-9548.10.3.237.
36. H. Xue, H. Khawaja, Investigation of Ice-PVC separation under flexural loading using FEM analysis, Int. J. Multiphys. 10 (3) (2016) 247–264, doi:10.21152/1750-9548.10.3.247.
37. H. Xue, H. Khawaja, Review of the phenomenon of ice shedding from wind turbine blades, Int. J. Multiphys. 10 (3) (2016) 265–276, doi:10.21152/1750-9548.10.3.265.
38. Z. Andleeb, et al., Thermoelastic Investigation of Carbon-Fiber-Reinforced Composites using Drop Weight Impact Test., Applied Sciences 11 (1) (2020) 207, doi:10.3390/app11010207.
39. S. Ludvigsen, et al., Multiphysics Analysis of Contact Pressure Profile of Airless tires as compared to Conventional Tires, The International Journal of Multiphysics 14 (4) (2020) 399–425, doi:10.21152/1750-9548.14.4.399.
40. Sushmit Dhar, Hassan Khawaja, Recognizing Potential of LiDAR for Comprehensive Measurement of Sea Spray Flux for Improving the Prediction of Marine Icing in Cold Conditions - A Review., Ocean Engineering 223 (2021) 108668, doi:10.1016/j.oceaneng.2021.108668.
41. Taimur Rashid, Hassan Khawaja, Kåre Edvardsen, Measuring Thickness of Marine Ice Using IR Thermography., Cold Regions Science and Technology 158 (2018) 221–229, doi:10.1016/j.coldregions.2018.08.025.
42. Taimur Rashid, Hassan Khawaja, Kåre Edvardsen, Determination of Thermal Properties of Fresh Water and Sea Water Ice using Multiphysics Analysis., The International Journal of Multiphysics 10 (3) (2016) 227–291, doi:10.21152/1750-9548.10.3.277.
43. Taimur Rashid, Hassan Khawaja, Kåre Edvardsen, Review of Marine Icing and Anti-/De-Icing Systems., The International Journal of Multiphysics 15 (2) (2016) 79–87, doi:10.1080/20464177.2016.1216734.

CHAPTER THREE

Characterizing and modeling of acoustical performance of natural sound absorber

Azma Putra[a], Thaynan Oliveira[b], Mhamed Souli[b], Dg Hafizah Kassim[a]
[a] Centre for Advanced Research on Energy, Universiti Teknikal Malaysia Melaka, Durian Tunggal, Melaka, Malaysia
[b] Unité de Mécanique de Lille, Université de Lille, Villeneuve-d'Ascq, Lille, France

3.1 Introduction

Consider a sound wave impinging upon a surface of a material. Some of the incident sound energy is reflected back, some energy is transmitted and the rest is absorbed by the material by energy conversion as illustrated in Fig. 3.1.

The role of an acoustic material or sound absorber is to maximize the absorption mechanism. For this purpose, they are usually constructed with porous and fibrous structures to trap the sound wave through the interconnected built-in pores.

Mechanism of sound absorptions in a porous or fibrous material can exist through three mechanisms, namely [1]

1. Viscous losses due to oscillatory flow in the pores.
2. Heat conduction through the pore structure.
3. Vibration of the solid skeleton of the porous materials.

Sound energy can also be absorbed through vibration of a solid impermeable panel or flexible panel, or in a structure with a narrow passage directly connected to a larger volume of space (Helmholtz resonator or microperforated panel).

Mineral and synthetic fibrous porous material, like glass wool, rock wool, and foam are typical commercial sound absorber materials which are still dominating the market until today. The applications range from acoustic materials in buildings to trim materials in vehicles.

The productions of these materials however, provide negative effect to environment according to the life cycle assessment in terms of the global warming potential as presented in [2], where the global warming potential

Multiphysics Simulations in Automotive and Aerospace Applications.
DOI: 10.1016/B978-0-12-817899-7.00010-1

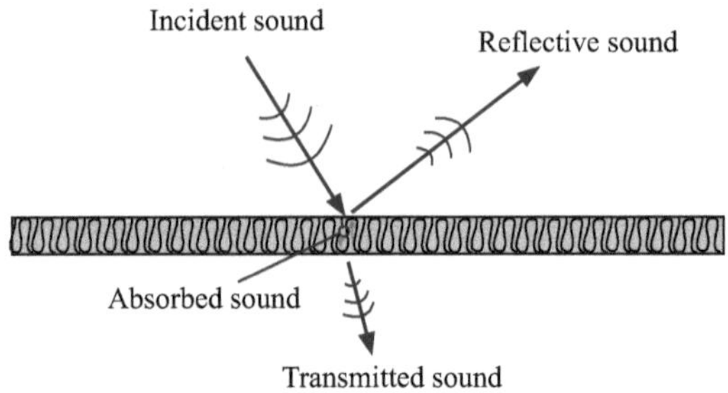

Figure 3.1 Illustration of sound impinging on a material.

represents the equivalent of kilogram CO_2 released to the air from the production to the installation of an acoustic material product [3]. This excludes the disposals of the mineral and synthetic materials which must also be regulated to minimize the environmental pollution.

The fibers from natural materials therefore become the strong candidate as the raw materials for the sound absorbers. The sources of the natural fibers are mainly from vegetables and animals. In particular, the vegetable fibers can be obtained from various parts of plants and can be categorized into: seed fiber (e.g., cotton and kapok); bast fiber or skin fiber (e.g., kenaf, flax, jute, hemp, ramie); leaf (or hard) fiber (e.g., palm, sisal, cocos, bromelia); wood fiber (e.g., straw of wheat and rice); and fruit fiber (e.g., coconut, oil palm) [4].

The natural fibers have been shown to possess good thermal conductivity, specific heat and density for applications in building insulation materials [5]. The natural fibers also have good structural properties as the composite materials in automotive application [6, 7], including for noise control materials in the automotive interior [8].

The most recent studies on natural fibers as sound absorbers include the oil-palm fruit bunch fibers [9], broom fibers [10], kenaf fibers [11, 12], pineapple leaf fibers [13], date palm fibers [14], and hemp fibers [15].

Several works of composite natural fibers for improving sound absorption have also been published. The most recent ones are the rice-husk grain and coir fibers [16], sisal and kenaf fibers [17], denim shoddy and jute fibers [18].

3.2 Kenaf fibers

Kenaf plant or *Hibiscus cannabinus* is closely related to cotton and okra as it belongs to the same Malvaceae family. Kenaf is entitled as the next industrial crop due to its various use as paper products, textiles, composites and building materials [31]. Its growth rate from seed to the height of

3.6–4.3 m can be within 5–6 months duration. The plant stalk has 40% usable parts and the bast obtained is processed in the same method as the coir fiber. India and China provide more than 75% of the worldwide total production, and in Malaysia the plantation is expanding as the replacement of tobacco [32, 33].

The kenaf fiber has diameter of 60–100 μm. Kenaf fiber also consists of cellulose (49–53.8%), hemicellulose (21.5–23%), lignin (15–19%), and ash (2–5%) [34]. Usually an alkaline treatment is used to enhance the surface of fiber to achieve a better adhesion with polymeric matrix. Even though the kenaf bast fiber has lumen inside, its unidirectional composites are able to reach an extreme fiber strength [35]. In this chapter, measurement and modeling the sound absorption of kenaf fibers are presented.

3.3 Measurement of sound absorption coefficient

Measuring the sound absorption coefficient of an acoustic material can be divided into two main methods, namely, the normal incidence method and the diffuse field method. In the first method, the material is excited by sound energy at angle normal to the surface of the material. For this purpose, the wavelength of the sound incident must be much greater to the diameter of the test material. Thus only relatively small size of material is required (usually circular shape with less than 100 mm in diameter) and sound energy is contained inside a circular tube, known as impedance tube. Meanwhile in the diffuse field method, the sound excites the material from any possible angles of incidence and thus closely resembles the situation in practice. However, a special, large room is required to enable the sound energy to be diffuse in the room to the lowest frequency possible, known as reverberation chamber. According to BS EN ISO-354 [19], a chamber with volume of at least 150 m^3 and test material with surface area of 10 m^2 are required to have an accurate result above 100 Hz. In this chapter, only the normal incidence method (or impedance tube method) is discussed and the experimental results are compared with the theoretical model.

3.3.1 Impedance tube method

Fig. 3.2 shows the measurement setup for the normal incidence absorption coefficient using the impedance tube. This setup using two acoustic microphones is based on BS EN ISO-10534 [20].

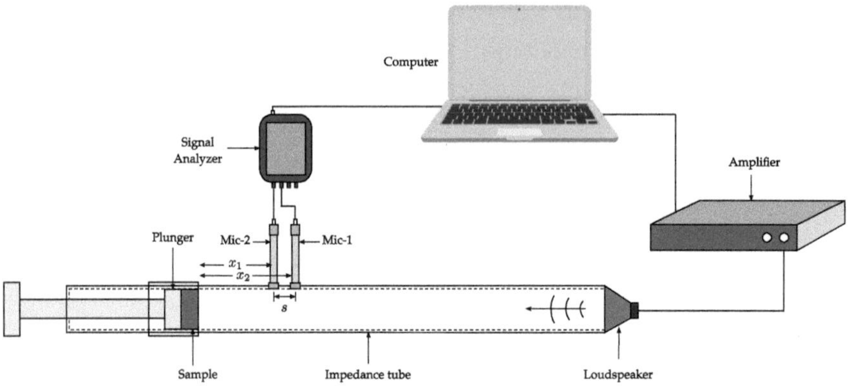

Figure 3.2 Measurement setup for sound absorption test using impedance tube method.

Random signal is generated by a loudspeaker and the two microphones record the incident and reflected waves. The incident and the reflected transfer functions are given by

$$H_I = e^{-\mathrm{j}\,k_0 s} \tag{3.1}$$

$$H_R = e^{\mathrm{j}\,k_0 s} \tag{3.2}$$

where $k_0 = \omega/c_0$ is the wavenumber, with ω the angular frequency, c_0 the speed of sound and $s = x_2 - x_1$ is the distance between the microphones and $\mathrm{j} = \sqrt{-1}$. The complex acoustic transfer functions is determined by

$$H_{12} = \frac{e^{\mathrm{j}\,k_0 x_2} + re^{-\mathrm{j}\,k_0 x_2}}{e^{\mathrm{j}\,k_0 x_1} + re^{-\mathrm{j}\,k_0 x_1}} \tag{3.3}$$

and the normal incidence reflection factor is

$$r = \frac{H_{12} - H_I}{H_R - H_{12}} e^{2\mathrm{j}\,k_0 x_1} \tag{3.4}$$

The specific acoustic impedance can be calculated as

$$Z_n = \rho_0 c_0 \left(\frac{1 + r}{1 - r} \right) \tag{3.5}$$

and the normal incidence sound absorption coefficient is

$$\alpha = 1 - |r|^2 = \frac{4\mathrm{Re}(z_n)}{[1 + \mathrm{Re}(z_n)]^2 + [\mathrm{Im}(z_n)]^2} \tag{3.6}$$

where $z_n = Z_n/\rho_0 c_0$.

Figure 3.3 Loose fibers of kenaf.

Table 3.1 Mass of fibers and thickness of test samples.

No.	Mass (g)	Thickness (mm)
1	1.5	10
2	3	20
3	4.6	30
4	6.1	40

3.3.2 Preparation of test samples

Fig. 3.3 shows the loose fiber from kenaf plant as the raw material to fabricate the test samples. The fibers were first weighted before they are inserted to a cylindrical shape mold. For the experiment, four samples with different thicknesses were prepared with the same bulk density of 178 kg/m^3. The sample must have a diameter of 33 mm to be fitted in the impedance tube. Table 3.1 lists the mass and thickness of the test samples for the measurement.

The fibers were then hot pressed approximately at 100°C using motorized hydraulic molding press machine as shown in Fig. 3.4 and Fig. 3.5 shows example of a test sample.

3.4 Measurement results and prediction

3.4.1 Acoustical parameters

The identification of macroscopic parameters of porous materials from measurements in an impedance tube is a widely used method concerning

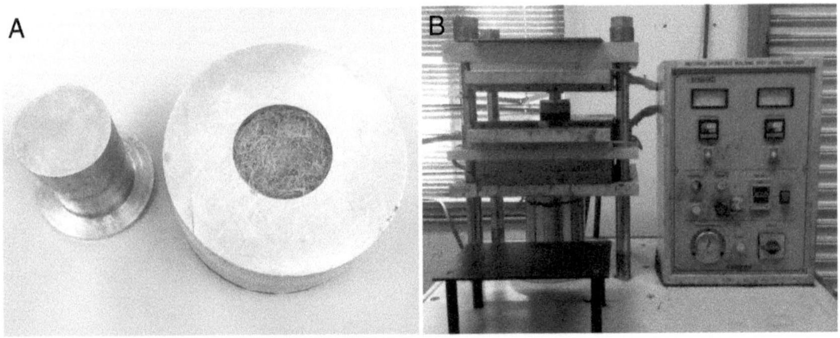

Figure 3.4 (A) Mold used to fabricate coir and kenaf samples and (B) motorized hydraulic test press machine.

Figure 3.5 Example of test sample with thickness of 20 mm.

the characterization [21, 22]. In this scenario, the method based on the adjustment of curves using a propagation model, usually for rigid or limp structure, stands out. The optimization method uses the measured surface impedance or absorption coefficient from the test sample as parameters to define the objective function.

The accuracy of the method will depend on the chosen model and the cost of the function used in the inverse algorithm. The model needs to represent the measured acoustic properties, that is, it should be sufficiently sensitive to the variations of the parameters. The most common

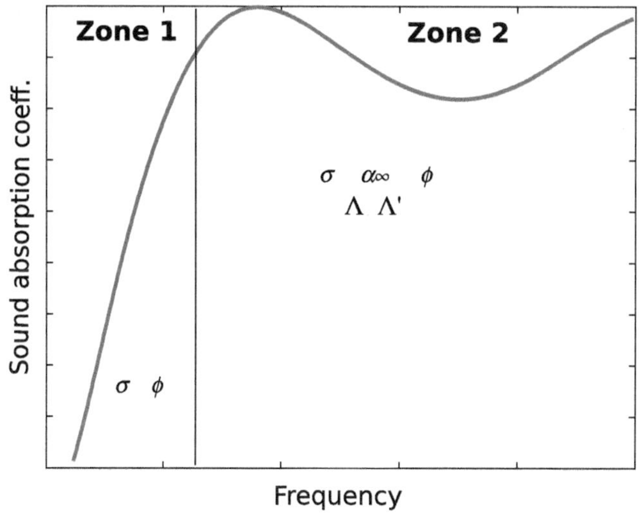

Figure 3.6 Importance of macroscopic parameters in the sound absorption coefficient.

phenomenological model considers the solid structure of the fiber as the rigid frame and incorporates five macroscopic fluid parameters, namely:

a. Flow resistivity, σ
b. Porosity, ϕ
c. Tortuosity, α_∞
d. Viscous characteristic length, Λ
e. Thermal characteristic length, Λ'

A sensitivity analysis of the absorption coefficient as a function of these five main macroscopic parameters is presented in Fig. 3.6.

Fig. 3.6 is an adaptation of a graphic based on simulation of various porous materials adopting a rigid structure model [23]. This demonstrates that the low frequency is sensitive to the flow resistivity σ and porosity ϕ, while the tortuosity α_∞, viscous length Λ, and thermal length Λ' affect the level of absorption coefficient at medium and high frequencies. The presentation of this trend is important for evaluation if the impedance tube used in the characterization has a proper cutoff frequency for computing the required variables. If the cutoff frequency is low, it is evident that the measured data can be insufficient in the spectral range in which these parameters have significant effect. According to Mareze [21], the determination of the characteristic lengths is quite complex and may be inaccurate using the genetic algorithm, but such a method becomes an important tool when direct measurements are not viable.

The objective function of the genetic algorithm used in this work is based on the comparison between the measured absorption coefficient $\alpha^{(\text{exp})}$ and the theoretical absorption coefficient $\alpha^{(\text{teo})}$ computed from the Johnson–Champoux–Allard (JCA) model [24, 25] using a rigid frame equivalent fluid approach. The objective function in this case is given by

$$\beta = \sum_{n=1}^{N} \left| \frac{\alpha_n^{(\text{exp})} - \alpha_n^{(\text{teo})}}{\alpha_n^{(\text{exp})}} \right|^2 \tag{3.7}$$

with n being the sample and N is the total number of samples in the measurement.

This technique was also applied by Pereira et al. [26] for porous concrete material characterization, where the concrete was modeled by using the Horoshenkov–Swift model [27].

The optimization code used in this work is based on the *ga* function in MATLAB/OCTAVE, that finds the minimum error to the objective function β in Eq. (3.7) based on genetic algorithms. Although direct measurement of each parameter is beneficial for a more robust characterization of the samples, the information of optimized parameters can be sufficient to represent the material in ordinary commercial codes of energy-based methods, like statistical energy analysis and transfer matrix method.

3.4.2 Governing mathematical model

The JCA model has been demonstrated to be a more sophisticated model compared to the Delany–Bazley model [28] as it incorporates the five acoustical parameters which determines the sound propagation within the porous material. Johnson [24] presents the effective density parameter ρ_{eff}, given by

$$\rho_{\text{eff}} = \rho_0 \left[\alpha_\infty + \frac{\nu\phi}{\mathrm{j}\,\omega q_0} \left(1 + \frac{4\,\alpha_\infty^2 q_0^2}{\Lambda^2 \phi^2 \nu} \mathrm{j}\,\omega \right)^{1/2} \right]. \tag{3.8}$$

where ρ_0 is the air density, $\nu = \eta/\rho_0$ is the kinematic viscosity of the medium, $\eta = 1.84 \times 10^{-5}$ [Pa s] is the static viscosity of the air, and the viscous permeability of the fiber q_0 may be approximated by $q_0 = \eta/\sigma$.

From Champoux and Allard [22,25], the effective bulk modulus K_{eff} can be computed as

$$K_{\text{eff}} = \gamma P_0 \left[\gamma - \frac{\gamma - 1}{1 + \dfrac{\nu' \phi}{\mathrm{j}\,\omega q_0'} \left(1 + \dfrac{\mathrm{j}\,\omega (\Lambda')^2}{16 \nu'} \right)^{1/2}} \right]^{-1} \tag{3.9}$$

where $\nu' = \nu/\mathrm{Pr}$ is the thermal viscosity, $\mathrm{Pr} = 0.77$ is the Prandtl number, q_0' is the thermal permeability which is given by $q_0' = \phi(\Lambda')^2/8$ under the hypothesis that the pores can be approximated by circular cross-shaped tubes.

From the equivalent fluid theory [29], it is possible to calculate the characteristic impedance $\tilde{Z}_c$ and the complex wave-number $\tilde{k}_c$ of the porous material normalized by the porosity [22], then

$$\tilde{Z}_c = \frac{\sqrt{\rho_{\text{eff}} K_{\text{eff}}}}{\phi} \tag{3.10}$$

and

$$\tilde{k}_c = \frac{\omega \sqrt{\rho_{\text{eff}} / K_{\text{eff}}}}{\phi} \tag{3.11}$$

By assuming the porous sample with thickness h over a rigid wall, the normalized surface impedance $\tilde{Z}_s$ of the system exited by a normal incidence plane wave, which is the assumption of the experimental setup in an impedance tube, is given by

$$\tilde{Z}_s = -\mathrm{j}\, \tilde{Z}_c \coth(\tilde{k}_c h) \tag{3.12}$$

From this surface impedance, the sound absorption coefficient is then calculated using Eq. (3.6) which gives

$$\alpha = \frac{4\mathrm{Re}(\tilde{z}_s)}{[1 + \mathrm{Re}(\tilde{z}_s)]^2 + [\mathrm{Im}(\tilde{z}_s)]^2} \tag{3.13}$$

where $\tilde{z}_s = \tilde{Z}_s / \rho_0 c_0$.

3.4.3 Sound absorption analysis

This section presents the optimized results from the model and their comparison to the experimental data. A set of measurements for different thicknesses of the kenaf fibers are evaluated. It is important to consider that the rigid frame equivalent fluid model does not consider the frame

Table 3.2 Algorithm input parameters.

	σ (kN/m^4)	ϕ (-)	α_∞ (-)	Λ (μm)	Λ' (μm)
Starting shot	20	0.75	1.02	100	150
Upper limit	50	0.80	1.06	180	400
Lower limit	10	0.65	1.00	60	100

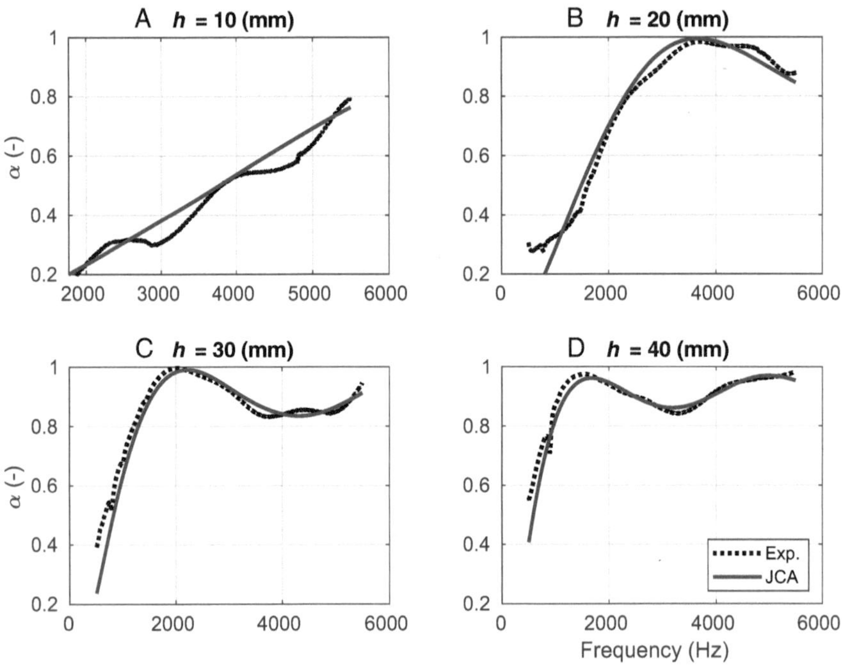

Figure 3.7 Comparison of optimized and experimental results for the kenaf fiber with different thicknesses (density 178 kg/m^3).

properties, which means that different acoustic parameters should be given for each structural configuration.

The initial parameter entries are informed as well as the intervals that limit the range of optimization of these variables. To choose the initial parameters, typical values of the macroscopic properties for mineral fibers are considered [30]. The input parameters used in this work are presented in Table 3.2.

In addition to the parameters reported in Table 3.2, it is evident that the sample thickness and the sound absorption coefficient of the sample

Table 3.3 Numerical macroscopic parameters for the kenaf fiber.

	σ (kN/m^4)	ϕ (-)	α_∞ (-)	Λ (μm)	Λ' (μm)
Kenaf fiber 178 (kg/m^3)	25.543	0.701	1.1	84.5	102.3

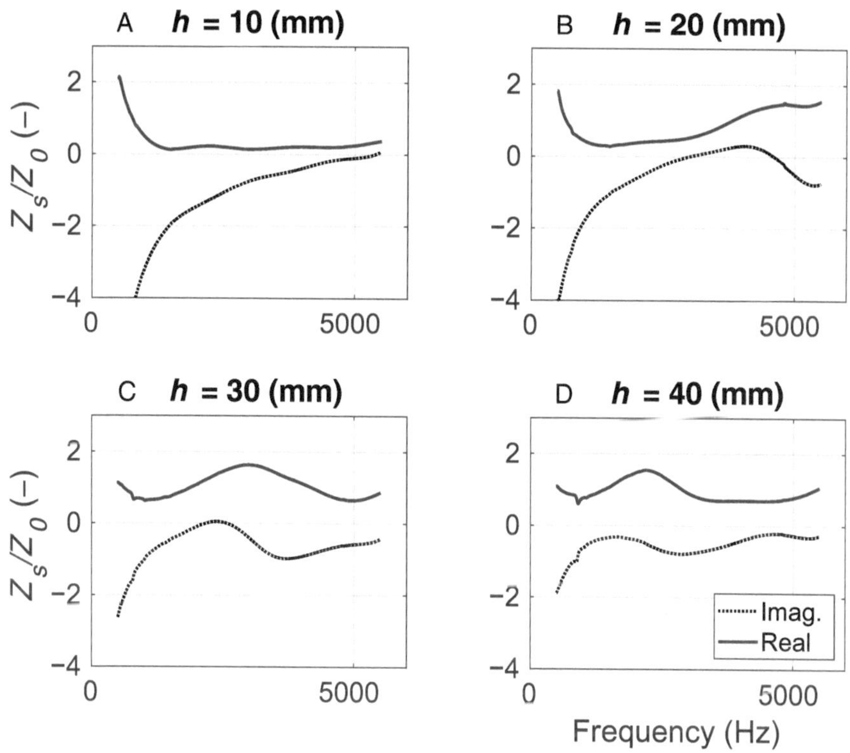

Figure 3.8 Experimental surface impedance normalized by $Z_0 = \rho_0 c_0$ for the kenaf fiber (density 178 kg/m^3).

obtained experimentally are necessary for the optimization of the analytical curve. The analysis was performed up to 5.5 kHz.

In this study, the samples were characterized taking into account the proposed model. Each simulation has an objective error value and a time limit for solution. Table 3.3 presents the mean obtained values from five runs of this optimization process, considering the kenaf samples.

The experimental results are compared with the numerical sound absorption coefficient in Fig. 3.7.

It can be seen in Fig. 3.7 that the optimized parameters shown in Table 3.3 represent well the sound absorption coefficient of the fiber for

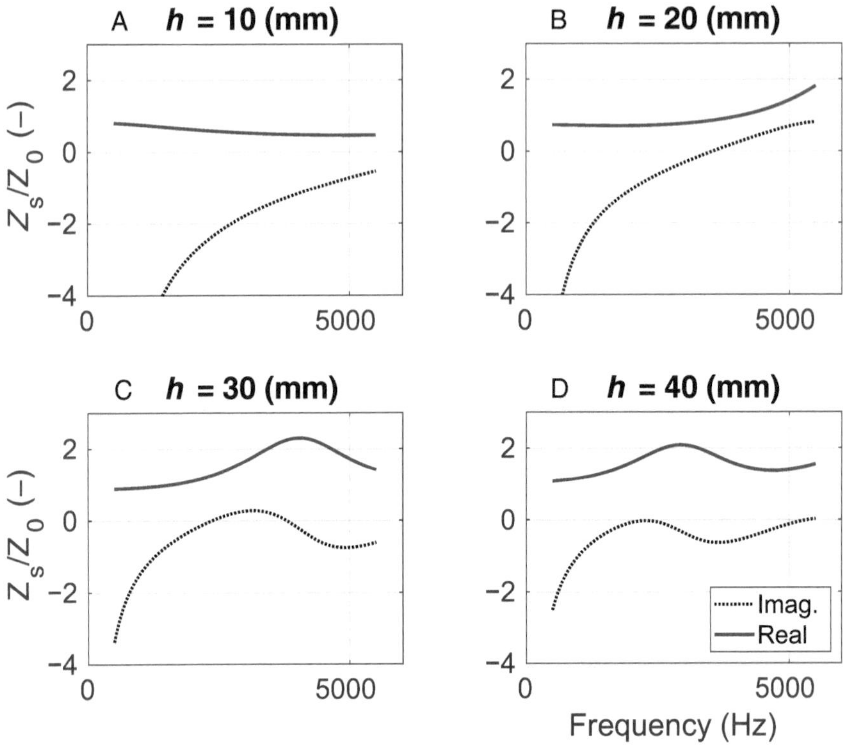

Figure 3.9 Numerical surface impedance normalized by $Z_0 = \rho_0 c_0$ for the kenaf fiber (density 178 kg/m^3).

the different values of thickness by using the JCA model. Notice that the lowest thickness of 10 mm in the experimental result in Fig. 3.7(A) has some antiresonances shown by the fluctuation of the absorption coefficient. It may be related to the frame vibration phenomena that cannot be modeled by rigid frame models, as well as the difficulty of experimental mounting conditions for very low thickness fibers in the impedance tube.

The normalized surface impedance $\tilde{Z}_s$ (as in Eq (3.12)) of the sample was also evaluated experimentally and numerically. Fig. 3.8 shows the real and imaginary part of the normalized impedance measured for four different thicknesses. Fig. 3.9 shows the results obtained from the inverse characterization. The same trend of the impedance can be observed in most parts of the spectrum, although discrepancy between experimental (Fig. 3.8) and theoretical (Fig. 3.9) results can be noticed in the real part of the impedance at very low frequencies.

It is important to clarify that this optimization process considers the porous materials as equivalent fluids. It means that the simulated acoustic properties should change for different solid frame properties, for example for different fiber density of the sample.

3.5 Summary

The proposed method offers a fast and practical solution for the characterization of fibrous acoustic materials in terms of its macroscopic properties. The model can be applied to similar type of fibers (such as coir fiber, palm oil fiber, jute fibers) and absorber samples with different densities. From the experimental test in the laboratory of a small test sample, the acoustical properties obtained can be used as the prediction data to simulate the sound absorption inside an enclosure (e.g., car cabin) or sound insulation performance of a partition with multilayer acoustic materials. Some commercial packages which include this acoustical characterization method in statistical energy analysis and transfer matrix method are available. To improve the accuracy of the prediction, the method can also be used together with measured porosity and air flow resistivity of the test sample and with the remaining three parameters (tortuosity, viscous and thermal characteristic lengths) to be obtained from the numerical simulation.

References

1. F. Fahy, D. Thompson, Fundamentals of Sound and Vibration, CRC Press, London, 2016.
2. F. Asdrubali, Survey on the acoustical properties of new sustainable materials for noise control, Acta Acustica United With Acustica 92 (2006) 89–98.
3. I.Z. Bribián, A.V. Capilla, A.A. Usón, Life cycle assessment of building materials: Comparative analysis of energy and environmental impacts and evaluation of the eco-efficiency improvement potential, Build. Environ. 46 (5) (2011) 1133–1140.
4. U. Berardi, G. Iannace, Acoustic characterization of natural fibers for sound absorption applications, Build. Environ. 94 (2015) 840–852.
5. F. Asdrubali, F. D'Alessandro, S. Schiavoni, A review of unconventional sustainable building insulation materials, Sustain. Mater. Technol. 4 (2015) 1–17.
6. F. Ahmad, H.S. Choi, M.K. Park, A review: natural fiber composites selection in view of mechanical, light weight, and economic properties, Macromol. Mater. Eng. 300 (1) (2015) 10–24.
7. R. Dunne, D. Desai, R. Sadiku, J. Jayaramudu, A review of natural fibers, their sustainability and automotive applications, J. Reinforced Plast. Compos. 35 (13) (2016) 1041–1050.
8. G. Thilagavathi, E. Pradeep, T. Kannaian, L. Sasikala, Development of natural fiber nonwovens for application as car interiors for noise control, J. Ind. Textiles 39 (3) (2010) 267–278.
9. K.H. Or, A. Putra, M.Z. Selamat, Oil palm empty fruit bunch fibres as sustainable acoustic absorber, Appl. Acoust. 119 (2017) 9–16.

10. U. Berardi, G. Iannace, M. Di Gabriele, The acoustic characterization of broom fibers, J. Nat. Fibers 14 (6) (2017) 858–863.
11. Z. Lim, A. Putra, M. Nor, M. Yaakob, Sound absorption performance of natural kenaf fibres, Appl. Acoust. 130 (2018) 107–114.
12. E. Taban, P. Soltani, U. Berardi, A. Putra, S.M. Mousavi, M. Faridan, S.E. Samaei, A. Khavanin, Measurement, modeling, and optimization of sound absorption performance of kenaf fibers for building applications, Build. Environ. 180 (2020) 107087.
13. A. Putra, K.H. Or, M.Z. Selamat, M.J.M. Nor, M.H. Hassan, I. Prasetiyo, Sound absorption of extracted pineapple-leaf fibres, Appl. Acoust. 136 (2018) 9–15.
14. E. Taban, A. Khavanin, A. Ohadi, A. Putra, A.J. Jafari, M. Faridan, A. Soleimanian, Study on the acoustic characteristics of natural date palm fibres: experimental and theoretical approaches, Build. Environ. 161 (2019) 106274.
15. A. Santoni, P. Bonfiglio, P. Fausti, C. Marescotti, V. Mazzanti, F. Mollica, F. Pompoli, Improving the sound absorption performance of sustainable thermal insulation materials: natural hemp fibres, Appl. Acoust. 150 (2019) 279–289.
16. H. Mamtaz, M.H. Fouladi, M.Z. Nuawi, S.N. Namasivayam, M. Ghassem, M. Al-Atabi, Acoustic absorption of fibro-granular composite with cylindrical grains, Appl. Acoust. 126 (2017) 58–67.
17. R. Dunne, D. Desai, R. Sadiku, Material characterization of blended sisal-kenaf composites with an ABS matrix, Appl. Acoust. 125 (2017) 184–193.
18. M. Raj, S. Fatima, N. Tandon, Recycled materials as a potential replacement to synthetic sound absorbers: a study on denim shoddy and waste jute fibers, Appl. Acoust. 159 (2020) 107070.
19. ISO-354:2003, Acoustics – Measurement of Sound Absorption in a Reverberation Room, Standard, International Organization for Standardization, 2003.
20. ISO-10534-2:1998(en), Acoustics – Determination of Sound Absorption Coefficient and Impedance in Impedance Tubes – Part 2: Transfer-Function Method, Standard, International Organization for Standardization, 2003.
21. P.H. Mareze, Análise Da Influência Da Microgeometria Na Absorção Sonora De Materiais Porosos De Estrutura Rígida (*in portuguese*), Ph.D. thesis, Universidade Federal de Santa Catarina, 2013.
22. J. Allard, N. Atalla, Propagation of Sound in Porous Media: Modelling Sound Absorbing Materials, second ed., Wiley, United Kingdon, 2009.
23. N. Atalla, Structural and acoustic noise control material modeling, in: S.A. Hambric, S.H. Sung, D.J. Nefske (Eds.), Engineering Vibroacoustic Analysis: Methods and Applications, Wiley, United Kingdom, 2017, pp. 230–265.
24. D.L. Johnson, J. Koplik, R. Dashen, Theory of dynamic permeability and tortuosity in fluid-saturated porous media, J. Fluid Mech. 176 (1987) 379–402.
25. J.-F. Allard, Y. Champoux, New empirical equations for sound propagation in rigid frame fibrous materials, J. Acoust. Soc. Am. 91 (6) (1992) 3346–3353.
26. M. Pereira, J. Carbajo San Martín, L. Godinho, P. Amado-Mendes, D. Mateus, J. Ramis-Soriano, Acoustic behavior of porous concrete. Characterization by experimental and inversion methods, Mater. Construc. 69 (336) (2019).
27. K.V. Horoshenkov, M. Swift, The acoustic properties of granular materials with pore size distribution close to log-normal, J. Acoust. Soc. Am. 110 (5) (2001) 2371–2378.
28. M. Delany, E. Bazley, Acoustical properties of fibrous absorbent materials, Appl. Acoust. 3 (2) (1970) 105–116.
29. C. Zwikker, C.W. Kosten, Sound Absorbing Materials, Elsevier, New York, 1949.
30. T.J. Cox, P. D'Antonio, Acoustic Absorbers and Diffusers: Theory, Design and Application, second, Taylor & Francis e-Library, New York, USA, 2009.
31. N. Saba, M.T. Paridah, M. Jawaid, Mechanical properties of kenaf fibre reinforced polymer composite: A Review, Construction and Building Materials 76 (2015) 87–96.

32. M.H. Hamidon, M.T.H. Sultan, A.H. Ariffin, A.U.M. Shah, Effects of fibre treatment on mechanical properties of kenaf fibre reinforced composites: A Review, Journal of Materials Research and Technology 8 (2019) 3327–3337.
33. H.P.S.A. Khalil, A.F.I. Yusra, A.H. Bhat, M. Jawaid, Cell wall ultrastructure, anatomy, lignin distribution, and chemical composition of Malaysian cultivated kenaf fiber, Industrial Crops and Products 1 (2010) 113–121.
34. O.E. Babatunde, J.M. Yatim, M.Y. Ishak, R. Masoud, R. Meisam, Potential of kenaf fiber in bio-composite production: A Review, Jurnal Teknologi 77 (12) (2015) 23–30.
35. H.T. Sreenivas, N. Krishnamurthy, G.R. Arpitha, A comprehensive review on light weight kenaf fiber for automobiles, International Journal of Lightweight Materials and Manufacture 3 (4) (2020) 328–337.

CHAPTER FOUR

Investigation of torsional vibration using weak FSI capabilities

Daniel Brunner[a,b], Gernot Boiger[a], Hassan Khawaja[b,c]
[a] Institute of Computational Physics, Zurich University of Applied Sciences, Winterthur, Switzerland
[b] UiT – The Arctic University of Norway, Tromsø, Norway
[c] Al Ghurair University, Dubai, United Arab Emirates

4.1 Introduction

Monitoring the physical properties of liquids, such as viscosity and density, is of great interest in many fields. In the automotive industry, monitoring the viscosity of lubricating oils is of great importance [21] because it can be related to its state of degradation [22]. Traditional rheological laboratory instruments are large, delicate, and expensive. Microelectromechanical system-based vibration sensors offer a more practical, cheaper solution to measure the oil viscosity within an automotive engine. There are different approached on how use these sensors to measure the lubricant's viscosity. Buhrdorf et al. presented a tuning fork-based sensor [6] that can be modularly integrated into an oil level sensor. This sensor can provide data for viscosity, density, permittivity, and temperature. Jakoby et al. [15] presented a sensor to measure the viscosity of engine oil via microsampling. Similarly, Ash et al. [1] presented a sensor to measure the viscosity, which only requires on droplet of the oil.

All these microelectromechanical sensors are mechanical resonators. The working principle of a resonant sensor is that a resonating element is brought into contact with a fluid. The fluid changes the resonant element's resonance frequency and damps its oscillation. Both the shift in the resonance frequency and damping can be measured and related to the fluid's properties. This chapter focuses on the modeling of torsional resonators using weak fluid–structure interaction (FSI) models. Traditionally, torsional resonators are purely cylindrical; therefore, they do not create mass displacements that are required to measure density. Cylindrical resonant elements emit only shear waves, which predominantly create viscous damping. This damping affects both the resonance frequency and bandwidth of the resonator. Brack

Multiphysics Simulations in Automotive and Aerospace Applications.
DOI: 10.1016/B978-0-12-817899-7.00007-1

et al. [2,3,23] and Valtorta and Mazza [20] analyzed both quantities, which enabled the measurement of effects such as viscoelasticity. It is also possible to measure the fluid density using torsional resonators if the oscillation has a normal velocity component on its surface. Torsional resonators are attractive because they are less prone to wall effects than transversely vibrating devices. Examples of torsional resonators include the symmetrical torsional resonator presented by Goodbread et al., [10] and the resonating paddle devised by Heinisch et al. [13].

To develop, and gain insights into the working principle of these sensors, analytical and numerical models are required. In most cases, the sensors can be modeled based on a simplified, single-degree-of-freedom analytical model, such as a mass–spring system. This approach was generically described by Heinisch et al. [12], who applied a reduced-order model based on lumped elements to describe a resonator's behavior. The behavior of various resonant elements, such as cylindrical torsional resonators [13], tuning forks [11], torsional paddles [18], cantilevers [19], and microchannels [17], can be modeled using similar mass–spring models, as will be discussed in the following chapter. In these models, the fluid interactions are typically described using separate damping and mass-loading terms. The mass-loading term increases the mass of the resonator, whereas the damping term damps the oscillation. The magnitude of the damping and mass-loading terms can be derived from the interaction of the resonant element with the fluid.

The flow field surrounding the resonant element is typically described by analytical approximations since in most cases, computational methods are feasible. To date, most analytical work on fluid models for resonant elements has been done on either flat in-plane oscillating plates or cylindrical geometries. In-plane oscillating planes are most relevant for thickness–shear resonators. Sensors of this type comprise a quartz disk that is excited by means of the piezoelectric effect; the quartz disk emits only shear waves similar to cylindrical torsional resonators. Johannsmann [16] reviewed the fundamentals of quartz disks, which included several fluid models for in-plane oscillating plates (e.g., Newtonian, viscoelastic, and non-Newtonian fluids); the interactions of a shear wave with a thin film and the deposits and fluids stratified in layers were also considered. For nonflat geometries, such as torsional resonators, the mathematical models reviewed by Johannsmann [16] can be used if the radius of the cylinder is much larger than the length scale of the fluid (i.e., the boundary-layer thickness). In case the boundary thickness increases, the curvature of the cylinder or tube must be considered [3,7–9,23].

For more complex geometries, analytical models become insufficient to describe either the viscous damping or the mass loading caused by the fluid. In such cases, the FSIs can be solved numerically with certain simplification. A fully, strongly coupled FSI simulation is in most cases not feasible. Using weak FSI models allows to decouple the structural domain from the fluid domain. This allows for solving both parts, fluid and structural independently at a much lower computational cost. A solution of the combined system can be obtained by coupling them with simplified single- or multiple-degrees-of-freedom models. This combined model approach allows to predict the resonator properties.

4.2 Methodology

The basic working principle of any resonance sensor is that the resonator's properties, namely, its resonance frequency and bandwidth, change as the resonator interacts with its surrounding fluid. The change in bandwidth and resonance frequency is then related to the fluid's viscosity and density. This relation can be described through models, which are discussed in the following sections.

Section 4.2.1 discusses the structural models/coupling models, which describe the structural part of the resonator in a simplified manner. First, we discuss the resonant element as a single-degree-of-freedom model (mass–spring system), then we discuss resonant elements with multiple degrees of freedom and at last, we consider continuous geometries, where the inertial mass is distributed and not allocated in discrete point.

In the second part, we discuss the interaction of the fluid with the resonator using weak FSI. At first, we derive a concept to incorporate the fluid forces (mass and damping terms) into the resonator model, which were discussed before. These mass and damping terms are computed based on the flow field around the resonator. Thus, the main part of this section is how to compute the flow field around a vibrating structure. First, the flow field is solved analytically for both a flat and cylindrical surface, then numerical for arbitrary geometries.

4.2.1 Analytical structural models

In most cases, the resonator can be considered as a linear continuous system like to a tube or a beam. In most applications, the sensors are designed to operate in one mode. To describe the behavior of this mode with respect to its resonance frequency and bandwidth, a single-degree-of-freedom model can be used. For structures in which either multiple modes are relevant, or the modes are changed by fluid interactions, we need models with multiple degrees of freedom. Continuous systems with infinite degrees of freedom are discussed last, where each mode (i.e., degree of freedom) can be described by a single-degree-of-freedom system.

4.2.1.1 Single-degree-of-freedom system

A single-degree-of-freedom system is a damped mass–spring system that has only one resonance frequency. This single-degree-of-freedom system describes most resonators in which only one resonance frequency is considered:

$$J\frac{\partial^2\varphi}{\partial t^2} + d\frac{\partial\varphi}{\partial t} + c\varphi = F(t) \tag{4.1}$$

where ϕ is the angular deflection in time t, J is an inertial mass, c is the spring constant, d is the damping term, and $F(t)$ is the transient excitation force.

Note that the damping term d and inertial mass J depend on the fluid properties and the resonator. These parameters are discussed at length in Section 4.2.2.1.

In more general terms, the resonator can be described as follows:

$$J\frac{\partial^2\varphi}{\partial t^2} + 2D\omega_0\frac{\partial\varphi}{\partial t} + \omega_0^2\varphi = F_N(t) \tag{4.2}$$

where D is the normalized damping term, ω_0 is the angular resonance frequency, $F_N(t)$ is the normalized forcing term, and f_0 is the resonance frequency, Eqs. (4.3)–(4.5).

$$D = \frac{d}{2\sqrt{cJ}} \tag{4.3}$$

$$\omega_0 = 2\pi f_0 = \sqrt{\frac{c}{J}} \tag{4.4}$$

$$F_N(t) = \frac{F(t)}{J} \tag{4.5}$$

The resonator's damping is typically expressed by a quality factor $Q = (2D)^{-1}$ or a bandwidth $\Gamma = f_0/Q$. The resonator system is underdamped if the normalized damping term D is less than 1. This is a necessary criterion when a resonator is used to measure the viscosity of a fluid. When the normalized damping factor $D = 1$, means critical damping. When the damping factor D is greater than one, the system does not oscillate but decays into its equilibrium state without overshooting, that is, oscillating.

If we assume that the force term $F(t) = 0$, then the fundamental solution is an exponentially decaying oscillation with a logarithmic decrement of $\tau_d = (D\omega_0)^{-1}$. Then, we have the following equality:

$$\varphi = \varphi_0 e^{i\omega_0 t} \tag{4.6}$$

where ω_0 is the angular resonance frequency, $i = \sqrt{-1}$, and ϕ_0 is the angular deflection at $t = 0$ s. The complex part of the angular resonance frequency ω_0 describes the exponential decay, that is, the logarithmic decrement $\tau_d = 1/\text{imag}(\omega_0)$.

In case the force term is a harmonic, we have

$$F(t) = \hat{F} e^{i\omega t} \tag{4.7}$$

where $\hat{F}$ is the amplitude of the force function $F(t)$ and ω is the angular frequency. Then, after the transient effects caused by the initial condition have phased out, the system eventually yields the quasi–steady-state solution $\phi_s(t)$ as follows:

$$\varphi_s(t) = \frac{\hat{F} e^{i\omega t}}{J(\omega_0^2 + 2iD\omega_0\omega - \omega^2)} \tag{4.8}$$

By convention, the transfer function X is the ratio between the quasi–steady-state angular deflection and the harmonic force term. By convention, this transfer function is formulated as follows:

$$X = \frac{\varphi_s(t)}{\hat{F} e^{i\omega t}} = \frac{1}{J(\omega_0^2 + 2iD\omega_0\omega - \omega^2)} \tag{4.9}$$

This transfer function X describes the amplification of the angular deflection and the phase shift with respect to the excitation. The amplification is defined by the Euclidean norm of X as

$$|X| = \frac{\hat{\varphi}_s}{\hat{F}} = \frac{1}{J}\sqrt{\frac{1}{\left(\omega_0^2 - \omega^2\right)^2 + (2D\omega_0\omega)^2}} \tag{4.10}$$

where $\hat{\varphi}_s$ is the amplitude of the term $\phi_s(t)$.

The phase shift $\Delta\phi_s$ between the force term $F(t)$ and the steady-state solution of the angular defection $\phi_s(t)$ is defined by the angle of X as follows:

$$\Delta\varphi_s = \text{ang}(X) = \arctan\left(\frac{2D\omega_0\omega}{\omega_0^2 - \omega^2}\right) \tag{4.11}$$

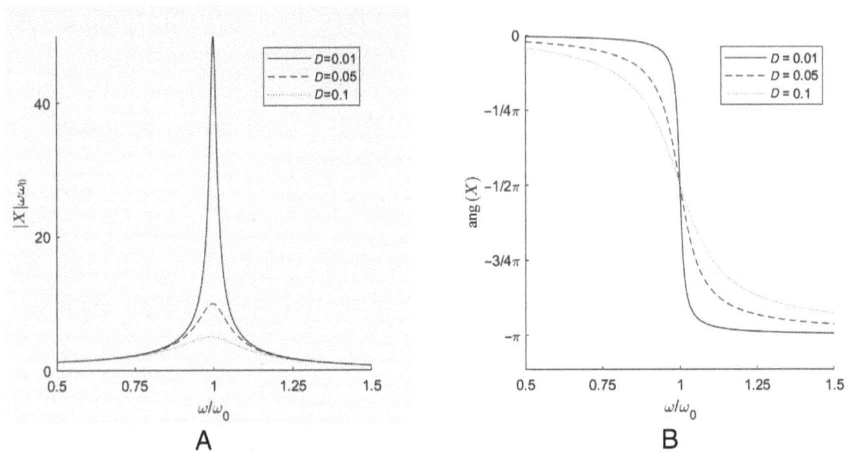

Figure 4.1 Amplitude (A) and phase (B) of a damped resonator at different normalized damping terms $D = [0.01\ 0.05\ 0.1]$, ω is the angular frequency, ω_0 is the angular resonance frequency, X is the transfer function, and ang(X) is the phase shift between the excitation term and angular deflection of the resonator.

Fig. 4.1 shows the amplification and phase shift for the different normalized damping values D. With increasing normalized damping values, the amplification decreases, and the phase shift increasingly spreads over a wide frequency range.

To experimentally characterize a single-degree-of-freedom system, the bandwidth Γ and resonance frequency f_0 needs to be determined. The bandwidth of the system can be computed using three methods; all these methods result in the same bandwidth value. The definitions of the three methods are given below.

In the first method, the bandwidth is the frequency difference between the phase shift curves when they cross $-\pi/4$ and $-3\pi/4$, as follows:

$$\Gamma = \frac{\omega\big(\mathrm{ang}(X(\omega)) = -\pi/4\big) - \omega\big(\mathrm{ang}(X(\omega)) = -3/4\pi\big)}{2\pi} \tag{4.12}$$

In the second method, the bandwidth is the frequency difference when the amplification $|X|$ is $-3dB$ of its maximum value. The $-3dB$ point is corresponding to $-3dB = 20 \cdot \log_{10}(1/\sqrt{2})$. Thus, the definition is given as follows:

$$\Gamma = \frac{\omega\left(\frac{|X(\omega)|}{\max(|X(\omega)|)} = -3dB\right)_{\omega > \omega_0} - \omega\left(\frac{|X(\omega)|}{\max(|X(\omega)|)} = -3\ dB\right)_{\omega < \omega_0}}{2\pi} \tag{4.13}$$

In the third method, when the excitation is turned off, the bandwidth is defined by the rate of exponential decay, as follows:

$$\Gamma = \frac{1}{\tau_d \pi} \tag{4.14}$$

Here, τ_d is the logarithmic decrement. The angular resonance frequency is defined by the phase shift $\Delta\phi_s$ between the forcing function and angular deflection at exactly $-\pi/2$:

$$\omega_0 = \omega\big(\text{ang}(X) = -\pi/2\big) \tag{4.15}$$

For more complex systems, a single-degree-of-freedom system will not capture all the effects. To account for either multiple frequencies or modal distortions, a model having multiple degrees of freedom is needed.

4.2.1.2 System with multiple degrees of freedom

When the resonator is too complex to be simplified to a single-degree-of-freedom model, it is possible to use models with multiple degrees of freedom. The likely scenarios include cases in which either multiple frequencies are of interest or the resonator itself is based on multiple masses and springs.

A model with multiple degrees of freedom is a linear combination of several single-degree-of-freedom systems. Such systems have been discussed in more detail by Irretier [14]. The dynamics of the system is given by the following system of differential equations:

$$J\frac{\partial^2 \vec{\varphi}}{\partial t^2} + D\frac{\partial \vec{\varphi}}{\partial t} + C\vec{\varphi} = \vec{F} \tag{4.16}$$

where the angular deflection of the torsional resonator $\vec{\varphi}$ is described by a vector and the inertial mass matrix J, the damping matrix D, the stiffness matrix C, and the force term $\vec{F}$.

The angular deflection vector $\vec{\varphi}$ has n elements, where each element represents one differential equation in the system. Note that $\vec{\varphi}$ is used here as a vector quantity, whereas ϕ in Section 4.2.1.1 was a scalar quantity.

For an angular displacement vector $\vec{\varphi}$ with n elements, matrices $\boldsymbol{J}$, $\boldsymbol{D}$, and $\boldsymbol{C}$ will be of size $n \times n$. The force term $\vec{F}$ has the same number of elements as $\vec{\varphi}$. In case the system of differential equations is homogeneous, that is, if $\vec{F} = \vec{0}$, there will be n fundamental solutions of the kind shown in the following equation:

$$\vec{\varphi}_m = \widehat{\vec{\varphi}}_m \cdot e^{i\omega_{0,m}t} \tag{4.17}$$

where $\omega_{0,m}$ is the angular resonance frequency of the angular deflection amplitude vector $\widehat{\vec{\varphi}}_m$ (mode shape), $i = \sqrt{-1}$, and t is the time.

The angular resonance frequencies and their corresponding mode shapes for the undamped system can be determined by solving the following eigenvalue problem:

$$(\boldsymbol{J}\lambda^2 + \boldsymbol{C})\vec{\varphi} = \vec{0} \tag{4.18}$$

where $\lambda = i\omega_{0,m}$ represents a solution of an eigenvalue and $\widehat{\vec{\varphi}}_m$ is its eigenvector. Therefore, each eigenvector and its corresponding eigenvalue build the fundamental solution shown in Eq. (4.17).

To solve the eigenvalue problem, the following condition must be satisfied:

$$\det(\lambda\boldsymbol{J} + \boldsymbol{C}) = 0 \tag{4.19}$$

When damping is included, the eigenvalue problem becomes more complex, Eq. (4.20).

$$(\boldsymbol{J}\lambda^2 + \boldsymbol{D}\lambda + \boldsymbol{C})\vec{\varphi} = \vec{0} \tag{4.20}$$

The eigenvalues λ of Eq. (4.20) are complex; the real part is the angular resonance frequency, and the complex part is the inverse of the logarithmic decrement. Thus, a multiple-degrees-of-freedom system, which consists of inertial masses and springs, can predict the angular resonance frequency and system bandwidth.

However, such multiple-degrees-of-freedom systems require the inertial mass is localized at points. For continuous resonators, the inertial mass is distributed, thus such resonators are described by continuous systems, see Section 4.2.1.3.

4.2.1.3 Continuous resonator

The mode structure of a tube-style resonator can be derived based on a linear mechanical system with infinite modes. The shape and resonance frequency of the modes is obtained by a modal analysis of the following differential equation:

$$\frac{\partial}{\partial x}\left(GI_p \cdot \frac{\partial \Psi}{\partial x}\right) = \rho_s I_p \frac{\partial^2}{\partial t^2}\Psi \tag{4.21}$$

where the angular deflection Ψ is a function of the position x and time t, G is the shear modulus, ρ_s is the density of the resonator, and I_p is the second moment of the area. Note that Ψ is used for continuous systems in the space domain, whereas ϕ and $\vec{\varphi}$ were used for discrete systems in Sections 4.2.1.1 and 4.2.1.2.

In case of a straight, homogeneous tube, neither the shear modulus nor the second moment of inertia depends on the axial position x; therefore, Eq. (4.21) is reduced to the following traditional wave equation:

$$GI_p \frac{\partial^2 \Psi}{\partial x^2} = \rho_s I_p \frac{\partial^2}{\partial t^2} \Psi \tag{4.22}$$

The solution to Eq. (4.22) with two nodal points as boundary conditions, Eqs. (4.23) and (4.24),

$$\Psi(x = 0) = 0 \tag{4.23}$$

$$\Psi(x = L) = 0 \tag{4.24}$$

is given as Eq. (4.25).

$$\Psi = A \cdot \sin\left(k\pi \frac{x}{L}\right) \cdot e^{i\omega_0 t} \tag{4.25}$$

Here, A is the complex wave amplitude, k is the wave number, ω_0 is the angular resonance frequency, and L is the distance between the nodal points $(x = 0)$ and $(x = L)$.

The angular frequency depends on the wave number k and is associated with its corresponding mode as follows:

$$\frac{G}{\rho_s}\left(\frac{k\pi}{L}\right)^2 = \omega_0^2 \tag{4.26}$$

To obtain the angular resonance frequency and the mode shape for a nonconstant tube radius, Eq. (4.21) is solved directly. The obtained solution can then be written as follows:

$$\Psi = \hat{\Psi} e^{i\omega_0 t} \tag{4.27}$$

where $\hat{\Psi}$ is the amplitude of the mode at the position x.

A method to obtain a solution is by solving Eq. (4.21) numerically under quasi–steady-state conditions, for example, by using MATLAB®. Fig. 4.4 shows an example of a solution of Eq. (4.21) for a tube having a thick central section with the boundary conditions given by Eqs. (4.23) and (4.24). It is possible to obtain different solutions for the mode shapes and their corresponding resonance frequency by choosing different initial conditions for the numerical solver. Fig. 4.2 shows the mode shapes at the three lowest

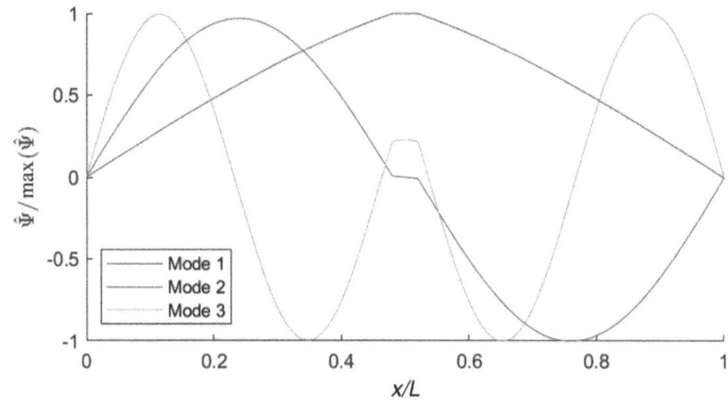

Figure 4.2 Normalized shapes of the three first modes of a torsional resonating tube with a thick central section, x is the axial position, L is the distance between the nodal points, and $\hat{\Psi}$ is the angular deflection amplitude.

angular resonance frequencies, that is, Modes 1–3.

Each mode can be described by a single-degree-of-freedom system for which the corresponding inertial mass J, the spring constant c, and the damping term d must be determined. The inertial mass J and damping term d depend on the fluid; the damping term d depends on the fluid properties and the intrinsic damping of the resonator, and J is a combination of the inertial mass of the fluid J_f and the inertial mass of the structure J_0. Modal analysis gives the values of the inertial mass of the structure J_0 and the spring constant c. The parameters d and J_f are discussed in Section 4.2.2.1.

The inertial mass of the structure J_0 for each mode is calculated by integrating the second moment of the area over the modal function:

$$J_0 = \int_{-\frac{l}{2}}^{-\frac{l}{2}} \hat{\Psi}(x) \cdot I_p(x) \rho_s \, dx \tag{4.28}$$

where the solution is normalized such that $\max(|\Psi(x)|) = 1$.

The spring constant c of the corresponding single-degree-of-freedom system is obtained as follows:

$$c = J_0 \cdot {\omega_0}^2 \tag{4.29}$$

where ω_0 is the angular resonance frequency of the corresponding mode.

In Section 4.1.1, the structural part of the resonator was discussed. To predict the resonance frequency and bandwidth based on the fluid properties, the fluid-dependent terms J and d need to be determined (see Section 4.2.2.1).

4.2.2 Analytical and numerical fluid models

To integrate the fluid interactions into the structural model with either a single degree of freedom or multiple degrees of freedom (see Sections 4.2.1.1–4.2.1.3), the damping and inertial mass caused by the fluid needs to be determined. The damping and inertial mass terms were derived based on the flow and pressure fields, see Section 4.2.1.

The Sections 4.2.2.2 up to 4.2.2.4. discuss different methods on computing the flow field around a torsionally vibrating structure. The simplest case, that is, an oscillating flat surface is discussed in Section 4.2.2.2. In case the surface is curved for a rod or a tube, this model may no longer suffice. Section 4.2.2.3 discusses an analytical model, which accounts for the curvature of, for example, a rod or a tube. If no analytical solution can be found, Computational Fluid Dynamics (CFD) provides the means to compute the flow and pressure fields numerically (see Section 4.2.2.4).

4.2.2.1 Fluid forces on a resonating body

When resonators are described structurally as mass–spring systems, the fluid's impact on the dynamics of the resonator must be accounted for in the inertial mass and damping terms (J and d, respectively). In single- or multiple-degree-of-freedom systems, solving for J and d through fluid models enables the coupling of the fluid domain with the structural domain. The spring constant c and the matrix $\mathbf{C}$ of a system with a single degree of freedom or multiple degrees of freedom, respectively, are given using resonator geometry. The total inertial mass J and the damping term d for a system with a single degree of freedom and the matrices $\boldsymbol{J}$ and $\boldsymbol{D}$ for a system with multiple degrees of freedom depend on the fluid interaction. Each position in the matrices $\boldsymbol{J}$ and $\boldsymbol{D}$ can be determined in the same manner as the inertial mass J and the damping term d for a system with a single degree of freedom. Therefore, hereinafter, we focus only on the terms J and d.

The inertial mass J is the sum of the inertial mass of the structure J_0 and the inertial mass of the fluid J_f. This total is given by the following equation:

$$J = J_0 + J_f \tag{4.30}$$

Similarly, the damping term is the sum of the intrinsic damping of the resonator d_0 and the fluid damping d_f.

$$d = d_0 + d_f \tag{4.31}$$

Note that the intrinsic damping d_0 is independent of the fluid interaction and is a superimposed damping effect that is experimentally determined see Section 4.2.3.3.

The two parameters d_f and J_f are real, positive values that depend on the fluid properties, sensor geometry, frequency of the resonator. Both parameters are derived based on the pressure and velocity fields surrounding the resonator. Therefore, the pressure-induced torque $M_{p,t}$ and the shear stress-induced torque $M_{v,t}$ acting on the fluid–structure boundary $d\Omega_r$ are computed as follows:

$$M_{v,t} = \eta \vec{e_i} \cdot \int_{d\Omega_r} (\vec{r} \times (\vec{n} \cdot \nabla \vec{u}))dA \tag{4.32}$$

$$M_{p,t} = \vec{e_i} \cdot \int_{d\Omega_r} \left(p\vec{n} \times \vec{r}\right)dA \tag{4.33}$$

where η is the dynamic viscosity of the fluid, ρ is the fluid density, $\hat{\varphi}$ is the angular deflection amplitude, ω is the angular frequency, $\vec{n}$ is the surface normal of the boundary, $\vec{r}$ is the distance from the axis, p is the pressure field, $\vec{u}$ is the velocity field, and $\vec{e_i}$ is the unity vector of the torsional axis.

Due to the harmonic oscillation of the system, the pressure and viscous amplitude are periodic if the system behaves linearly. Note that for nonperiodic solutions, strong fluid–structure coupling may have to be considered; therefore, this method may not be suitable.

The amplitudes of the pressure and shear stress-induced torque are obtained by Fourier analysis. Due to the linearity of the system, the amplitude is zero for any frequency other than the frequency of the excitation. Therefore, only the amplitude at the angular frequency ω of the excitation is determined, see Eqs. (4.34) and (4.35):

$$\hat{M}_v = \mathcal{F}\left(M_{v,t}\right) \tag{4.34}$$

$$\hat{M}_p = \mathcal{F}\left(M_{p,t}\right) \tag{4.35}$$

where $\hat{M}_v$ is the amplitude of the complex shear stress torque, $\hat{M}_p$ is the amplitude of the complex pressure torque, and $\mathcal{F}()$ is the operator of the Fourier analysis.

The inertial mass of the fluid J_f and the fluid damping term d_f are obtained from the real and imaginary parts of the sum of the amplitudes of the pressure and shear stress torque, Eqs. (4.36) and (4.37).

$$J_f = \Re\left(\frac{\hat{M}_p + \hat{M}_v}{\omega^2\hat{\varphi}}\right) \tag{4.36}$$

$$d_f = \mathcal{J}\left(\frac{\hat{M}_p + \hat{M}_v}{\omega\hat{\varphi}}\right) \tag{4.37}$$

Thus, the damping term d and inertial mass J can be computed by using Eqs. (4.30) and (4.31) based on the properties of the fluid and resonator, which allows the prediction of the resonance frequency and bandwidth through the previously described structural models, see Section 4.1.

However, to compute the terms $\hat{M}_p$ and $\hat{M}_v$, the solution of the velocity and pressure fields needs to be known. The solution method also depends on the geometry, as discussed in Sections 4.2.2.2–4.2.2.4.

4.2.2.2 In-plane oscillating plate

The in-plane oscillating plate is a fluid model that is commonly used; it is especially suitable for shear wave-emitting piezoelectric sensors, such as thickness–shear resonators or cylindrical torsional resonators. Johannsmann [16] investigated many different fluid and solid constellations of an in-plane oscillating plate, ranging from single to multiple fluid/solid layers with either Newtonian or viscoelastic properties. In this section, only Newtonian fluids in a semi-infinite domain are considered.

In this approach, the flow u_x over the surface is modeled under the assumption that the flow is unidirectional (especially in the x direction), and the flow changes only in the z direction, normal to the flow direction u_x. Therefore, we have

$$\frac{\partial u_x}{\partial t} = \frac{\eta}{\rho}\frac{\partial^2 u_x}{\partial z^2} \tag{4.38}$$

where t is the time, η is the dynamic viscosity, ρ is the fluid density, and z is the distance from the surface.

Eq. (4.38) can be solved under the assumption of a periodic solution as follows:

$$\hat{u}_x i\omega = \frac{\eta}{\rho}\frac{\partial^2 \hat{u}_x}{\partial z^2} \tag{4.39}$$

where ω is the angular velocity and $i = \sqrt{-1}$.

The boundary conditions are imposed on the surface of the plate and at an infinite distance to obtain Eqs. (4.40) and (4.41),

$$\hat{u}_x(z = 0) = \omega\hat{\Psi}\hat{\varphi}e^{i\omega t} \tag{4.40}$$

$$\hat{u}_x\left(\lim_{z\to\infty} z\right) = 0 \tag{4.41}$$

where $\hat{\varphi}$ is the amplitude of the oscillation and $\hat{\Psi}$ is the value of the modal function.

The solution to Eq. (4.39) is given as follows:

$$\hat{u}_x(z, t) = -\omega\hat{\Psi}\hat{\varphi}e^{\frac{z}{\delta(1+i)}}e^{i\omega t} \tag{4.42}$$

where δ is the thickness of the boundary layer and represents the length scale of the flow; this thickness is the distance from the wall where the velocity amplitude is e^{-1} times its original velocity amplitude as follows:

$$\delta = \sqrt{2\frac{\eta}{\rho\omega}} \tag{4.43}$$

The boundary-layer thickness δ is independent of the wall velocity and depends only on the dynamic viscosity η, fluid density ρ, and the angular velocity ω.

The in-plane oscillating plate model can be used when the length scale of the flow δ is much smaller than the scale of geometrical features of the resonating structure (e.g., cylinder), and the radius is much larger than the flow structure. If the boundary-layer thickness is in the same order of magnitude than the radius, then the curvature of the tube or cylinder would influence the flow field. This case has been discussed in Section 4.2.2.3.

4.2.2.3 Torsional oscillating tube

The model described in Section 4.2.2 neglects the curvature of the structure's surface (e.g., the tube of the tubular viscosity sensor). This approach is reasonably accurate when the radius of the tube is much larger than the boundary-layer thickness. As the boundary-layer thickness increases, the curvature must be included in the fluid model. For this, the equations of fluid mechanics are solved using cylindrical coordinates under the assumption that there is no axial flow, no azimuthal change, and no radial flow, as shown in the following equations:

$$\frac{\partial u_\alpha}{\partial t} = \frac{\eta}{\rho}\left(\frac{1}{r}\frac{\partial u_\alpha}{\partial r} + \frac{\partial^2 u_\alpha}{\partial r^2} + \frac{u_\alpha}{r^2}\right) \tag{4.44}$$

$$\frac{{u_\alpha}^2}{r} = \rho\frac{\partial p}{\partial r} \tag{4.45}$$

where u_α is the azimuthal velocity, r is the radial coordinate, ρ is the fluid density, η is the dynamic viscosity, and p is the pressure.

Assuming temporal periodicity, we transfer the equation from the time into the frequency domain equation (4.45):

$$i\omega\hat{u}_\alpha = \frac{\eta}{\rho}\left(\frac{1}{r}\frac{\partial \hat{u}_\alpha}{\partial r} + \frac{\partial^2 \hat{u}_\alpha}{\partial r^2} - \frac{\hat{u}_\alpha}{r^2}\right) \tag{4.46}$$

where ω is the angular frequency and $i = \sqrt{-1}$.

The solution for Eq. (4.46) is given as follows:

$$\hat{u}_\alpha(r) = c_1 J_1\left((-1)^{\frac{3}{4}} r\sqrt{\frac{\omega\eta}{\rho}}\right) + c_2 Y_1\left(-(-1)^{\frac{3}{4}} r\sqrt{\frac{\omega\eta}{\rho}}\right) \tag{4.47}$$

where J_1 is the Bessel function of the first kind and first order, Y_1 is the Bessel function of the second kind and first order, and c_1 and c_2 are constants that are determined by the following boundary conditions:

$$\hat{u}_\alpha(r = 0) = 0 \tag{4.48}$$

$$\hat{u}_\alpha(r = R) = \omega\hat{\varphi}\hat{\Psi}R \tag{4.49}$$

where R is the inner radius of the tube, $\hat{\varphi}$ is the angular deflection amplitude, and $\hat{\Psi}$ is the value of the modal function.

Note that for computing the damping term d_f and the inertial mass J_f in Eqs. (4.30) and (4.31), respectively, the mode shape described by the modal function $\hat{\Psi}$ needs to be considered. The mode shape $\hat{\Psi}$ for a tube has been discussed in Section 4.2.1.3.

A solution of Eq. (4.46) is shown in Fig. 4.3 for four fluids with different viscosities at a fluid density of 1000 kg/m^3. When the viscosity increases, the velocity field $\hat{u}_\alpha(r)$ penetrates further into the tube.

Analytical solutions have been presented for either flat or circular geometries. However, there is no guarantee that analytical solutions can be found for more complex geometries. Section 4.2.2.4 introduces a numerical approach to compute the flow field. This approach is suitable for arbitrary geometries.

4.2.2.4 Computational fluid dynamics

A more general approach to obtain the velocity and pressure fields around a resonating structure is by numerically solving the fluid mechanics equations. For coupling the flow field with a resonating structure, we are only interested in the quasi–steady-state solution. In most cases, we can also

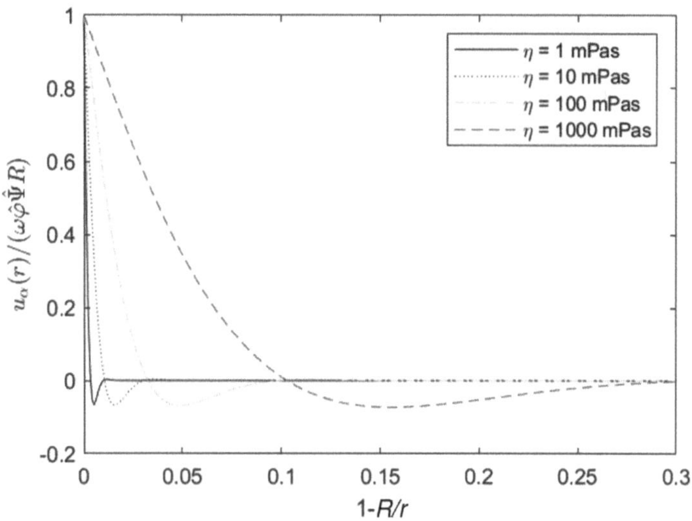

Figure 4.3 Flow field in a tube for viscosities η of four fluids at a fluid density of 1000 kg/m^3. Here, $u_a/(\omega\hat{\varphi}\hat{\psi}R)$ is the normalized azimuthal velocity, r is the radial position, and R is the inner radius of the tube.

assume the fluid to be incompressible, thermally uniform, homogeneous, and Newtonian. For this case, the equations of fluid mechanics are given as follows:

$$\nabla \cdot \vec{u} = 0 \tag{4.50}$$

$$\rho\left(\frac{\partial \vec{u}}{\partial t} + (\vec{u} \cdot \nabla)\vec{u}\right) = -\nabla p + \eta \Delta \vec{u} + \vec{F} \tag{4.51}$$

Here, ρ is the fluid density, $\vec{u}$ is the velocity field, p is the pressure field, $\vec{F}$ is a force term, η is the dynamic viscosity, and t is the time.

Most resonant sensors create microvibrations, where the vibration amplitude is orders of magnitude smaller than the structural features. Therefore, the flow tends to behave linear and the convective term of the fluid mechanics equations $(\vec{u} \cdot \nabla)\vec{u}$ is negligible, as shown in Eq. (4.52). Neglecting the convective term reduces the computational effort. However, the validity of this assumption may still be case specific; therefore, it needs to be treated as such.

$$\rho\left(\frac{\partial \vec{u}}{\partial t}\right) = -\nabla p + \eta \Delta \vec{u} + \vec{F} \tag{4.52}$$

In Sections 4.2.2.2 and 4.2.2.3, the flow field is analytically computed by assuming temporal periodicity and a simple geometry. For more complex geometries, it is not possible to simplify the fluid mechanics equations and the boundary conditions to a degree for which analytical solution can be found. Thus, to obtain a solution of the velocity and pressure fields, Eqs. (4.50) and (4.51) or Eqs. (4.50) and (4.52) are solved numerically.

There are two main approaches for numerically solving the fluid mechanics equations: the finite volume method and the finite element method.

The difference between the finite volume and finite element methods lies in how the equations are discretized. In the simplest terms, the finite volume method solves for the fluxes between the cells. The main advantage of this method is that the mass flux is inherently conserved, that is, the mass flux that is transferred from one cell to another cell has the same value. Therefore, even if the flux has a numerical error from the solver, the mass is neither created nor lost.

The finite element method is different; it uses the shape functions for the derivatives. The advantage of using shape functions is that higher-order discretization schemes can be used unlike the finite volume method. However, the numerical method does not inherently conserve mass flux. Thus, there might be some discrepancy in the flux from one cell to another. However, for many numerical applications, this error is negligible.

4.2.2.5 Reduced-order modeling and boundary conditions

A challenge in the fluid mechanics application of resonant sensors is the difference of scales. As previously mentioned, the length scale of the flow and the geometrical features can vary by orders of magnitude. Since both large geometrical features and small-scale flow features need to be resolved, a high computational is invadable unless reduced-order modeling is applied. In case of torsional resonators, the shapes tend to be long and continuous, thus, geometries can potentially be reduced from three-dimensional domain into a two-dimensional domain. In many cases, simulation may need to be restricted to two dimensions because the resources for full three-dimensional simulations are not available.

Brunner et al. [5] conducted a study called "*CFD modeling of pressure and shear rate in torsionally vibrating structures using ANSYS CFX and COMSOL Multiphysics*," which compared different numerical methods of computing the flow field around a torsionally oscillating structure. The geometry used in the study was a cylinder with four attached fins. The simulations were conducted in a two-dimensional domain in order to reduce the numerical requirements. The mesh of the two-dimensional domain is shown in Fig. 4.4. The mesh needs to be able to resolve both the structural component

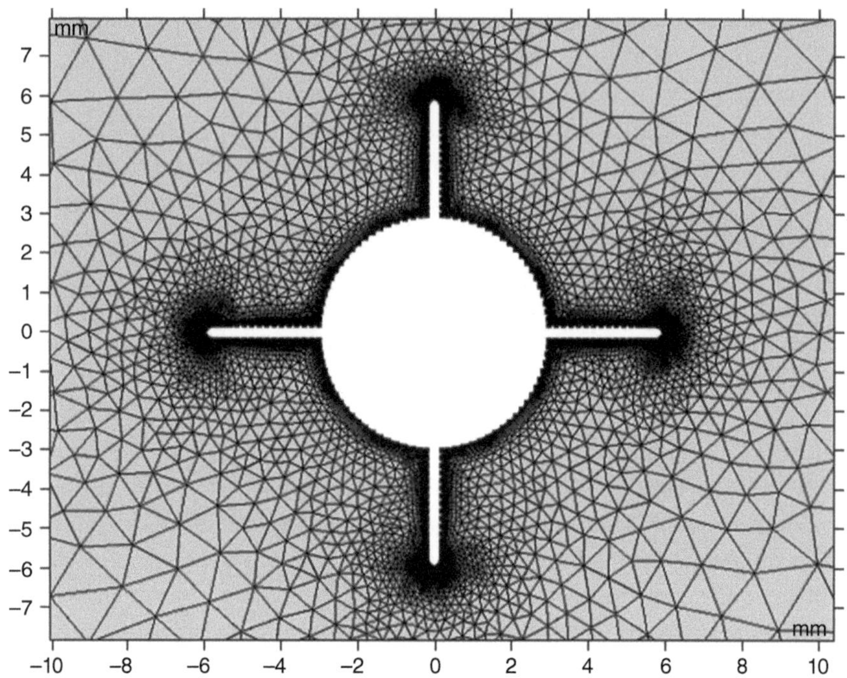

Figure 4.4 Two-dimensional mesh of the computational domain in COMSOL Multiphysics® with 26,558 elements [5].

as well as the flow in the vicinity of the geometry. In this case, the structure vibrates at a frequency of 7500 Hz. This lead to a boundary-layer thickness of 6 μm for a fluid with the viscosity and density of water. The mesh needs to be able to resolve the flow within this micrometer scale near the wall (see boundary layers) as well as structure itself (12 mm in outer diameter). Therefore, to compute a mesh-independent solution, a large number of cells in needed, especially in the vicinity of the boundary (as shown in Fig. 4.4).

The main subject of the study presented by Brunner et al. [5] was the comparison between different numerical methods and different boundary conditions. This study aimed to show that a solution of the flow field can be obtained that is invariant of numerical method and boundary condition. The methods used in the study were the following:

- *Moving wall approach*

  In the moving wall approach, the torsional vibration is modeled by the boundary condition of the oscillating structure. The torsional oscillation is around the axis $\vec{e_i}$, which is normal to the computational domain (i.e., the cross section). The velocity at the boundary of the resonator $d\Omega_r$ is defined by Eq. (4.53) as

$$\vec{u}(d\Omega_r) = \omega X \cdot \cos(\omega t)(\vec{r} \times \vec{e_i}) \tag{4.53}$$

where $\vec{u}$ is the velocity field, ω is the angular frequency, $\vec{r}$ is the distance from the axis $\vec{e_i}$ to the boundary, and X is the oscillation amplitude.

- *Frame of reference*

  Instead of inducing the torsional vibration on the wall, the frame of reference approach induces the motion in the domain. Thereby, the force term $\vec{F}$ is replaced by the centrifugal acceleration $\vec{\Omega} \times (\vec{\Omega} \times \vec{r})$, Coriolis acceleration $2\vec{\Omega} \times \vec{u}$, and Euler acceleration $(\partial\vec{\Omega})/(\partial t \times \vec{r})$, as follows:

$$\rho\left(\frac{\partial \vec{u}}{\partial t} + \vec{u} \cdot \nabla \vec{u} + 2\vec{\Omega} \times \vec{u} + \vec{\Omega} \times \left(\vec{\Omega} \times \vec{r}\right) + \frac{\partial \vec{\Omega}}{\partial t} \times \vec{r}\right) = -\nabla p + \eta \Delta \vec{u} \tag{4.54}$$

  where $\vec{\Omega}$ is the rotation vector of the domain. The motion of the domain is described by $\vec{\Omega}$ as,

$$\vec{\Omega} = \overrightarrow{e_i} X \cdot \sin(\omega t) \tag{4.55}$$

  where ω is the angular frequency, $\overrightarrow{e_i}$ is the unity vector of the torsional axis, and X is the oscillation amplitude.

The solutions of the shear rates and pressures on the boundaries were compared. These quantities are important because they are needed to compute the torque on the structure, that is, resonant element.

The shear rates and pressure values computed by the different methods were compared under quasi–steady-state conditions at three geometrical sections. These geometrical sections are the radial section (the base cylinder in the center), the fin, and the tip of the fin. All three sections are labeled in Fig. 4.5.

Since the pressure and shear rate a harmonic, they were compared by their amplitude and phase. Fig. 4.6 shows the shear rate amplitude for two fluid viscosities and three numerical methods. The solutions for all the methods were found to be in good agreement.

Fig. 4.7 shows the pressure amplitude, and Fig. 4.8 shows the phase of the pressure for the radial section. The pressure and phase were only plotted for a dynamic viscosity of 2 mPas since the solutions were almost identical to those at 20 mPas. For all the methods, the pressure amplitude was reduced

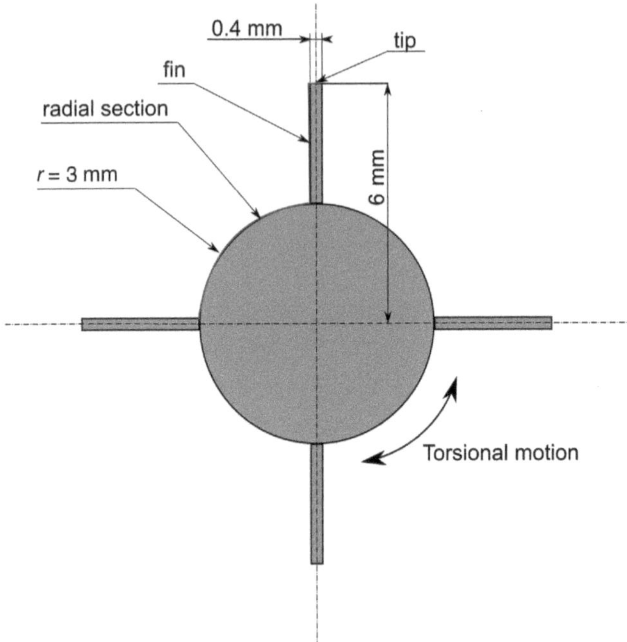

Figure 4.5 Geometry of the sensor tip of a viscosity–density sensor [5].

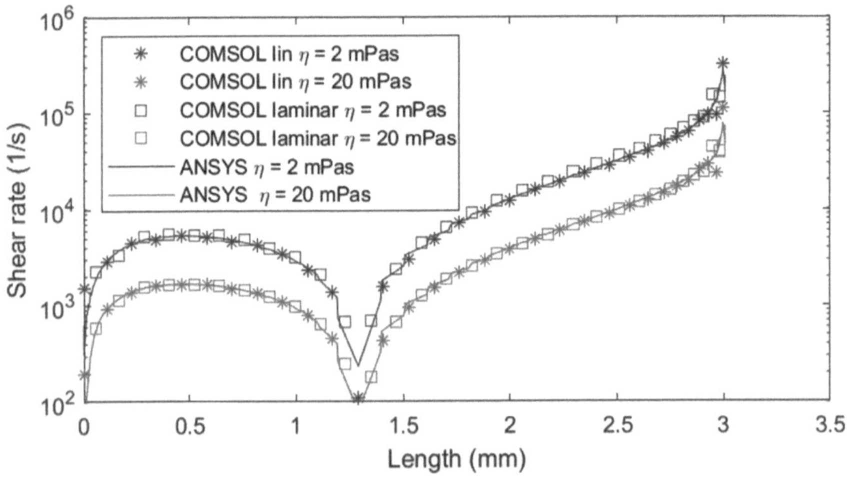

Figure 4.6 Comparison of shear rate amplitudes over the fin between COMSOL Multiphysics® (full and linearized version of fluid mechanics equations) and ANSYS CFX [5].

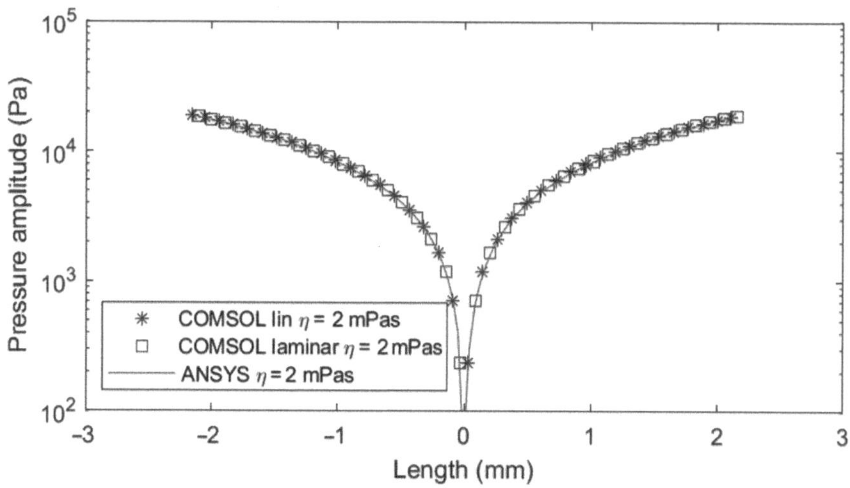

Figure 4.7 Comparison of the pressure amplitude over the radial section between COMSOL Multiphysics (full and linearized version of the Navier–Stokes equation) and ANSYS CFX [5].

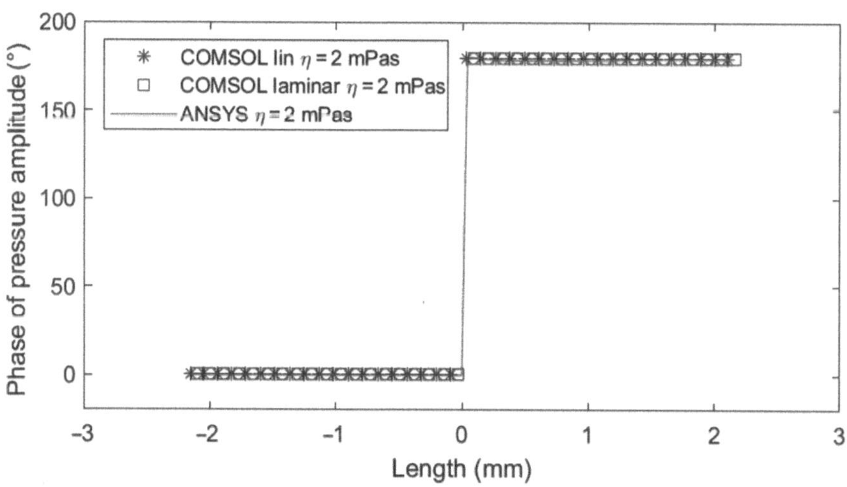

Figure 4.8 Comparison of the phase of the pressure over the radial section between COMSOL Multiphysics (full and linearized version of the Navier–Stokes equation) and ANSYS CFX [5].

strongly in the center of the radial section. At the same location, the pressure phase switched from slightly above 0° to almost 180°, which indicated that the pressure field was asymmetric. For both the phase and amplitude, all the three methods were in good agreement.

The result of the study was that all the numerical models yielded the same solution; therefore, the solution was independent of the numerical method and how the motion was applied. This underlines the potential of CFD for investigating the fluid mechanism around a torsional resonator.

4.3 Example of a torsional tubular resonator

In this section, we discuss how to apply the methods described on how to compute the bandwidth and resonator of a sensor. The presented example is a study by Brunner et al. [4]. They proposed a new tubular sensor that can measure the product of the fluid viscosity and density under conditions of internal flow. The sensor comprised of a straight, thin walled, stainless-steel tubular body; the flow passes through the sensor body without any interruption. This allowed the tubular sensor to be directly integrated into a process line.

The working principle of the tubular sensor is based on torsional resonance. The first torsional mode of the resonator was excited at a frequency close to its natural frequency. The excited resonance created motion in the fluid. The shear stresses caused by the fluid motion induced a torque on the sensor, which dampened the resonator's oscillation. The damping of the oscillation can be measured by a bandwidth and related to the viscosity–density product of the fluid.

Fig. 4.9 shows the schematic of the tubular sensor. The resonator consists of a tube and two large decoupling disks mounted onto the outer diameter. The first torsional mode was excited between the two masses via two permanent magnets that were mounted onto the tube. These magnets were driven by electromagnets, which produced an oscillating torque near the natural frequency of the first torsional mode. This driving torque was turned off after sufficient energy had been provided to the resonator. Then, the decay in the torsional oscillation was measured using electromagnets. Based on the measured signal, the bandwidth Γ and resonance frequency f_0 were computed.

Appropriate models are required to numerically predict the bandwidth and resonance frequency. The model used to compute the bandwidth and resonance frequency of the sensor comprises of different parts that were discussed previously. The resonator itself, that is, the tube, is modeled as a continuous resonator. This is done to compute the shape of the mode, the inertial mass of the resonator as well as the resonance frequency, see

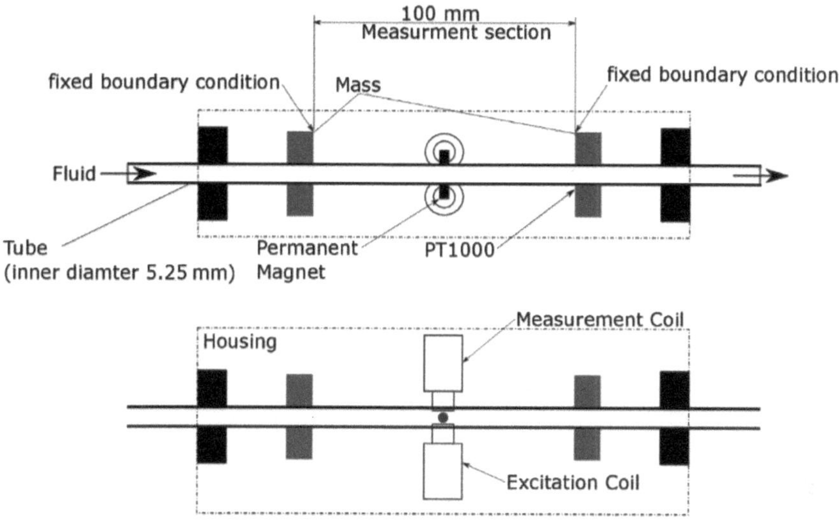

Figure 4.9 Experimental setup of the tubular sensor [4].

Section 4.2.2.3. Based on the inertial mass and resonance frequency of the mode, the resonator can be modeled as a single-degree-of-freedom resonator, see Section 4.2.1.1. To include the fluid interaction, the inertial mass of the fluid J_f and the fluid damping d_f must be computed. The value of these two parameters can be computed based on the flow field within the tube according to Section 4.2.1. The flow field can be solved analytically for this circular geometry as described in Section 4.2.2.3.

The approach taken in the study was slightly different. They did not describe the fluid forces by the inertial mass of the fluid J_f and the fluid damping d_f but by a complex amplitude. The end result, however, is the same.

In the study, experiments were performed to validate the predictions of the numerical models. Fig. 4.10 shows the product of fluid viscosity and density ($\rho\eta$) versus the normalized bandwidth Γ/f_0. The colormap shows the temperatures at which the $\rho\eta$-values were measured. The black line shows the prediction of the model. The model was in good agreement with the experimental result both quantitatively and qualitatively. Hence, this underlines the capability of modeling to predict the sensitivity of a sensor prior to its manufacturing.

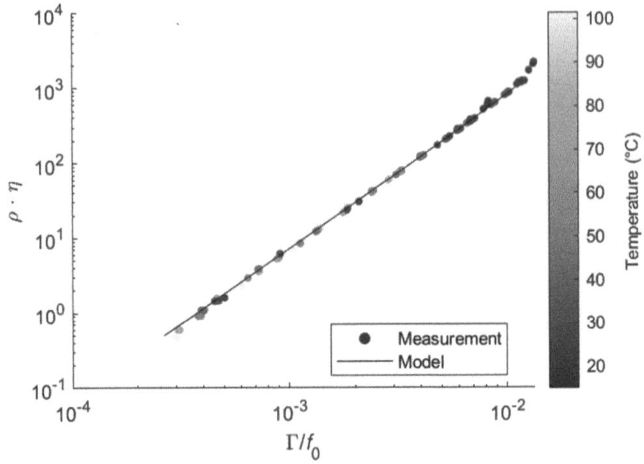

Figure 4.10 Comparison between the numerical prediction and experimental results for different fluids and temperatures.

4.4 Conclusions

This chapter presents the methodology on how to predict the properties of torsional resonators, namely, their bandwidth and resonance frequency with respect to the fluid interaction. The chapter covers resonant elements with a single degree of freedom, multiple degrees of freedom, as well as continuous resonators. All these different models can be weakly couple with a solution of the flow field, surrounding the resonator. Different methods of computing the solution of a flow field were discussed, ranging from analytical models for flat and cylindrical surfaces to numerical methods by which the flow around arbitrary geometries can be computed. For validation of the numerical method, the results of the study "*CFD modeling of pressure and shear rate in torsionally vibrating structures using ANSYS CFX and COMSOL Multiphysics*" were discussed, which demonstrated the a solution of the flow field that is independent of numerical method can be obtained. As a modeling example, the paper "*Analysis of a Tubular Torsionally Resonating Viscosity–Density Sensor*" was discussed, which describes and validates a continuous resonator, that was reduced to a single-degree-of-freedom resonator and coupled with an analytical fluid model.

Within this chapter, a variety of modeling schemes were presented that can be applied for modeling viscosity and density sensors based on torsional resonance. Such predictive models allow for model-based development and may accelerate the development time and reduce the development cost of, for example, new sensor generations.

References

[1] D.C. Ash, M.J. Joyce, C. Barnes, C.J. Booth, A.C. Jefferies, Viscosity measurement of industrial oils using the droplet quartz crystal microbalance, Meas. Sci. Technol. 14 (2003) 1955–1962.
[2] T. Brack, Multi-Frequency Phase Control of a Torsional Oscillator for Applications in Dynamic Fluid Sensing, ETH Zürich, Switzerland, 2017.
[3] T. Brack, S. Bolisetty, J. Dual, Simultaneous and continuous measurement of shear elasticity and viscosity of liquids at multiple discrete frequencies, Rheol. Acta 57 (2018) 415–428.
[4] D. Brunner, J. Goodbread, K. Häusler, S. Kumar, G. Boiger, H.A. Khawaja, Analysis of a tubular torsionally resonating viscosity–density sensor, Sensors 20 (2020) 3036.
[5] D. Brunner, H. Khawaja, M. Moatamedi, G. Boiger, CFD modelling of pressure and shear rate in torsionally vibrating structures using ANSYS CFX and COMSOL multiphysics, Int. J. Multiphys. 12 (2018) 349–358.
[6] A. Buhrdorf, H. Dobrinski, O. Lüdtke, J. Bennett, L. Matsiev, M. Uhrich, O. Kolosov, Multiparameteric oil condition sensor based on the tuning fork technology for automotive applications, Adv. Microsyst. Automot. Appl. 2005 (2005) 289–298.
[7] S. Clara, F. Feichtinger, T. Voglhuber-Brunnmaier, A.O. Niedermayer, A. Tröls, B. Jakoby, Balanced torsionally oscillating pipe used as a viscosity sensor, Meas. Sci. Technol. 30 (2019) 015101, doi:10.1088/1361-6501/aae755.
[8] J. Dual, Experimental Methods in Wave Propagation in Solids and Dynamic Viscometry, ETH Zürich, Switzerland, 1989.
[9] M. Fuchs, W. Drahm, C. Matt, A. Wenger, A Coriolis meter with direct viscosity measurement, Flomeko 14 (4) (2003) 1–5, doi:10.1049/cce:20030413.
[10] J. Goodbread, J. Dual, Resonant conductor measurement system and method, US 8,752,716 B2, 2014.
[11] M. Heinisch, S. Clara, I. Dufour, B. Jakoby, A spiral spring resonator for mass density and viscosity measurements, Proc. Eng. 87 (2014) 1143–1146.
[12] M. Heinisch, T. Voglhuber-Brunnmaier, E.K. Reichel, I. Dufour, B. Jakoby, Reduced order models for resonant viscosity and mass density sensors, Sens. Actuat. A Phys. 220 (2014) 76–84.
[13] M. Heinisch, T. Voglhuber-Brunnmaier, E.K. Reichel, I. Dufour, B. Jakoby, Electromagnetically driven torsional resonators for viscosity and mass density sensing applications, Sens. Actuat. A Phys. 229 (2015) 182–191.
[14] H. Irretier, Grundlagen der Schwingungstechnik 2, Friedr. Vieweg & Sohn Verlagsgesellschaft mbH, BraunschweigjWiesbaden, 2001.
[15] B. Jakoby, M. Scherer, M. Buskies, H. Eisenschmid, An automotive engine oil viscosity sensor, IEEE Sens. J. 3 (2003) 562–568.
[16] D. Johannsmann, The Quartz Crystal Microbalance in Soft Matter Research: Fundamentals and Modeling, Springer, Cham Heidelberg New York Dordrecht London, 2014.
[17] M.F. Khan, S. Schmid, P.E. Larsen, Z.J. Davis, W. Yan, E.H. Stenby, A. Boisen, Online measurement of mass density and viscosity of pL fluid samples with suspended microchannel resonator, Sens. Actuat. B Chem. 185 (2013) 456–461.
[18] H. Li, J. Wang, X. Li, D. Chen, Viscosity–density sensor with resonant torsional paddle for direct detection in liquid, IET Nanobiotechnol. 5 (2011) 121–125.
[19] N. McLoughlin, S.L. Lee, G. Hähner, Simultaneous determination of density and viscosity of liquids based on resonance curves of uncalibrated microcantilevers, Appl. Phys. Lett. 89 (2006) 184106.

[20] D. Valtorta, E. Mazza, Dynamic measurement of soft tissue viscoelastic properties with a torsional resonator device, Med. Image Anal. 9 (2005) 481–490.
[21] L. Vinet, A. Zhedanov, Advanced Microsystems for Automotive Applications 2005, Springer, Berlin, Heidelberg, Berlin, Heidelberg, 2005.
[22] O.O Amund, A.G. Adebiyi, Effect of viscosity on the biodegradability of automotive lubricating oils, Tribology International 24 (4) (1991) 235–357, doi:10.1016/0301-679X(91)90049-F.
[23] T. Brack, J. Dual, Multimodal torsional vibrations for the characterization of complex fluids, WIT Transactions on the Built Environment 129 (2013), doi:10.2495/FSI130171.

CHAPTER FIVE

Noise control treatments modeling using statistical energy analysis and the transfer matrix method

Thaynan A. Oliveira
Univ. Lille, ULR 7512 - Unité de Mécanique de Lille - Joseph Boussinesq (UML), F-59000 Lille, France

5.1 Introduction

Acoustical and structural systems are often found in engineering, whose importance of dynamic responses to random excitation is extended to high frequencies. Examples of such situations include:

- The calculation of sound pressure into a car;
- The structural response of an aerospace payload in flight;
- The vibration result of a structure under diffuse field conditions;
- The dynamical analysis of an aircraft structure (fuselage) due to jet noise and other dynamical sources;

Classical numerical methods, as the finite element method and modal analysis, are suitable in the area of vibration and acoustics at low frequencies. For high frequency, these tools cannot be used due to the small size of the wavelength, which requires a large number of degrees of freedom in the model and the high computational cost for obtaining the solution [1]. For solving dynamical problems at high frequencies, statistical approaches have been developed. Statistical energy analysis (SEA) is one of them, which has proved to be a promising tool, mainly for solving acoustic and vibration problems in complex structures.

The SEA method allows the user to model a given system as an interaction of modes associated with isolated parts of the system (subsystems), considering each mode group as a single-degree-of-freedom oscillator described by his average energy response. The responses are obtained from the analysis of the energy distribution through the system, excited by a power source. This technique is based on the power balance equations for each mode group [2].

Multiphysics Simulations in Automotive and Aerospace Applications.
DOI: 10.1016/B978-0-12-817899-7.00003-4

Some hypotheses are assumed in the determination of a model in SEA. The most important is that the vibrational energy varies from a subsystem to another with a rate proportional to the difference between the modal energy and the dissipated energy into the subsystem. Another important hypothesis assumes the weak coupling condition, based on the magnitude of the coupling loss factors and the dissipation loss factors of the coupled subsystems, presented in the following sections [1]. The weak coupling condition is valid if the internal dissipation of the subsystem is more important than the coupling path in the total loss of energy. This makes the ratio between the coupling factors and the internal dissipation factors considerably less than 1. Thus, the energy flow between the subsystems may be based on the difference between the energies of the decoupled subsystems [3].

Additionally, it is important to ensure that the input powers are uncorrelated, and they can be simply summed. It guarantees the linearity of the equations that govern the SEA problem. Natural frequencies must also be distributed within the frequency range of interest, as well as the energy distribution is assumed to be egalitarian for all modes of a subsystem present in a given frequency band.

5.1.1 Noise control treatments in the industrial context

Passive noise control treatments (NCTs) are widely used in aerospace and automotive industries whose main function is the dissipation of vibration (or acoustic) energy of a vibroacoustic system. In this context, sound packages containing a sequence of high dissipative porous materials and thin solid layers are generally employed to combine the high energy loss into a poroelastic media with the sound insulation capacity of double-wall systems.

For high frequency, analytical methods are generally used for the performance assessment of a NCT. The transfer matrix method (TMM) is a useful tool capable of characterizing a multilayer treatment from the resolution of a simple linear system. Computational efficiency makes this method extensively used in industrial applications for design purposes [4, 5].

The main contribution of this chapter is an overview of these high-frequency modeling approaches and the coupling between the TMM and SEA for solving acousticâ¬"structural problems with trimmed subsystems. The procedure for obtaining the SEA parameters was presented with coding examples. Practical results obtained from different techniques were also compared.

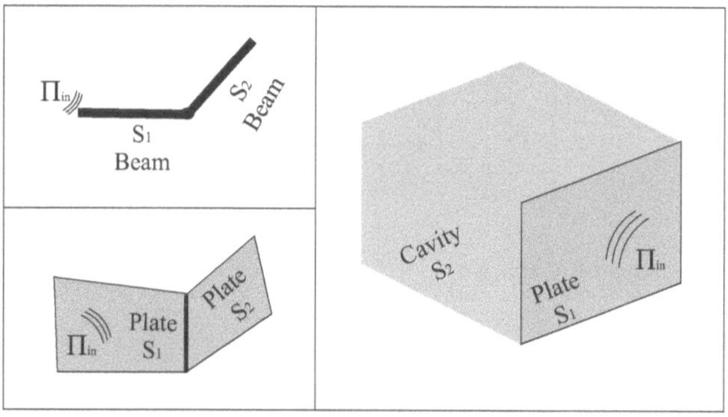

Figure 5.1 Examples of simple SEA problems.

5.2 SEA equations

As it was previously mentioned, the system is divided in SEA into physical elements in which the vibroacoustic characteristics, such as damping, excitation, and coupling properties, can be considered similar. Subsystems directly excited by external sources will transmit energy to the others by means of coupling characteristics. The main mechanism in SEA is based on the relationship between power inputs and outputs from each subsystem. Fig. 5.1 shows some simple problems that can be considered using SEA.

Fig. 5.1 shows the coupling of two SEA subsystems. However, it is common to have more robust models with a high number of elements. Considering a subsystem i coupled to a number of n subsystems, the power flow is

$$\Pi_i^{in} = \Pi_i^{\text{absorbed}} + \sum_{j \neq i}^{n} \Pi_{i;j}^{\text{transmited}}, \tag{5.1}$$

where Π represents the power in Watts. Eq. (5.1) shows that there are a dissipated portion of the input power into a subsystem (controlled by the energy dissipation capacity of the media) and a transmitted portion of that, controlled by the coupling nature. If considering the simple problems exposed in Fig. 5.1, the power flow can be illustrated by Fig. 5.2.

It is possible to obtain the equations of the power flow illustrated in the Fig. 5.2, for each subsystem, as

$$\text{Subsystem 1:}\quad \Pi_{in} + \Pi_{21,trans} = \Pi_{1,diss} + \Pi_{12,trans} \tag{5.2}$$

$$\text{Subsystem 2:}\quad \Pi_{12,trans} = \Pi_{2,diss} + \Pi_{21,trans}. \tag{5.3}$$

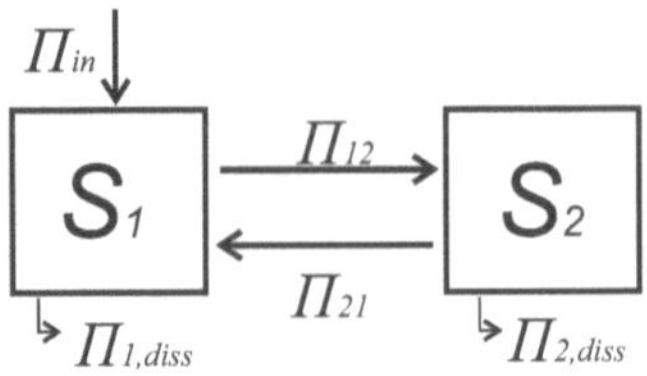

Figure 5.2 Two subsystems SEA flow.

The power Π is related to the energy from

$$\Pi_i = \omega \eta_i E_i, \tag{5.4}$$

where ω is the angular frequency, η_i is the loss factor and E_i is the energy of a subsystem i. The dissipated and the transmitted power will be given by

$$\Pi_{i,diss} = \omega \eta_{i,diss} E_i \qquad [i = 1, 2]\ ; \tag{5.5}$$

$$\Pi_{ij,trans} = \omega \eta_{ij} E_i \qquad [i \neq j]. \tag{5.6}$$

with η_{ij} being the coupling loss factor that control the energy exchanging between two subsystems. Substituting Eq. (5.5) and (5.6) into (5.2), considering the subsystems 1 and 2, the following linear system of equations is obtained

$$\omega(\eta_{1,diss} E_1 + \eta_{12} E_1 - \eta_{21} E_2) = \Pi_{in}\ , \tag{5.7}$$

$$\omega(-\eta_{12} E_1 + \eta_{21} E_2 + \eta_{2,diss} E_2) = 0. \tag{5.8}$$

It can be observed that the SEA method is based on the resolution of linear systems for calculating the energy in each subsystem. In the matrix form, the energy balance will be

$$\omega \begin{bmatrix} \eta_{1,diss} + \eta_{12} & -\eta_{12} \\ -\eta_{21} & \eta_{2,diss} + \eta_{21} \end{bmatrix} \begin{Bmatrix} E_1 \\ E_2 \end{Bmatrix} = \begin{Bmatrix} \Pi_{in} \\ 0 \end{Bmatrix}. \tag{5.9}$$

The determination of damping variables for complex subsystems is based on empirical/theoretical models or it comes from direct experimental data. For an area connection (plate-cavity), the transmitted powers by coupling between subsystems will depend on the radiation characteristics and differences between the environments. If the modal densities of the subsystems involved in the problem are known, the coupling loss factors in one direction is required, and the following consistency ratio can be used to complement the damping matrix by assuming the symmetric SEA approach

$$n_i \eta_{ij} = n_j \eta_{ji}, \tag{5.10}$$

where n_i e n_j are the modal densities of the subsystems i and j, respectively. Therefore, it will be necessary to know the following parameters to solve the linear system exposed in Eqs. (5.9) and (5.10):

- The modal densities of subsystems;
- The dissipation loss factors;
- The coupling loss factors;
- The input power.

Next sections are briefly presenting the parameters related to area connections, which is the most important junctions concerning the application of NCTs in SEA.

5.2.1 Modal densities for structural plates and acoustic cavities

The mode count is very important in the SEA method, since it is responsible for representing the capacity to receive and store energy in the subsystem through the resonant modes. In this context, it is possible to highlight three main variables, as the number of modes N in a given frequency band and the modal density n which represents the changing rate of the number of modes as a function of frequency, that is [2]

$$n(\omega) = \frac{\mathrm{d}N}{\mathrm{d}\omega} \; . \tag{5.11}$$

The greater number of modes present in a frequency band will result in the creation of a larger number of stationary waves in the domain when excited by a broadband noise (having all frequencies), so it is evident that such parameters govern the energy storage in the concerned subsystem.

Among the SEA parameters, the modal density is usually the most common to be determined using analytical or experimental methods. In experimental terms, such parameter can be extracted from the real part of the mechanical mobility in the case of structural subsystems, or from the real part of the average acoustic impedance to an acoustic subsystem. However, several analytical models are presented in theory for calculating modal density. For an acoustic cavity, as shown in Fig. 5.1, the modal density will be given by

$$n_{cav}(\omega) = \frac{\omega^2 V}{2\pi^2 c^3}, \tag{5.12}$$

where V is the volume and c the fluid sound velocity.

For a two-dimensional nonstiffened structure, there are also studies that identify its modal density commonly used in SEA codes. This parameter is given by

$$n_s(\omega) = \frac{Sk}{2\pi c_g}, \tag{5.13}$$

where k is the wavenumber and c_g is the group velocity for bending, longitudinal, or shear waves [6].

An important factor related to the modal characteristics of the subsystem is the *Modal Overlap*. This parameter considers not only the modal density, but also the subsystem damping. It is known that the statistical analysis becomes adequate from the occurrence of a sufficient number of modes in the spectral band, however, a highly damped system may lose some resonant characteristics. The consideration of the damping, besides the modal density, is a qualitative tool to obtain the appropriate response range of analysis. The modal overlap can be calculated as [3]

$$M(\omega) = \frac{\pi}{2}\omega n(\omega)\eta_T, \tag{5.14}$$

where the η_T is the total loss factor of the subsystem. High values of modal overlap result in small values of variance and narrow confidence intervals, and the mean response may be considered adequate. As a general rule, it is preferable that the parameter be greater than 1.

5.2.2 Coupling factors for cavity-plate area connections

The coupling between acoustic and structural subsystems Fig. 5.1 is controlled by different propagation paths. The most important ones are controlled by the resonant and nonresonant sound transmission. The first one is affected by the radiation capacity of the structural system and is important in frequencies above the coincidence spectral region of the plate. On the other hand, the nonresonant path is controlled by the mass of the structure and it is important for global responses below the coincidence frequency [7]. The radiation loss factor that models the connection between plate s and cavity a is given by

$$\eta_{sa} = \eta_{rad} = \frac{\rho_a c_a S_s}{\omega M_s}\sigma_{rad}, \tag{5.15}$$

with ρ_a being the fluid density, c_a the fluid sound velocity, S_s the area of the partition, M_s the structural mass, and σ_{rad} the radiation efficiency of the structural system that can be recovered in theoretical models as that one proposed by Leppington [8]. If a plate is modeled as the partition between two cavities, the nonresonant loss factor that models the coupling between the two cavities i and j is

$$\eta_{ij} = \eta_{mass} = \frac{c_i A_i \tau_m}{4\,\omega\, V_i}, \tag{5.16}$$

where A is the total surface area, V is the volume of the cavity, and τ_m is the mass controlled transmission coefficient of the system. The field incidence transmission coefficient can be used in this modeling strategy [7].

5.2.3 Dissipation loss factor

The DLF is a dimensionless quantity that is usually related to other energy loss variables, depending on the physics involved in the system. By definition, it is given by the ratio between the dissipated power and the input power of the system, or

$$\eta_{diss} = \frac{\Pi_{diss}}{\Pi_{input}}. \tag{5.17}$$

In structural dynamics, one can highlight the relationship with the viscous damping factor $\zeta = 0.5\eta$. In acoustics, the loss factor can be related to the reverberation time of a subsystem, which is directly related to the energy decay as a function of time, or more common from the mean surface absorption of the subsystem [9].

The representation of dissipative quantities is necessary as the experimental methods for characterization of systems are expanded and more natural forms of representing the energy loss arise. On acoustics, the reverberation time T_R can be measured directly from the average impulse response decay of a room. For a structure, it is very common to obtain the modal loss factor from the half-power bandwidth method from a measured frequency response function. For simple structures, there are empirical data available in the literature that indicate the loss factor in certain systems. The dissipation loss factor of a solid plate can vary between 0.2% and 3% [10].

5.2.4 Energy variables

With the parameters that govern the SEA equations, one can solve the linear system, as exemplified in (5.9), to obtain the average energy in each SEA subsystem. However, it is often necessary to present the results in function of representative dynamic variables. The energy response of a structural system in terms of the *rms* velocity, which is represented by [2, 6]

$$< v_{rms}^2 >= \frac{E}{M}, \tag{5.18}$$

where M is the total mass of a subsystem (admitting uniform distribution).

For an acoustic subsystem, it is interesting to convert the average energy of the subsystem into sound pressure. This conversion will depend on the characteristic impedance of the medium and also on the volume of the cavity, so that [6]

$$p_{rms}^2 = \frac{E}{V}\rho_a c_a^2. \tag{5.19}$$

It is also common to present this result in terms of the sound pressure level defined as

$$L_p = 10\log(p_{rms}^2/p_{\text{ref}^2}), \tag{5.20}$$

where $p_{\text{ref}} = 2.10^{-5}$ in [Pa].

5.3 Noise control treatments: characterization of a porous sample

In this section, the theory of sound propagation through porous media is discussed. The concepts of acoustic impedance are presented and the importance of that parameter to describe more usual indicators, like the absorption and reflection coefficients, is discussed. Some semiphenomenological models of porous media are mentioned, as well as the macroscopic properties that constitute the fluid phase of a porous material. From these theoretical models, it is possible to characterize a medium that can be used to approximate the indicators of a SEA subsystem.

5.3.1 Acoustic impedance

The concept of impedance appears in several areas of engineering. It is given by the ratio between a potential quantity and a kinetic quantity. In acoustics, the variable that can be associated with potential energy is the sound pressure and the one associated with the kinetic energy is the particle velocity or volume velocity. There are different types of acoustic impedance. The *point* or *specific impedance* is given by the ratio between effective sound pressure and particle velocity (both in frequency domain) at a single point located inside an acoustic medium [11], that is

$$\tilde{Z}(f) = \frac{\tilde{p}(f)}{\tilde{u}(f)} \qquad \text{Ns/m}^3 \text{ or Rayls}, \tag{5.21}$$

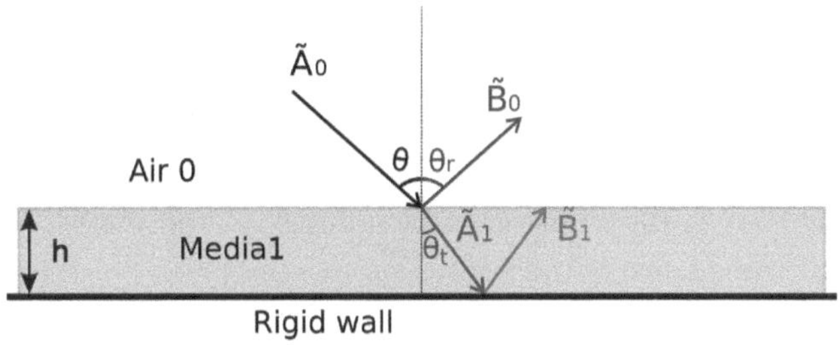

Figure 5.3 Sample between the fluid and the rigid wall.

where $\tilde{Z}(f)$ is the complex specific impedance, $\tilde{p}(f)$ is the sound pressure, and $\tilde{u}(f)$ is the particle velocity at the point.

The *characteristic impedance* Z_c is general quantity related to the acoustic medium. That one represents the ratio pressure/velocity when we consider free field propagation of plane waves inside the fluid medium [11]. The relationship between Z_c and the physical properties of the fluid is given by

$$Z_c = \rho\, c \;\; [rayls], \tag{5.22}$$

with ρ being the density function and c the sound velocity in the propagating medium.

Surface impedance

The surface impedance is the most important parameter for the acoustic characterization of a porous sample. From this parameter, it is possible to calculate the reflection and the absorption coefficient and, in some cases, the transmission loss (TL) of a porous treatment [12]. This quantity is given by the ratio of the sound pressure to the normal particle velocity at the surface between two medias. An example is given by considering a porous interface between a fluid and a rigid wall, represented by the Fig. 5.3.

The Fig. 5.3 shows a fluid, for example, the air (0), in contact with a porous material with thickness h_1 (media 1) backed by a rigid surface. The variable $\tilde{A}_0$ is the complex wave amplitude that insides on the material with angle θ relative to the normal direction in the air 0. Part of the wave is reflected to 0, with amplitude B_0 and $\theta_r = \theta$ if assuming smooth surface. A portion of energy is refracted to the media 1, with A_1 and θ_t being the amplitude and the angle of this wave. The rigid wall is perfectly reflective and B_1 is the amplitude of the reflected wave inside the material. The surface impedance for this problem is given by

$$\tilde{Z}_s(\omega) = -\mathrm{j}\frac{\tilde{Z}_c}{\cos\theta_t}\cot(\tilde{k}_c\cos(\theta_t)h), \tag{5.23}$$

where $\tilde{k}_c = \omega/\tilde{c}$ is the complex wavenumber in the media 1, $\tilde{Z}_c$ is the characteristic impedance of the porous sample, and $\mathrm{j} = \sqrt{-1}$ is the complex operator.

5.3.2 Reflection and absorption coefficients

The reflection coefficient R is a quantity that contains the relevant acoustic properties to express how much the surface reflects the sound wave. When $|R| = 1$, the surface is perfectly reflective like a rigid wall [13]. Considering the situation presented in Fig. 5.3, this parameter can be computed as

$$R = \frac{\tilde{Z}_s\cos\theta_i - \rho_0 c_0}{\tilde{Z}_s\cos\theta_i + \rho_0 c_0} \tag{5.24}$$

with $\tilde{Z}_s$ the surface impedance of the sample [12].

Another important quantity to be presented in this section is the sound absorption coefficient. This amount represents the fraction of incident energy that is not reflected by the sample, related to the energy dissipation into the material or the transmission of the sound wave through the media [13]. Considering the sample between air and rigid surface (5.3), the absorption coefficient α will be

$$\alpha = 1 - |R|^2. \tag{5.25}$$

In this case, all the energy that is not reflected to the medium 0 is dissipated by the material, since it is on a rigid wall. This is one of the mechanisms that provide mathematical support for impedance tube measurements.

5.3.3 Classical models of fibrous materials

There are several semiphenomenological models designed to characterize the acoustic behavior of porous materials. These models take into account macroscopic properties, and consideration of parameters varies from different theories. In this section, we will present the semiempirical model of Delany-Bazley, which is one of the simplest in terms of calculation, and also the Jhonson-Champoux-Allard model, which considers five macroscopic parameters to characterize the sound propagation in the porous media: the flow resistivity σ, porosity ϕ, tortuosity α_∞, dissipation characteristic length Δ, and thermal characteristic length Δ'.

Delany-Bazley model

In the Delany-Bazley model, the acoustic characteristics of the sample depend entirely on the σ [14] flow resistivity. The characteristic impedance Z_c can be obtained from Eq. ((5.26))

$$Z_c = \rho_0 c_0 \left[1 + 0,057 \left(\frac{\rho_0 f}{\sigma} \right)^{-0,754} - \mathrm{j}\,0,087 \left(\frac{\rho_0 f}{\sigma} \right)^{-0,732} \right], \tag{5.26}$$

as well as the characteristic wavenumber of the sample k_c

$$\frac{k_c}{k_0} = 1 + 0,0978 \left(\frac{\rho_0 f}{\sigma} \right)^{-0,700} - \mathrm{j}\,0,189 \left(\frac{\rho_0 f}{\sigma} \right)^{-0,595}, \tag{5.27}$$

where σ is the flow resistivity.

Johnson-Champoux-Allard model

The Johnson-Champoux-Allard model is more sophisticated compared to the Delany-Bazley model as it incorporates the five parameters which are able to influence the sound propagation within the porous material. Johnson et al. [15] present the effective density parameter ρ_{ef}, given by

$$\rho_{ef} = \rho_0 \left[\alpha_\infty + \frac{\nu\phi}{\mathrm{j}\,\omega q_0} \left(1 + \frac{4\,\alpha_\infty^2 q_0^2}{\Lambda^2 \phi^2 \nu} \mathrm{j}\omega \right)^{1/2} \right]. \tag{5.28}$$

where $\nu = \eta/\rho_0$ is the kinematic viscosity of the medium, $\eta = 1.8410^-5$ [Pa s] is the static viscosity of the air, the viscous permeability of the fiber q_0 may be approximated by $q_0 = \eta/\sigma$. Champoux and Allard [16] present the effective bulk modulus K_{ef}, given by

$$K_{ef} = \gamma P_0 / \left[\gamma - \frac{\gamma - 1}{1 + \frac{\nu'\phi}{\mathrm{j}\omega q_0'} \left(1 + \frac{\mathrm{j}\omega (\Lambda')^2}{16\nu'} \right)^{1/2}} \right], \tag{5.29}$$

where $\nu' = \nu/\mathrm{Pr}$ is the thermal viscosity, $\mathrm{Pr} = 0.77$ is the Prandtl number, q_0' is the thermal permeability which is given by $q_0' = \phi(\Delta')^2/8$ under the hypothesis that the pores can be approximated by circular cross-shaped tubes.

The characteristic impedance and the complex wavenumber of a porous sample is computed from the equivalent fluid theory as [17]

$$\tilde{Z}_c = \frac{\sqrt{\rho_{ef} K_{ef}}}{\phi}, \tag{5.30}$$

$$\tilde{k}_c = \frac{\omega\sqrt{\rho_{ef}/K_{ef}}}{\phi}. \tag{5.31}$$

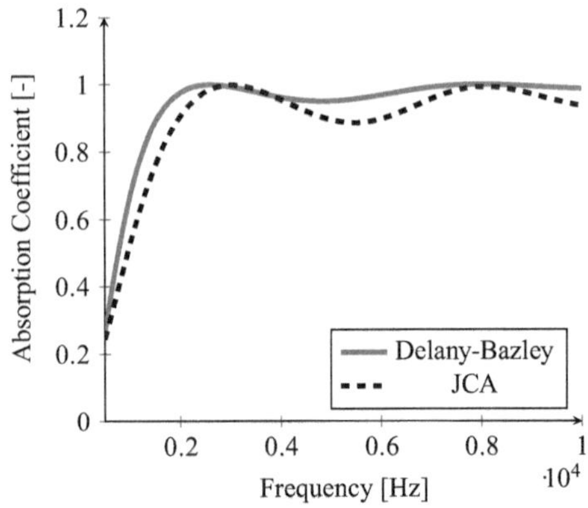

Figure 5.4 Sound absorption coefficient computed from the Delanyâ¬"Bazley model and Johnsonâ¬"Champouxâ¬"Allard model for a 30 mm thick glass fiber.

Table 5.1 Material properties.

	σ	ϕ (%)	α_∞ (-)	Λ (μm)	Λ' (μm)
Glass fiber	22000,0	98,0	1,1	105,0	340,0

Fig. 5.4 shows a comparison of the normal incidence absorption coefficient ($\theta_i = 0$) for a fibrous material, calculated from the models presented in this section. The material properties considered for this simulation is given by the Table 5.1.

As it can be seen in the Fig. 5.4, there are some important differences between the two models. However, if considering a statistical point of view, the result shown a similar tendency for both curves. The Delany-Bazley approach can be seen as a simplest but useful tool for representing fiber materials in some applications on SEA.

5.4 The Transfer matrix method

Multilayer panels are often used in the context of noise control. The mechanisms that involves the application of a NCT on a face of a structural system are: the sound insertion loss (IL), which can change the acoustic radiation and the mass law transmission of a structure, the sound absorption seen by the acoustic medium in contact with the surface and the added structural damping. The methodology of the TMM [12] is based on the theory of plane wave propagation in a given layer. The material of each layer is assumed homogeneous, isotropic, and the surface of the treatment

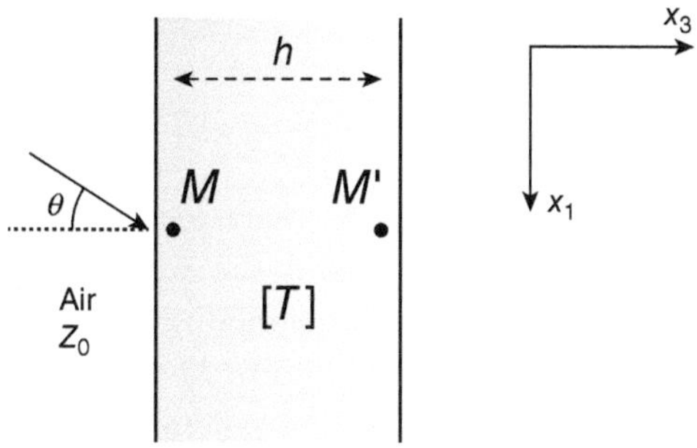

Figure 5.5 Illustration of an infinite extent layer on the transfer matrix method.

is considered infinite in this formulation. Fig. 5.5 shows an incident plane acoustic wave upon a media with thickness h, at an incident angle θ.

By considering oblique incidence, the x_1 component of the propagating wavenumber, also denoted as trace wavenumber k_t, is given by

$$k_t = k_0 \sin(\theta) \tag{5.32}$$

Note that k_0 represents the wavenumber in the surrounding fluid. The sound propagation inside the layer is represented by a transfer matrix T

$$V(M) = [T]V(M'), \tag{5.33}$$

where V is a vector containing the response for each degree of freedom of the media at the points M and M' set close to the forward and the backward of the layer [12]. Thomson [18] suggested the transmission matrices representing the propagation of plane elastic waves through stratified mediums. In this study, solid and fluid domains were initially approached, but the final equations in this formulation were limited to some special cases. The boundary conditions based on continuity of stresses and velocities proposed by Thompson was also used by Folds and Loggins in 1977 [19], who derived the transfer matrices of parallel solid plates excited by oblique incidence plane waves.

5.4.1 Transfer matrices

The most important definition of this method is the transfer matrix of the layer media, which can be seen as a frequency response function of the linear system. In this section, the fluid, solid and poroelastic matrices are presented, as well as the continuity conditions between them. These parameters are used for assembling the global matrix and allowing the resolution of the linear problem.

5.4.1.1 Fluid layer

A linear acoustic fluid domain can be characterized by two independent degrees of freedom: the pressure p and the particle velocity v_3^f in the longitudinal direction x at a point into the medium. In the notation exposed in Eq. (5.33), the response vector at the point M will be given as

$$V^f(M) = [p(M)\ v_3^f(M)]^{\mathrm{T}}, \tag{5.34}$$

where the index 3 represents the longitudinal direction of propagation [12] and T indicates the transposed operation. The longitudinal propagating wavenumber into the fluid is given as

$$k_3 = \sqrt{k_{eq}^2 - k_t^2}; \tag{5.35}$$

where k_{eq} is the characteristic wavenumber of the medium, which can be complex if the layer is a porous material represented by an equivalent fluid (see Eq. (5.31)).

The transfer matrix of a fluid is derivated from the solution of the 1D wave equation, by calculating the pressure and velocity field at the points M and M'. If considering $M' = 0$, the distance between two points will be represented by the thickness h of the layer and the position of $M = -h$. In this context, the transfer matrix $[T^f]$ will be given as

$$[T^f] = \begin{bmatrix} \cos(k_3 h) & \mathrm{j}\frac{\omega\rho_0}{k_3}\sin(k_3 h) \\ \mathrm{j}\frac{k_3}{\omega\rho_0}\sin(k_3 h) & \cos(k_3 h) \end{bmatrix}, \tag{5.36}$$

For more details on the solution of the acoustic wave equation, references [11] and [13] can be consulted.

5.4.1.2 Solid layer

According to [20] four waves can propagate in a complete description of a solid media: two longitudinal and two shear waves (incident and reflected). For an isotropic solid, the propagating wavenumbers associated with each direction depends on the Lam coefficients, which is given as

$$\lambda = \frac{E(1+\mathrm{j}\eta)\,\nu}{(1+\nu)(1-2\nu)}; \qquad \mu = \frac{E(1+\mathrm{j}\eta)}{2(1+\nu)}, \tag{5.37}$$

being E the Young's modulus, ν the Poisson's ratio, η the dissipation loss factor, λ and μ the first and second Lam coefficients of the plate, respectively. Both parameters are used for calculating the characteristic wavenumber in the longitudinal δ_1 and shear δ_2 directions

$$\delta_1 = \sqrt{\frac{\omega^2 \rho}{\lambda + 2\mu}}$$

$$\delta_3 = \sqrt{\frac{\omega^2 \rho}{\mu}} \tag{5.38}$$

where ρ is the density of the solid elastic material.

Folds et al. [19] presented the velocities v_1, v_3 and stress components σ_{33}^s and σ_{13}^s in function of displacement potentials of the longitudinal and shear waves. The response vector of a structural layer at the point M becomes

$$V^s = [v_1^s(M)\ v_3^s(M)\ \sigma_{33}^s\ \sigma_{13}^s]^{\mathrm{T}} \tag{5.39}$$

In order to obtain the transfer matrix of the elastic solid, the vector V at the points M and M' should be connected with the four by four system that depends on the four amplitudes of the propagating waves cited before, which can be expressed as

$$V^s(M) = [\Gamma(M)]W \tag{5.40}$$

$$V^s(M') = [\Gamma(M')]W \tag{5.41}$$

where W is the amplitude vector extracted from the potential relations and Γ is a position-dependent matrix given as

$$[\Gamma(x_3)] = \begin{bmatrix} \omega k_t cos(k_{13}x_3) & -j\omega k_t \sin(k_{13}x_3) & j\omega k_{33}\sin(k_{33}x_3) & -\omega k_{33}cos(k_{33}x_3) \\ -j\omega k_{13}\sin(k_{13}x_3) & \omega k_{13}cos(k_{13}x_3) & \omega k_t cos(k_{33}x_3) & -j\omega k_t \sin(k_{33}x_3) \\ -D_1 cos(k_{13}x_3) & jD_1\sin(k_{13}x_3) & jD_2 k_{33}\sin(k_{33}x_3) & -D_2 k_{33}cos(k_{33}x_3) \\ jD_2 k_{13}\sin(k_{13}x_3) & -D_2 k_{13}cos(k_{13}x_3) & D_1 cos(k_{33}x_3) & -jD_1\sin(k_{33}x_3) \end{bmatrix}, \tag{5.42}$$

where $k_{13} = \sqrt{\delta_1^2 - k_t^2}$ is the shear propagating wavenumber, $k_{33} = \sqrt{\delta_3^2 - k_t^2}$ is the normal wavenumber on the x_3 direction, $D_1 = \lambda(k_t^2 + k_{13}^2) + 2\mu k_{13}^2$ and $D_2 = 2\mu k_t$ are elastic coefficients [12].

By considering the origin point $M' = 0$ and $M = -h$ (see Fig. 5.5), the transfer matrix of an elastic solid can be calculated from the ratio between

Eqs. (5.40) and (5.41), which is

$$V^s(-h) = [T^s]V^s(0) = [\Gamma(-h)][\Gamma(0)]^{-1}V^s(0) \tag{5.43}$$

$$T^s = [\Gamma(-h)][\Gamma(0)]^{-1}, \tag{5.44}$$

where $[T^s]$ is the transfer matrix of the elastic solid layer. This approach fixes the inversion of Γ at the point zero, which allows to obtain an inverse matrix independent of the thickness h from a simple symbolic calculation,

$$[\Gamma(0)]^{-1} = \begin{bmatrix} \frac{D_2}{\omega(D_1+D_2k_t)} & 0 & \frac{-1}{D_1+D_2k_t} & 0 \\ 0 & \frac{D_1}{k_{13}(D_1+D_2k_t)} & 0 & \frac{-k_t}{k_{13}(D_1+D_2k_t)} \\ 0 & \frac{D_2}{\omega(D_1+D_2k_t)} & 0 & \frac{1}{D_1+D_2k_t} \\ \frac{-D_1}{\omega(D_1k_{33}+D_2k_tk_{33})} & 0 & \frac{-k_t}{D_1k_{33}+D_2k_tk_{33}} & 0 \end{bmatrix}, \tag{5.45}$$

This chapter focuses in practical computational aspects of the proposed methods and the connection between the two approaches. The derivation of $\Gamma(x3)$ from the potential displacements are detailed in references [12] and [19].

5.4.1.3 Septum and thin plate layers

Some modifications of the elastic solid can be useful for more simple modeling strategies. An interesting alternative is to model the plate as a thin elastic solid in bending. It means that the thickness of the layer is sufficiently small and normal waves are much more important than shear waves.

Even simpler, a septum layer is a limp thin layer. It means that the layer is assumed to have no stiffness and can be fully described by the mass per unit area of the material. The form of the equation of motion of this kind of media is given by

$$Z_s(\omega)v_3(M') = \sigma_{33}(M') - \sigma_{33}(M) \tag{5.46}$$

where $Z_s(\omega)$ is the mechanical impedance of the panel, with

$$Z_s = j\omega m'', \quad \text{for a septum}$$
$$Z_s = j\omega m''\left(1 - \frac{Dk_t^4}{\omega^2 m''}\right), \quad \text{for a thin plate} \tag{5.47}$$

with D being the bending stiffness and m'' the mass per unit area of the plate. The 2 degrees of freedom response vector of the mechanical field in a point

M is $V^t(M) = [\sigma_{33}(M)v_3(M)]^{\mathrm{T}}$, and the transfer matrix for this media is given by

$$[T^t] = \begin{bmatrix} 1 & -Z_s \\ 0 & 1 \end{bmatrix}. \tag{5.48}$$

If the septum is considered massless, the mechanical impedance becomes zero and the transfer matrix reduces to identity. This case can simulate a thin impervious screen used to protect a porous material in practical meanings.

5.4.1.4 Poroelastic layer

According to the Biot's theory of propagation of elastic waves in a fluid saturated porous solid [21], two compressional waves and shear wave can propagate in a porous material. These wave equations were used by Lauriks et al. [20] to evaluate the transfer matrix of a poroelastic material containing 6 degrees of freedom, with response vector given by

$$V^P(M) = [v_1^s\ v_3^s(M)\ v_3^f(M)\ \sigma_{33}^s(M)\ \sigma_{13}^s(M)\ \sigma_{33}^f(M)]^{\mathrm{T}} \tag{5.49}$$

where v_1^s and v_3^s are the x_1 and x_3 components of the solid frame velocity, v_3^f is the fluid particle velocity component, σ_{13}^s and $\sigma_3 3^s$ are the shear and normal components of the solid stress tensor and σ_{33}^f the longitudinal component in the fluid stress.

For the problem illustrated in Fig. 5.5, the x_3 component of the propagating wavenumber is given by

$$k_{i3} - \sqrt{(\delta_i^2 - k_1^2)} \quad i = [1, 2, 3] \tag{5.50}$$

with δ_i being the wavenumbers for each wavetype i.

Similar to the solid layer procedure, the transfer matrix of a poroelastic media is evaluated from the auxiliar matrix $\Gamma(x_3)$ which relates the degrees of freedom at a point with the incident and reflected amplitudes of the wave potential relations [12]. Analogous to Eq. (5.45), the TM of a poroelastic material is computed by

$$[T^P] = [\Gamma(-h)][\Gamma(0)]^{-1}, \tag{5.51}$$

The six columns of $\Gamma(x)$ are given in Table 5.2, where D_i and E_i are elastic functions that depends on the elasticity coefficients Biot parameters, presented with details in references [20] and [12].

Table 5.2 Auxiliar matrix $\Gamma(x)$ for a poroelastic material [12].

The three first columns		
$\omega k_t \cos(k_{13}x_3)$	$-j\omega k_t \sin(k_{13}x_3)$	$\omega k_t \cos(k_{23}x_3)$
$-j\omega k_{13} \sin(k_{13}x_3)$	$\omega k_{13} \cos(k_{13}x_3)$	$-j\omega k_{23} \sin(k_{23}x_3)$
$-j\omega k_{13}\mu_1 \sin(k_{13}x_3)$	$\omega k_{13}\mu_1 \cos(k_{13}x_3)$	$-j\omega k_{23}\mu_2 \sin(k_{23}x_3)$
$-D_1 \cos(k_{13}x_3)$	$jD_1 \sin(k_{13}x_3)$	$-D_2 \cos(k_{23}x_3)$
$j2Nk_t k_{13} \sin(k_{13}x_3)$	$-2Nk_t k_{13} \cos(k_{13}x_3)$	$j2Nk_t k_{23} \sin(k_{23}x_3)$
$-E_1 \cos(k_{13}x_3)$	$jE_1 \sin(k_{13}x_3)$	$-E_2 \cos(k_{23}x_3)$
The three last columns		
$-j\omega k_t \sin(k_{23}x_3)$	$j\omega k_{33} \sin(k_{33}x_3)$	$-\omega k_{33} \cos(k_{33}x_3)$
$\omega k_{23} \cos(k_{23}x_3)$	$\omega k_t \cos(k_{33}x_3)$	$-j\omega k_t \sin(k_{33}x_3)$
$\omega k_{23}\mu_2 \cos(k_{23}x_3)$	$\omega k_t \mu_3 \cos(k_{33}x_3)$	$-j\omega k_t \mu_3 \sin(k_{33}x_3)$
$jD_2 \sin(k_{23}x_3)$	$2jNk_t k_{33} \sin(k_{33}x_3)$	$-2Nk_{33}k_t \cos(k_{33}x_3)$
$-2Nk_t k_{23} \cos(k_{23}x_3)$	$N(k_{33}^2 - k_t^2)\cos(k_{33}x_3)$	$-jN(k_{33}^2 - k_t^2)\sin(k_{33}x_3)$
$jE_2 \sin(k_{23}x_3)$	0	0

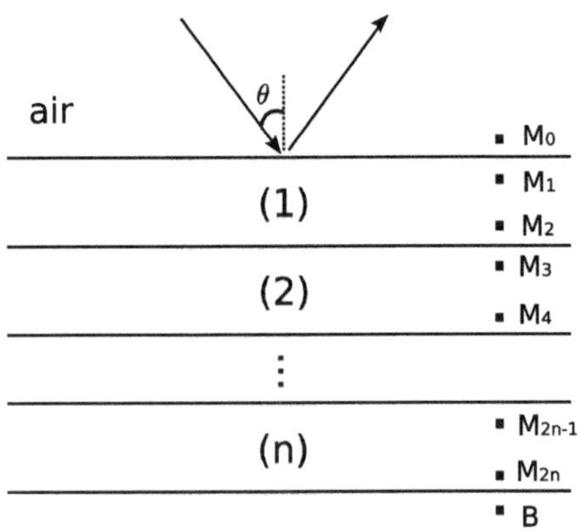

Figure 5.6 Illustration of a multilayer NCT with n layers and $2n$ continuity interfaces.

5.5 Numerical procedure and the SEA-TMM coupling

When a multilayered treatment is modeled, the transfer matrices presented in Section 5.4.1 as well as the continuity conditions between each layer should be considered. In this method, these conditions are given in terms of coupling matrices that simplifies the assembling process of a global transfer matrix. Fig. 5.6 illustrates a multilayer NCT with n layers and $2n$ interface surfaces.

If considering the connection between layer (1) and (2), the following relation is established

$$V^{(1)}(M_2) = V^{(2)}(M_3) = [T]^{(2)} V^{(2)}(M_4) \tag{5.52}$$

with $[T]^{(2)}$ being the transfer matrix of the layer 2. At the continuity surfaces, the relationship between degrees of freedom of each layer should be explained. This relationship is given by interface matrices $[I]$ and $[J]$ that imposes the conditions in which Eq. (5.52) can be valid. Taking into account these parameters, Eq. (5.52) can be rewritten as

$$[I_{(1)(2)}]V^{(1)}(M_2) + [J_{(1)(2)}][T^{(2)}]V^{(2)}(M_4) = 0 \tag{5.53}$$

Where $[I_{(1)(2)}]$ and $[J_{(1)(2)}]$ are the interface matrices between layer (1) and (2).

5.5.1 Assembling of the global transfer matrix

The process of assembling the global transfer matrix is similar to a finite element process. As the interface matrices are not square, a simple vectorial product cannot be used to make it. If considering Fig. 5.6, the generalization of Eq. 5.53 is given as follows

$$I_{(n)(n+1)}V^{n}(M_{2(n)}) + J_{(n)(n+1)}T^{(n+1)}V^{(n+1)}(M)_{2(n+1)} = 0. \tag{5.54}$$

The incident media (air) is assumed as layer zero (0) in this generalization. By following Eq. 5.54, using the local transfer matrices and the interface matrices presented in this section, the global matrix GM_0 is assembled as follows

$$[TG_0] =$$
$$\begin{bmatrix} [I_{(01)}] & [J_{(01)}][T^{(1)}] & [0] & \cdots & [0] & [0] \\ [0] & [I_{(12)}] & [J_{(12)}][T^{(2)}] & \cdots & [0] & [0] \\ \vdots & \vdots & \vdots & \ddots & \vdots & \vdots \\ [0] & [0] & [0] & \cdots & [I_{(n-1)(n)}] & [J_{(n-1)(n)}][T^{(n)}] \end{bmatrix} \tag{5.55}$$

and the right-hand side

$$V_0 = [V^{(0)}(A)\ V^{(1)}(M_2)\ V^{(2)}(M_4)\ \ldots\ V^{(n)}(M_{2n})]^{\mathrm{T}}. \tag{5.56}$$

The global transfer matrix TG_0 is not square and do not have same number of lines as V_0, but it follows the dimension of the continuity matrices presented in the following section. If N is the number of degrees of freedom of the problem, TG_0 in this step has $N-4$ lines if the last layer is a porous, $N-3$ if it is a solid and $N-2$ if the last media is a fluid. It means that extra equations are necessary to complete the linear problem. These equations are given by the boundary and initial conditions.

Table 5.3 Continuity matrices.

$I_{ff} = \begin{bmatrix} 1 & 0 \\ 0 & 1 \end{bmatrix}$	$J_{ff} = -\begin{bmatrix} 1 & 0 \\ 0 & 1 \end{bmatrix}$
$I_{ss} = \begin{bmatrix} 1 & 0 & 0 & 0 \\ 0 & 1 & 0 & 0 \\ 0 & 0 & 1 & 0 \\ 0 & 0 & 0 & 1 \end{bmatrix}$	$J_{ss} = -\begin{bmatrix} 1 & 0 & 0 & 0 \\ 0 & 1 & 0 & 0 \\ 0 & 0 & 1 & 0 \\ 0 & 0 & 0 & 1 \end{bmatrix}$
$[I_{pp}] = \begin{bmatrix} 1 & 0 & 0 & 0 & 0 & 0 \\ 0 & 1 & 0 & 0 & 0 & 0 \\ 0 & 1-\frac{\phi_2}{\phi_1} & \frac{\phi_2}{\phi_1} & 0 & 0 & 0 \\ 0 & 0 & 0 & 1 & 0 & 1-\frac{\phi_2}{\phi_1} \\ 0 & 0 & 0 & 0 & 1 & 0 \\ 0 & 0 & 0 & 0 & 0 & \frac{\phi_2}{\phi_1} \end{bmatrix}$	$[J_{pp}] = -\begin{bmatrix} 1 & 0 & 0 & 0 & 0 & 0 \\ 0 & 1 & 0 & 0 & 0 & 0 \\ 0 & 0 & 1 & 0 & 0 & 0 \\ 0 & 0 & 0 & 1 & 0 & 0 \\ 0 & 0 & 0 & 0 & 1 & 0 \\ 0 & 0 & 0 & 0 & 0 & 1 \end{bmatrix}$
$I_{sf} = \begin{bmatrix} 0 & 1 & 0 & 0 \\ 0 & 0 & 1 & 0 \\ 0 & 0 & 0 & 1 \end{bmatrix}$	$J_{sf} = \begin{bmatrix} 0 & -1 \\ 1 & 0 \\ 0 & 0 \end{bmatrix}$
$I_{pf} = \begin{bmatrix} 0 & (1-\phi) & \phi & 0 & 0 & 0 \\ 0 & 0 & 0 & 1 & 0 & 0 \\ 0 & 0 & 0 & 0 & 1 & 0 \\ 0 & 0 & 0 & 0 & 0 & 1 \end{bmatrix}$	$J_{pf} = \begin{bmatrix} 0 & -1 \\ (1-\phi) & 0 \\ 0 & 0 \\ \phi & 0 \end{bmatrix}$
$I_{sp} = \begin{bmatrix} 1 & 0 & 0 & 0 \\ 0 & 1 & 0 & 0 \\ 0 & 1 & 0 & 0 \\ 0 & 0 & 1 & 0 \\ 0 & 0 & 0 & 1 \end{bmatrix}$	$J_{sp} = -\begin{bmatrix} 1 & 0 & 0 & 0 & 0 & 0 \\ 0 & 1 & 0 & 0 & 0 & 0 \\ 0 & 0 & 1 & 0 & 0 & 0 \\ 0 & 0 & 0 & 1 & 0 & 1 \\ 0 & 0 & 0 & 0 & 1 & 0 \end{bmatrix}$

5.5.1.1 Continuity between layers

A compilation of the continuity matrices for different connection types is presented in Table 5.3. The most important matrices were extracted from reference [12]. It is used the index f for fluid, s for solid, and p for porous. With layers of different nature, the matrices should be permuted if the reverse order is required, it means

$$[I_{ij}] = [J_{ji}] \qquad i, j = [f\ s\ p]. \tag{5.57}$$

Septum and thin plate layers can be considered as equivalent fluids with similar continuity conditions of a fluid layer, but the acoustic pressure is equivalent to the negative value of normal stresses $p(M) = -\sigma_{33}(M')$, which demand a sign changing in the column of $[I]$ and $[J]$ relative to the normal stresses.

It is also possible to do a simple multiplication of matrices in same type layers connection, which means that a single block with equivalent

Table 5.4 Rigid wall boundary arrays

$[Y^p] = \begin{bmatrix} 1 & 0 & 0 & 0 & 0 & 0 \\ 0 & 1 & 0 & 0 & 0 & 0 \\ 0 & 0 & 1 & 0 & 0 & 0 \end{bmatrix}$	$[Y^s] = \begin{bmatrix} 1 & 0 & 0 & 0 \\ 0 & 1 & 0 & 0 \end{bmatrix}$	$[Y^f] = \begin{bmatrix} 1 & 0 \end{bmatrix}$

properties will be considered for a group of layers of same nature connected simultaneously. However, in case of porous-porous junction, the difference of porosities should be taken into account. The $[I]$ matrices are used to consider this condition and the equivalent matrix is given by

$$[T_{eq}^{(n)}] = [T^{(n-1)}][I_{(n-1)(n)}][T^{(n)}]. \tag{5.58}$$

Eq. (5.58) is valid only for layers of same nature.

5.5.2 Applying the boundary conditions

A NCT in SEA analysis can change three main parameters: the damping of a cavity from the surface absorption of the NCT, the radiation impedance of a plate in contact with the air and the damping of a structural system. Different boundary conditions are appropriated for computing each parameter, based on typical SEA simulation approaches [22].

5.5.2.1 Rigid wall or pressure release terminations

If the multilayer is backed by a rigid wall, the velocities at the last interface are equal to zero, which characterize the infinite impedance condition at the boundary. The number of equations will depend on the number of velocity degrees of freedom of the last layer. For poroelastic materials, three conditions are necessary. If the last layer is a solid, two equations should be added. Finally, one equation for the normal velocity is requested if the last layer is a fluid.

These conditions can be rearranged for being presented in a matrix form. In this context, a boundary matrix $[Y^{(n)}]$ can be found as

$$[Y^{(n)}]V^{(n)}(M_{2n}) = 0. \tag{5.59}$$

Eq. (5.59) needs to indicate that the velocity components in the vector $V^{(n)}$ are zero. The nonzero term of the matrix $[Y^{(n)}]$ should be given in according to the position of that components at the degrees of freedom vector. Table 5.4 shows a compilation of the boundary matrix for three cases of termination, where the index p, s and f represents poroelastic, solid, and fluid terminations, respectively.

Table 5.5 Pressure release boundary arrays

$[Y^p] = \begin{bmatrix} 0 & 0 & 0 & 1 & 0 & 0 \\ 0 & 0 & 0 & 0 & 1 & 0 \\ 0 & 0 & 0 & 0 & 0 & 1 \end{bmatrix}$	$[Y^s] = \begin{bmatrix} 0 & 0 & 1 & 0 \\ 0 & 0 & 0 & 1 \end{bmatrix}$	$[Y^f] = \begin{bmatrix} 0 & 1 \end{bmatrix}$

A similar approach is used for the pressure release condition, which means that the pressure/stress components are equal to zero at the termination, characterizing the state of infinite admittance at the boundary. Table 5.5 shows the boundary matrices if considering this case.

These matrices will compose the last lines of the global matrix $[TG]$, at the termination columns (related to the last degrees of freedom of the vector $V^{(n)}$), which is

$$[TG] = \begin{bmatrix} [\ldots & \ldots & TG_0 & \ldots & \ldots] \\ [0] & \ldots & [0] & \ldots & [Y^{(n)}] \end{bmatrix}. \tag{5.60}$$

The terms not related with the last layer are completed with zeros in the global transfer matrix. The global response vector in this case is composed by the original degrees of freedom of the problem $V_G = V_0$ (Eq. 5.56).

5.5.2.2 *Semi-infinite fluid termination*

For semi-infinite fluid termination, an extra air layer with a characteristic impedance $Z_0 \cos(\theta)$ is added at the end of the global matrix and two more degrees of freedom $V(B) = [p(B)\, v(B)]^{\mathrm{T}}$ will be added to the response vector V_0. First, the continuity equations will be applied as an extension of the Eq. (5.54) for the connection between the last layer and the semi-infinite fluid, then

$$[I_{(n)f}]V^{(n)}(M_{2n}) + [J_{(n)f}]V^{f}(B) = 0. \tag{5.61}$$

Where n can be p, s, or f according to the nature of the last layer. When the continuity is done, the impedance condition is applied in the fluid B by means of a boundary vector $[Y^{(sif)}]$ given by

$$[Y^{(sif)}] = [-1 \quad Z_B \cos\theta], \tag{5.62}$$

As expected, this vector has the same size as the $[Y^f]$ presented in the Table 5.4. The proper application of the characteristic impedance can be evaluated by checking the relation

$$[Y^{(sif)}]V^{f}(B) = 0. \tag{5.63}$$

In terms of global matrix, the continuity matrices and the boundary vector are applied as follows

$$[T_G] = \begin{bmatrix} \cdots & & & [0] \\ & D_0 & & [0] \\ \cdots & & & [0] \\ \cdots & & I_{(n)f} & J_{(n)f} \\ \cdots & & [0] & [Y^{(sif)}] \end{bmatrix} \tag{5.64}$$

with the new global response vector being $V_G = [V_0 \quad V_B]^{\mathrm{T}}$.

5.5.3 Direct solution: initial condition

If executing all the steps presented in the previous sections, one line will be missing for the global transfer matrix to become square. It is possible to extract the acoustical parameters from determinants of the global matrix, but this approach is computationally inefficient and it will not be presented in this section. A more efficient way includes an input excitation (velocity or pressure), which allows calculating a direct solution of a global linear system.

If fixing an arbitrary value for the initial velocity and eliminating the second column of T_G and second line of V_G, a constant velocity piston excitation is considered. This excitation should be informed in the right-hand side of the system of equations, which means that the eliminated column becomes an excitation vector P, that is

$$[T_G]^F V_G^F = -v_0 P, \tag{5.65}$$

where $[T_G]^F$ and V_G^F are the global matrix and response vector after eliminating the excitation degree of freedom, respectively, and v_0 is the amplitude of the input velocity. In Eq. (5.65), $[T_G]^F$ becomes square and a simple linear system can be solved for each frequency and incident angle, by using traditional numerical methods,

5.5.4 SEA indicators

Two acoustic indicators are important for the application studied in this chapter: the acoustic surface impedance Z_s of the multilayer and the transmission factor T. The acoustic impedance for a given frequency and incident angle is given by

$$Z_s(\omega, \phi) = \frac{V^F(1)}{v_0}, \tag{5.66}$$

where $V^F(1)$ is the first degree of freedom in the global response vector, which is the acoustic pressure at the surface in contact with the incident medium. The transmission factor is related to the sound insulation capacity of the treatment, that is

$$T(\omega,\theta) = (1+R)\frac{V_1(n-1)}{V_1(1)}, \tag{5.67}$$

where $R = (Z_s\cos(\theta) - Z_0)/(Z_s\cos(\theta) + Z_0)$ is the reflection coefficient of the NCT. As it is a transmission problem, Eq. (5.67) assumes that there is an acoustic medium in the reception side, it means that a semi-infinite fluid is considered as the termination condition. For absorption problems, a rigid wall termination is more representative for practical experiments. More information about these aspects is given in next sections.

5.5.4.1 Sound absorption from the TMM

The first important quantity to be presented in this section is the sound absorption coefficient. This amount represents the fraction of incident energy that is not reflected by the sample, related to the dissipation of energy into the NCT. The angle-dependent absorption coefficient α for a sample backed by a rigid wall (see Eq. (5.25)) can be also presented as [23]

$$\alpha(\omega,\theta) = 1 - |R(\theta,\omega)|^2 = \frac{4\text{Re}(Z_A)/\cos(\theta)}{|Z_A + 1/\cos(\theta)|^2}. \tag{5.68}$$

where $Z_A = \tilde{Z}_s(\omega,\theta)/Z_0$ is the normalized surface impedance of the treatment. If considering random incidence, the sound absorption coefficient should be integrated over the angle being calculated as follows

$$\alpha_d(\omega) = \frac{\int_0^{\theta_f} \alpha(\omega,\theta)\sin(\theta)\cos(\theta)d\theta}{\int_0^{\theta_f} \sin(\theta)\cos(\theta)d\theta}. \tag{5.69}$$

where θ_f is the final angle value of the integration range. According to Beranek [7], $\theta_f = 78°$ represents the field incidence, more indicated for comparing with reverberant chamber data. The Gauss' integration can be applied in Eq. 5.69 for improving the computational efficiency.

The surface sound absorption is then used in the dissipation loss factor of the SEA cavity in contact with the surface of the treatment. This parameter is calculated by the following equation

$$\eta = \frac{cA_t\overline{\alpha}}{4\pi fV}, \tag{5.70}$$

where c is the fluid sound velocity, A_t is the total surface area of the cavity, V is the volume, and $\overline{\alpha}$ is the surface mean absorption, given by

$$\overline{\alpha} = \frac{\sum A_i \alpha_d^{(i)}}{A_t}. \tag{5.71}$$

where A_i is the absorbing surface area i (e.g., the application area of the NCT) with absorption coefficient $\alpha_d^{(i)}$.

5.5.4.2 Insertion Loss from the TMM

The IL is defined as the ratio between the acoustic TL of a panel with the attached treatment over the TL of the untreated panel. This parameter can be calculated from the TMM simulation if considering a semi-infinite fluid as the boundary condition (fluid in both sides). The insertion coefficient IC will be given by

$$IC(\omega) = \frac{\int_0^{\theta_f} |T_{NCT}(\omega, \theta)|^2 \sin(\theta)\cos(\theta) \mathrm{d}\theta}{\int_0^{\theta_f} |T_{bare}(\omega, \theta)|^2 \sin(\theta)\cos(\theta) \mathrm{d}\theta}, \tag{5.72}$$

where T_{NCT} and T_{bare} are the transmission coefficients of the treated and untreated systems, respectively. T_{bare} is computed by running the TMM problem with a single plate layer, while T_{NCT} is solved considering plate + treatment. The IL $IL(\omega) = 10\log_{10}(IC(\omega))$ is recovered from the insertion coefficient. Finally, this parameter is applied as a scaling factor to the radiation and the mass law coupling loss factors in the SEA problem [24], then

$$\eta_{rad} = \eta_{rad}\, IC(\omega) = \eta_{rad}(10^{-IL(\omega)/10})\ ; \tag{5.73}$$

$$\eta_{mass} = \eta_{mass}\, IC(\omega) = \eta_{mass}(10^{-IL(\omega)/10})\ . \tag{5.74}$$

Fig. 5.7 shows the calculated IL for single- and double-wall configurations by using a bare plate (Plate 1) and an equivalent fluid porous model. An extra plate (Plate 2) is added at the extremity of the porous material to simulate the double-wall setup. The simplified MATLAB/OCTAVE code for resolution of this problem is presented in Section 5.6.1. Material properties are given in Table 5.6.

It can be noticed in Fig. 5.7 that the formulation of transfer matrices is capable of capturing the poor performance of the double-wall configuration in frequencies below the resonance of the spring-mass system ($f \approx 300$ Hz), represented by a negative value of the IL. An example code of these IL results applied to a simple SEA simulation is presented in Section 5.6.2.

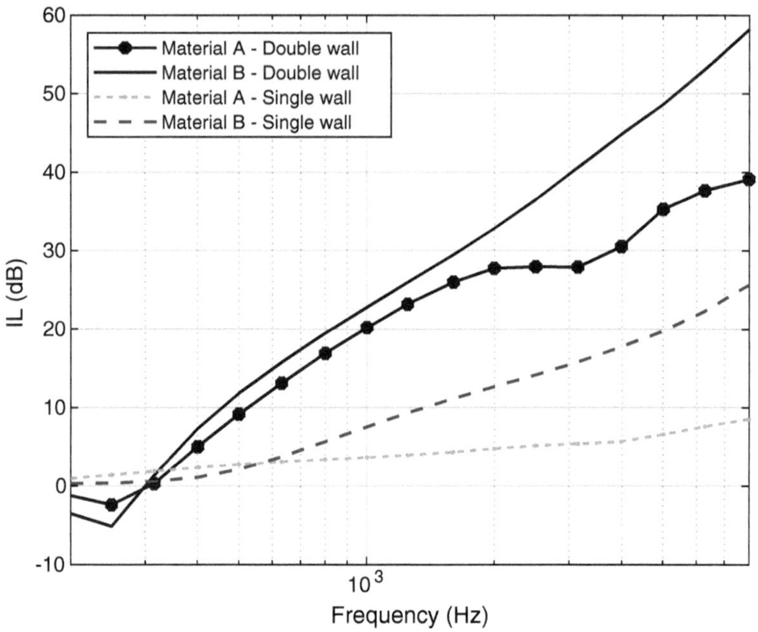

Figure 5.7 Insertion loss *IL* obtained from the TMM solver for single- and double-wall configurations [25].

Table 5.6 Material properties of the setups considered in this simulation [25].

	Solid			Porous	
	Plate 1		**Plate 2**	**Material A**	**Material B**
Thickness h (mm)	1.0		0.5	50	58
Young's modulus E (Pa)		71E9		-	-
Poisson's ratio ν (-)		0.33		-	-
Density ρ (kg/m^3)		2700		27	9.6
Flow resistivity σ (kN/m^4)		-		5	38.72
Porosity ϕ (-)		-		0.89	0.92
Tortuosity α_∞ (-)		-		1.37	1.06
Viscous charac. length Δ (μm)		-		213	37
Thermal charac. length Δ' (μm)		-		288	148

5.5.4.3 Damping loss factor from TMM

When the subsystem is excited by structural forces, it is also possible to apply a NCT in the simulation by using transfer matrices by assuming light coupling between structure and the sound package. In this scenario, the response of the treatment will be modeled as an equivalent loss factor that should be added to the dissipation factor of the solid plate [26].

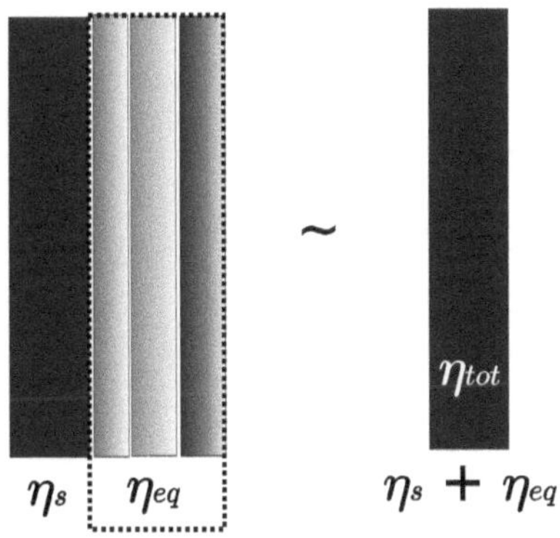

Figure 5.8 Illustration of a treated panel represented by structure and equivalent loss factors [26].

The TMM simulation of this problem consists in a bare plate + NCT system backed by a pressure release boundary condition as illustrated in Fig. 5.8. The propagating wavenumber in this case is not angle dependent by assuming normal structural excitation, then $k_t = k_b$, where k_b is the bending wavenumber of the structural system [24]. A total loss factor $\eta_{tot} = \eta_s + \eta_{eq}$ represents the equivalent treated SEA subsystem, where η_s is the loss factor of the single solid plate.

The equivalent loss factor is computed from the ratio between the input power and dissipated powers of the plate which can be extracted from classical SEA power balance equations (see Eq. (5.4)). This energy parameter can be wrote as

$$\eta_{eq} = \frac{\Pi_{input} S}{\omega M < v^2 >} \tag{5.75}$$

where S is the surface area, M is the total mass, and $< v^2 >= (v_0/2)^2$ is the mean square velocity of the SEA bare plate (also applied to the TMM simulation). The Π_{input} is computed from

$$\Pi_{input} = \frac{\mathrm{Re}(p(\omega)\ v0^*(\omega))}{2} \tag{5.76}$$

where p represents the normal pressure at the bare panel rear side, which is obtained after the solving of a full TMM linear system, and v_0 is the input amplitude velocity applied at the front side. As the input excitation is a normal incidence wave, the pressure and velocity parameters do not depend on the angle in this situation. Fig. 5.9 shows the velocity level of a Plate-FOAM system when the foam is modeled as (a) an equivalent fluid

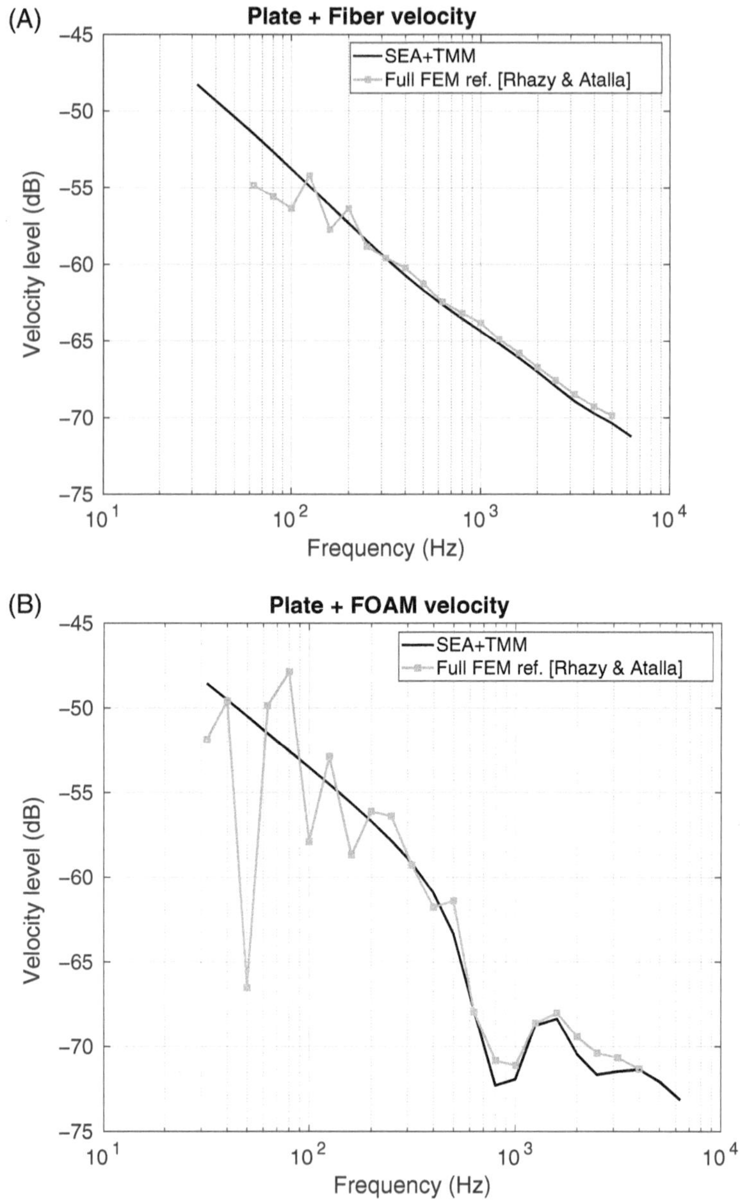

Figure 5.9 Velocity level of a plate + NCT system: comparison between SEA/TMM and full FEM analysis from [26].

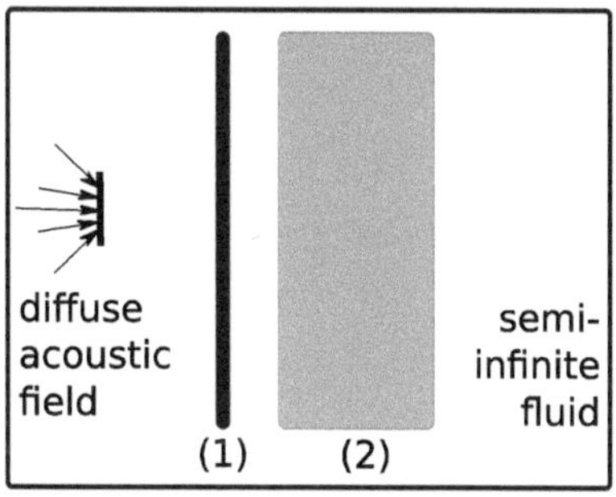

Figure 5.10 TMM problem composed by a (1) bare panel layer and a (2) equivalent fluid (fiber) material.

and (b) a poroelastic material. The SEA simulation is compared to full FEM simulation extracted from [26].

It can be seen in Fig. 5.9 that the statistical method is not capable of capturing the modal effects in the velocity response at low frequencies. At high frequency, the method shows good agreement with the deterministic result.

5.6 Numerical examples in OCTAVE/MATLAB

This section brings some application codes that simulates a treated plate system that can be solved by using the proposed energy approaches. The TL problem was used for exemplifying the capabilities of the SEA method together with the TMM formulation. Section 5.6.1 solves a thin plate + equivalent fluid model in terms of the IL parameter. This quantity is applied to the SEA code for a TL problem numerically exemplified in Section 5.6.2.

5.6.1 Simple TMM plate-fiber problem

Fig. 5.10 illustrates the problem approached in the following code. The physical properties of the Plate 1 for the bare panel and Material B for the equivalent fluid material were considered in this example. All the properties are presented in Table 5.1.

TMM code

```
%% TRANSFER MATRIX METHOD
%%% This code simulates the insertion loss of a plate-porous system by using
%%% the Transfer Matrix Method
```

```
clear all; close all;

v0=1; % input velocity
Ntheta = 1000; % number of angles
theta_final = 78; % THETA in degrees -field incidence

f = [250 315 400 500 630 800 1000 1250 1600 2000 2500 ...
   3150 4000 5000 6300 8000]; % third octave freq vector
j = sqrt(-1); % complex operator

  %% Air props
   Po = 101320; % atm pressure
   mu = 1.83e-5; % dynamic viscosity
   gama = 1.4; % cp/cv
   Pr = 0.71; % Prandtl number
   c0 = 343; % Sound velocity
   rho0 = 1.21; % Fluid density

  %% MATERIAL B props
   rho_1 = 9.6 ; % fiber density
   sigma = 3.872e+004; % flow resistivity
   por = 0.92; % porosity
   tor = 1.06; % tortuosity
   vis_length = 3.7e-005; % viscous length
   ter_length = 0.000148; % thermal length
   h_fib = 0.050; % thickness

  %% SOLID properties
   hplate = 0.001; rho_pla = 2700; Es = 71e9;
   eta1 = 0.01; Es = Es*(1+j*eta1); poiss = 0.33;
   D_bend = Es*hplate^3/(12*(1-poiss^2)); % bending stiffness

%% Layers [type     h ]
layers = [1      h_fib       % fluid layer
          2      hplate    ];      % thin plate

num_layers = length (layers (: ,1));
TM_fluid = zeros (2 ,2);
TM_solid = zeros (2 ,2);
```

```
Z0 = rho0*c0;

%% from eq.fluid simulation - complex wavenumber and equivalent density
   %%%%             % keq                          rho_eq
JCA_simulation = [1.4518e+01 - 2.7297e+00i 9.3692e+00 - 3.0720e+00i
                  1.7704e+01 - 4.0580e+00i 8.7653e+00 - 3.5359e+00i
                  2.1442e+01 - 5.9306e+00i 7.9679e+00 - 3.9285e+00i
                  2.5286e+01 - 8.1614e+00i 7.0980e+00 - 4.1453e+00i
                  2.9580e+01 - 1.0897e+01i 6.1452e+00 - 4.1676e+00i
                  3.4393e+01 - 1.4013e+01i 5.2060e+00 - 3.9731e+00i
                  3.9408e+01 - 1.6994e+01i 4.4473e+00 - 3.6380e+00i
                  4.5237e+01 - 1.9885e+01i 3.8382e+00 - 3.2219e+00i
                  5.3147e+01 - 2.2887e+01i 3.3357e+00 - 2.7416e+00i
                  6.2163e+01 - 2.5439e+01i 3.0101e+00 - 2.3353e+00i
                  7.3522e+01 - 2.7930e+01i 2.7750e+00 - 1.9782e+00i
                  8.8403e+01 - 3.0588e+01i 2.5970e+00 - 1.6666e+00i
                  1.0791e+02 - 3.3594e+01i 2.4580e+00 - 1.4034e+00i
                  1.3077e+02 - 3.6793e+01i 2.3539e+00 - 1.2038e+00i
                  1.6024e+02 - 4.0647e+01i 2.2622e+00 - 1.0355e+00i
                  1.9830e+02 - 4.5338e+01i 2.1785e+00 - 8.9362e-01i
                  2.4248e+02 - 5.0477e+01i 2.1076e+00 - 7.8405e-01i];

for ifreq=1: length(f) % loop frequency

freq = f (ifreq);
omega = 2*pi*freq;

for itheta = 1:Ntheta % loop theta

dtheta = (theta_final/90)*(pi/2)/(Ntheta-1); % trapezoidal integration
theta = (itheta - 1)*dtheta;
kt = (omega/c0)*sin(theta);

   for l=1: length(layers (:,1)) % loop layers

    if layers(l,1)==1 % equivalent fluid

       h=layers (l ,2);
       ka = JCA_simulation(ifreq ,1);
```

```
        rho= JCA_simulation(ifreq ,2);

        k3 = sqrt(ka^2 - (kt)^2); % this is k3

        TM_fluid = [       cos(k3*h)        j*omega*rho*sin(k3*h)/k3
                          j*k3*sin(k3*h)/(omega*rho)
cos(k3*h)];
    else % solid thin plate

    m = rho_pla*hplate; % surface mass
    Z = j*omega*m*(1 - D_bend*kt^4/(omega^2*m));
    Ts11 = 1; Ts12 = -Z; Ts21 = 0; Ts22 = 1;

    TM_solid = [Ts11 Ts12
                Ts21        Ts22 ] ;
   end % end if fluid

end % end loop layers

%% ASSEMBLING

Ift = eye(2,2); Jft = [-1 0; 0 1]; % fluid - thin plate/septum
Itf = Jft; Jtf = Ift;
D0_11 = Ift;
D0_12 = Jft*TM_solid;
D0_22 = Itf;
D0_23 = Jtf*TM_fluid;
D0_21 = zeros(size(D0_11));
D0_13 = zeros(size(D0_23));

D0 = [D0_11 D0_12 D0_13
    D0_21 D0_22 D0_23];

%%%%%%%%%%%%%%%%%%%%%%%%%%%%%%%%%%%%%%%%%%%%%
%%%% Applying boundary conditions – rigid wall for last layer

    Zc = Z0/cos(theta); % sif impedance
    bcY = zeros(1,length(D0(1,:))); % zeros for the boundary line
```

```
   bcY(end-1) = -1; bcY(end) = Zc; % [-1 Zc] for sif
  %bcY(end-1) = 0; bcY(end) = 0; % [0 1] for rigid
      D = [D0
         bcY];
%%%%%%%%%%%%%%%%%%%%%%%%%%%%%%%%%%%%%%%%
%%%% Piston excitation
F = -D(:,2)*v0;% initial velocity degree of freedom - 2nd column
D_final =[D(:,1) D(:,3:end)]; % eliminating 2nd column from D
V2 = D_final_F; % solving linear system

%%%%%%%%%%%%%%%%%%%%%%%%%%%%%%%%%%%%%%%%
%%% Acoustic indicators

Zs = V2(1)/v0; % surface impedance
R = (Zs*cos(theta) - Z0)./(Zs*cos(theta) + Z0); % reflection coeff.

T1 = (1+R).*V2(end-1)/V2(1);% transmission factor
T1sq(itheta) = abs(T1)$^2$*sin(theta)*cos(theta); % square for integration

 %% BARE PANEL response
  sigma33_bare = -1/(1-(Z/Zc));
  va_bare = -sigma33_bare/Zc;
  Zs_bare = 1/va_bare;
  R_bare = (Zs_bare*cos(theta) - Z0)./(Zs_bare*cos(theta) + Z0);

  T_bare = (1+R_bare)*(-sigma33_bare);
  Tbare_sq(itheta) = abs(T_bare)$^2$*cos(theta)*sin(theta);

end % end loop theta

  TA = trapz(T1sq*dtheta); % trapezoidal integration treated plate
  Tbare_int =trapz(Tbare_sq*dtheta); % trapezoidal integration bare plate
  IC(ifreq) = TA/Tbare_int; % insertion coefficient

end % end loop freq

IL = -10*log10(IC); % Insertion loss
      format shortE
IL_show = [f' IL']
```

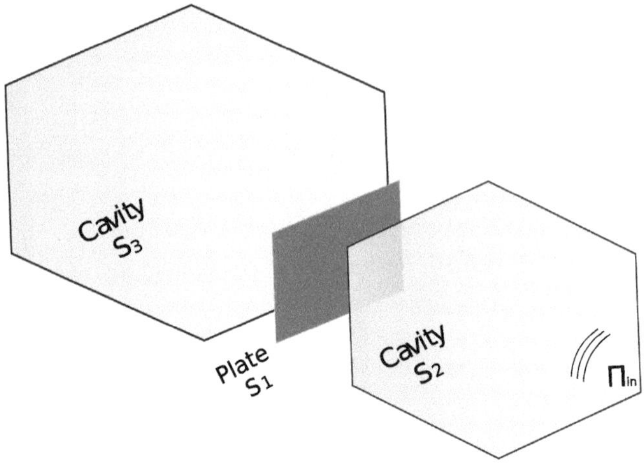

Figure 5.11 Example of a sound transmission loss problem.

The result from this simulation is the treatment's IL parameter. The obtained value for the studied system can be applied to the SEA simulation, as explained in the next section.

5.6.2 Importing the Insertion Loss to the SEA model

The obtained IL can be applied to a structural system used in automotive or aerospace industry to simulate the NCT performance in the design phase. A SEA model cavity-plate-cavity is approached in this section and the numerical results are compared with experimental data extracted from [27]. The SEA problem is composed by the flexural energy of the plate and longitudinal energy of the cavities. Fig. 5.11 illustrates the proposed example. Fig. 5.12 shows the Sound Transmission Loss of the proposed system, by considering the Insertion Loss extracted from Example 5.6.1.

The problem shown in Fig. 5.11 can be represented by the following linear system

$$\omega\begin{bmatrix} \eta_{1,diss}+\eta_{12}+\eta_{13} & -\eta_{12} & -\eta_{13} \\ -\eta_{21} & \eta_{2,diss}+\eta_{21}+\eta_{23} & -\eta_{23} \\ -\eta_{31} & -\eta_{32} & \eta_{3,diss}+\eta_{31}+\eta_{32} \end{bmatrix}\begin{Bmatrix} E_1 \\ E_2 \\ E_3 \end{Bmatrix}=\begin{Bmatrix} 0 \\ \Pi_{in} \\ 0 \end{Bmatrix}. \tag{5.77}$$

The mean square value of the pressure inside the cavities $p_i^2 = E_i\rho_i c_i^2/V_i$ can be recovered from the solution, with ρ_i being the fluid density, c_i the sound velocity, and V_i the volume of the cavity $i = 2, 3$. The TL of the partition

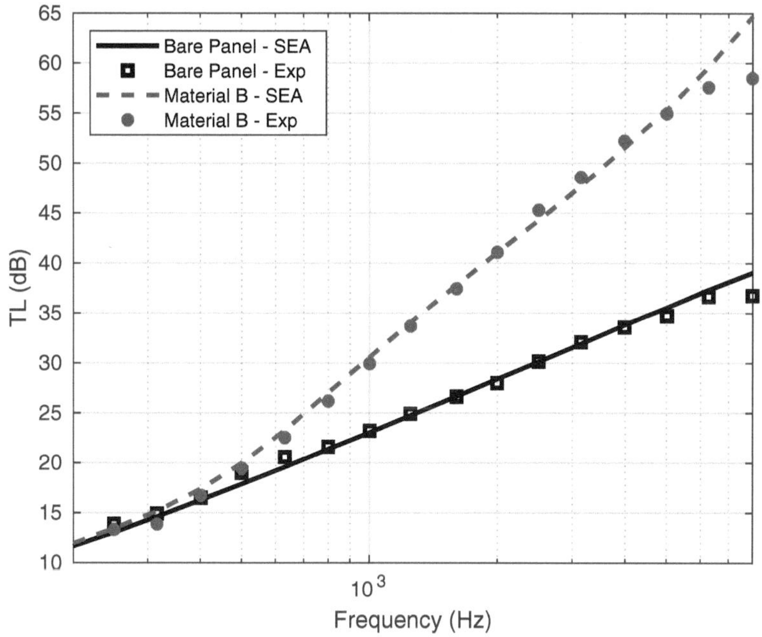

Figure 5.12 Sound transmission loss obtained for the single-wall configuration Plate 1 + Material B. Experimental results are extracted from [27].

can be calculated as [28]

$$TL = 10\log_{10}\left(\frac{p_2^2}{p_3^2}\right) + 10\log_{10}\left(\frac{S}{A_{eq}}\right), \tag{5.78}$$

where $A_{eq} = 4\omega V_3 \eta_{3,diss}/c_3$ is the equivalent absorption area of the reception cavity and S is the area of the panel, being V_3 the volume and c_3 the sound velocity of subsystem 3.

```
clear all; close all
% FREQUENCY 1/3 octave
freq = [250 315 400 500 630 800 1000 1250 1600 2000 2500 3150 ...
   4000 5000 6300 8000];

   % Input Insertion Loss from TMM - Plate + Material B
insertion_tmm = [4.4799e-01 6.3228e-01 1.1387e+00 2.1244e+00
3.7099e+00...
   5.6754e+00 7.5447e+00 9.3200e+00 1.1134e+01 1.2660e+01
1.4151e+01...
   1.5792e+01 1.7730e+01 1.9778e+01 2.2270e+01 2.5598e+01];
```

```
% Experimental data from the bibliography
Exp_TL = [13.4139 15.1725 17.3777 19.9746 23.2683 27.0320 30.5854
34.0621...
    37.7966 41.0511 44.2793 47.7323 51.5522 55.3433 59.6216 64.6599];
Exp_bare_panel=[13.886 14.910 16.504 19.008 20.601 21.626 ...
    23.219 24.926 26.634 28.000 30.162 32.097 33.577 34.715 36.650
36.764];
nfreqs = length(freq);

%%% SEA subsystems properties
    plate_dim = [1.006 1 0.001]; % [l b h]
    plate_props = [2700 71e9 0.33]; %[rho Youngs poisson]
    eta_diss_plate = 0.01; % diss. loss factor

    cav_dim = [70 190 147.54
               72 209 199];
    cav_props = [1.21 343
                 1.21 343];
    eta_diss_cav = 0.01;
    inp_pow = 1;

    n11 = eta_diss_plate;
    n22 = eta_diss_cav; n33=n22;

  %%% coupling factors from classical models of sig rad and tau
  %%% n12 n13 n21 n23 n31 n32(Section 1.2.2)
coup_fac=[...
1.6652e-04    1.6652e-04    1.5126e-05    3.1329e-05    1.1619e-05
2.3228e-05
1.4931e-04    1.4931e-04    8.8963e-06    1.6776e-05    6.7893e-06
1.2438e-05
1.3363e-04    1.3363e-04    5.1081e-06    8.6818e-06    3.8764e-06
6.4367e-06
1.2074e-04    1.2074e-04    3.0303e-06    4.6454e-06    2.2897e-06
3.4441e-06
1.0900e-04    1.0900e-04    1.7605e-06    2.4101e-06    1.3253e-06
1.7869e-06
9.8447e-05    9.8447e-05    1.0037e-06    1.2139e-06    7.5329e-07
9.0001e-07
```

```
8.9930e-05    8.9930e-05    5.9467e-07    6.3611e-07    4.4524e-07
4.7162e-07
8.2633e-05    8.2633e-05    3.5347e-07    3.3185e-07    2.6415e-07
2.4604e-07
7.5933e-05    7.5933e-05    2.0013e-07    1.6088e-07    1.4930e-07
1.1928e-07
7.1127e-05    7.1127e-05    1.2079e-07    8.3357e-08    9.0003e-08
6.1801e-08
6.7594e-05    6.7594e-05    7.3865e-08    4.3089e-08    5.4984e-08
3.1947e-08
6.5503e-05    6.5503e-05    4.5290e-08    2.1712e-08    3.3686e-08
1.6098e-08
6.5581e-05    6.5581e-05    2.8225e-08    1.0673e-08    2.0979e-08
7.9130e-09
6.8901e-05    6.8901e-05    1.9030e-08    5.4911e-09    1.4138e-08
4.0711c-09
7.8878e-05    7.8878e-05    1.3754e-08    2.7560e-09    1.0213e-08
2.0433e-09
1.0882e-04    1.0882e-04    1.1789e-08    1.3504e-09    8.7513e-09
1.0012e-09];


%% SEA EQUATIONS
for ifreq = 1:nfreqs

    f=freq(ifreq);
    w = 2*pi*f; % circular frequency omega

n12 = coup_fac(ifreq,1); n13 =coup_fac(ifreq,2);
n21 = coup_fac(ifreq,3); n23 = coup_fac(ifreq,4);
n31 = coup_fac(ifreq,5); n32 = coup_fac(ifreq,6);

  for ind_il = 1:2 % 1 for bare panel | 2 for treated panel

    if ind_il == 2 % it is treated panel
    IL = insertion_tmm(ifreq);
    n13 = n12*(10$^(-IL/10)$); n31 = n21*10$^(-IL/10)$; % plate-cav
    n23 = n23*10$^(-IL/10)$; n32 = n32*10$^(-IL/10)$; % cav-cav
    end
```

```
SEA_mat = [n11+n12+n13-n12 -n13
          -n21n22+n21+n23-n23
          -n31-n32n31+n32+n33];
Pow = [0;1;0]; % input power in the cavity
E = linsolve(SEA_mat,Pow/w);

for icav = 1:2

V = cav_dim(icav,3);
rho = cav_props(icav,1);
c =cav_props(icav,2);

p_sq(icav) = E(icav+1)*rho*c/V;
   if icav==2 % reception
   Aeq = 4*w*V*eta_diss_cav/c;
     end
end

S = plate_dim(1)*plate_dim(2); % surface area of the panel
TL(ifreq, ind_il) =10*log10(p_sq(1)/p_sq(2)) + 10*log10(S/Aeq);

   end %loop IL
end % loop freq

a = colormap(gray);
figure()
set(gca, 'fontsize', 15)
semilogx(freq,TL(:,1), 'color', a(1,:), 'linewidth', 2 ); hold on
semilogx(freq,Exp_bare_panel, 's', 'color', a(1,:), 'linewidth', 2); hold on
semilogx(freq,TL(:,2),':.','color',a(140,:), 'linewidth', 3); hold on
semilogx(freq,Exp_TL, 's','color',a(140,:), 'linewidth', 2); hold on
xlim([200 8000])
ylim([10 65])
ylabel('TL [dB]')
xlabel('Frequency [Hz]')
grid on
legend('Bare Panel - SEA', 'Bare Panel - Exp', 'Treated Panel - SEA',...
   'Treated Panel - Exp','Location', 'Northwest')
```

5.7 Summary

This chapter presented two popular modeling techniques in the automotive and aerospace industries for high-frequency analysis. The SEA, a classical method based on the energy exchanging between subsystems, was primarily introduced and discussed. The second part of this chapter was composed of a brief introduction of porous materials modeling and the presentation of the TMM, which is a wave-based approach used to evaluate the performance of a multilayer NCT.

Three indicators are involved in the application of an NCT at a surface: the IL, the sound absorption coefficient, and the damping loss factor. The first one is important when a trim is applied at a structural partition since this parameter models the sound TL capacity of the system. The second mechanism is related to the damping of the acoustic media in contact with the surface of the NCT. The application of this kind of treatment in a small room, for example, tends to increase the mean absorption that changes the reverberating characteristics of the acoustic environment. Finally, when the treatment covers the surface of a vibrating solid, the structural damping loss factor is also increased by the equivalent damping of the trim.

The main contribution of this chapter is the presentation of a modeling approach that provides the coupling between the TMM method and SEA for solving simple acoustic-structural problems with trimmed subsystems. The procedure for obtaining the mentioned indicators was presented with coding examples. Practical results obtained from different techniques were compared.

References

1. F.J. Fahy, Statistical energy analysis: a critical overview, Philos. Trans. R. Soc. Lond. Ser. A: Phys. Eng. Sci. 346 (1681) (1994) 431–447.
2. R.H.Lyon, R.G.DeJong, M. Heckl, Theory and application of statistical energy analysis, 1995.
3. K.D. Langhe, High frequency vibrations: contributions to expiremental and computational sea parameter identification techniques, Ph.D. thesis, Katholieke Universiteit Leuven, 1996.
4. J.A. Cordioli, S.N. Gerges, A.K. Pererira, M. Carmo, C. Grandi, Vibro-acoustic modeling of aircrafts using statistical energy analysis, in: Proceedings of the Brazilian Congress on Mechanical Engineering, Natal, RN, Brazil, 2004.
5. D. Blanchet, M. Feist-Muench, Innovative acoustic material concept integration into vehicle design process, in: SAE Technical Paper, SAE International, 2020. 10.4271/2020-01-1527
6. D. Johansson, P. Comnell, Statistial Energy Analysis Software., Development and Implementation of an Open Source Code in Matlab/Octave (2010).

7. L.L. Beranek, I.L. Ver, Noise and Vibration Control Engineering: Principles and Applications, Wiley, Cambridge, USA, 1992, p. 814..
8. F.G. Leppington, E.G. Broadbent, K. Heron, The acoustic radiation efficiency of rectangular panels, Proc. R. Soc. Lond. A. Math. Phys. Sci. 382 (1783) (1982) 245–271.
9. J.A.Cockburn, A.C. Jolly, Structural-acoustic response, noise transmission losses and interior noise levels of an aircraft fuselage excited by random pressure fields, Technical Report, Wyle Labs Inc Huntsville al Testing Div, 1968.
10. J.J. Wijker, Random Vibrations in Spacecraft Structures Design: Theory and Applications, 165, Springer Science & Business Media, Netherlands, 2009.
11. L.L. Beranek, Acoustics, Acoustical Society of America, New York, USA, 1954.
12. J. Allard, N. Atalla, Propagation of Sound in Porous Media: Modelling Sound Absorbing Materials, second ed., Wiley, United Kingdom, 2009.
13. H. Kuttruff, Acoustics: An Introduction, Taylor & Francis e-Library, New York, USA, 2007.
14. M. Delany, E. Bazley, Acoustical Properties of Fibrous Absorbent Materials, Appl. Acoust. 3 (2) (1970) 105–116.
15. D.L. Johnson, J. Koplik, R. Dashen, Theory of dynamic permeability and tortuosity in fluid-saturated porous media, J. Fluid Mech. 176 (1987) 379–402.
16. J.-F. Allard, Y. Champoux, New empirical equations for sound propagation in rigid frame fibrous materials, J. Acoust. Soc. Am. 91 (6) (1992) 3346–3353.
17. C. Zwikker, C.W. Kosten, Sound Absorbing Materials, Elsevier, Netherlands, 1949.
18. W.T. Thomson, Transmission of elastic waves through a stratified solid medium, J. Appl. Phys. 21 (2) (1950) 89–93.
19. D. Folds, C. Loggins, Transmission and reflection of ultrasonic waves in layered media, J. Acoust. Soc. Am. 62 (5) (1977) 1102–1109.
20. W. Lauriks, P. Mees, J.F. Allard, The acoustic transmission through layered systems, J. Sound Vibr. 155 (1) (1992) 125–132.
21. M.A. Biot, Theory of propagation of elastic waves in a fluid-saturated porous solid. II. Higher frequency range, J. Acoust. Soc. Am. 28 (2) (1956) 179–191.
22. N. Atalla, F. Sgard, Finite element and boundary methods in structural acoustics and vibration, CRC Press, Florida, USA, 2015.
23. S.-I. Thomasson, On the absorption coefficient, Acta Acust. United Acust. 44 (4) (1980) 265–273.
24. N. Atalla, Structural and acoustic noise control material modeling, in: S.A. Hambric, S.H. Sung, D.J. Nefske (Eds.), Engineering Vibroacoustic Analysis: Methods and Applications, Wiley, United Kingdom, 2017, pp. 230–265.
25. T. Oliveira, Numerical evaluation of the sound transmission of a structural system with attached noise control treatments using statistical energy analysis, in: Proceedings of International Conference in Structural Dynamics (EURODYN), Athens, Greece, 2020.
26. D. Rhazi, N. Atalla, Transfer matrix modeling of the vibroacoustic response of multi-materials structures under mechanical excitation, J. Sound Vibr. 329 (13) (2010) 2532–2546.
27. F.L.V.Q. Kulakauskas, Avalição do Uso de Materiais Porosos na Perda de Transmissão de Painéis Duplos (in portuguese), Master Thesis, Universidade Federal de Santa Catarina, 2016.
28. ISO:10140-2, Laboratory Measurement of Sound Insulation of Building Elements: *Measurement of Air-borne Sound Insulation*. International Standard Organization, 2010.

CHAPTER SIX

Failure and fatigue life due to random vibration in aircraft applications

Essam A. Al-Bahkali[a], Hisham Elkenani[a], Mhamed Souli[b]
[a]Department of Mechanical Engineering, King Saud University, Riyadh, Saudi Arabia
[b]Laboratoire de Mécanique, de Lille, UMR CNRS, Villeneuve d'Ascq, France

6.1 Introduction

Most real fatigue loadings are random processes in respect of frequency and stress amplitude. Estimating fatigue damage with power spectrum density (PSD) was first proposed by Rice [1] in 1954. The stress PSD represents the frequency domain approach input into the fatigue. This is a scalar function that describes how the power of the time signal is distributed among frequencies [2]. Mathematically, this function can be obtained by using a Fourier transform (FT) of the stress time history's autocorrelation function, and its area represents the signal's standard deviation. It is clear that PSD is the most complete and concise representation of a random process [3]. Finite element analysis is very efficient in taking the PSD of applied loads and determining the PSD of the resulting stresses at various points in the structure. A method of taking the PSD of stress and calculating fatigue lives therefore has attractions. Much of the early work on fatigue analysis from PSD's was carried out by NASA in order to determine the fatigue damage caused by vibration and buffeting of space vehicles.

6.2 Random response analysis

Random response linear dynamic analysis is used to predict the response of a structure subjected to a nondeterministic continuous excitation that is expressed in a statistical sense by a cross-spectral density matrix. The random response procedure uses the set of eigenmodes extracted in a previous eigenfrequency step to calculate the corresponding PSDs of response variables (stresses, strains, displacements, etc.) and, hence—if required—the variance and root mean square (RMS) values of these same

Multiphysics Simulations in Automotive and Aerospace Applications.
DOI: 10.1016/B978-0-12-817899-7.00009-5

variables. This section provides brief definitions and explanations of the terms used in this type of analysis based on the book by Clough and Penzien [4].

Examples of random response analysis are the study of the response of an airplane to turbulence; the response of a car to road surface imperfections; the response of a structure to noise, such as the "jet noise" emitted by a jet engine; and the response of a building to an earthquake.

Since the loading is nondeterministic, it can be characterized only in a statistical sense. We need some assumptions to make this characterization possible. Although the excitation varies in time, in some sense it must be stationary—its statistical properties must not vary with time. Thus, if $x(t)$ is the variable being considered (such as the height of the road surface in the case of a car driving down a rough road), then any statistical function of $x, f(x)$, must have the same value regardless of what time origin we use to compute f:

$$f(x(t)) = f(x(t+\tau)) \text{ for any } \tau. \tag{6.1}$$

We also need the excitation to be ergodic. This term means that, if we take several samples of the excitation, the time average of each sample is the same. These restrictions ensure that the excitation is statistically constant.

The following subsections explain some concepts used in random response analysis.

6.2.1 Statistical measures

The mean value of a random variable $x(t)$ over an interval T sec is:

$$E(x) = \lim_{T\to\infty} \frac{1}{T} \int_{-\frac{T}{2}}^{\frac{T}{2}} x(t)\, dt. \tag{6.2}$$

Since the dynamic response is computed about a static equilibrium configuration, the mean value of any dynamic input or response variable will always be zero:

$$E(x) = 0. \tag{6.3}$$

The variance of a random variable measures the average square difference between the point value of the variable and its mean:

$$\sigma_r^2(x) = \lim_{T\to\infty} \frac{1}{T} \int_{-\frac{T}{2}}^{\frac{T}{2}} (x(t) - E(x))^2\, dt. \tag{6.4}$$

Since $E(x) = 0$ for our applications, the variance is the same as the mean square value:

$$E\left(x^2\right) = \lim_{T\to\infty} \frac{1}{T} \int_{-\frac{T}{2}}^{\frac{T}{2}} x^2(t)\, dt. \tag{6.5}$$

The units of variance are (amplitude)^2, and are expressed as RMS values:

$$\sigma_r(x) = \sqrt{\sigma_r^2}. \tag{6.6}$$

6.2.2 Correlation

Correlation measures the similarity between two variables. Thus, the cross-correlation between two random functions is the integration of the product of the two variables x_1 and x_2, with one of them shifted in time by some fixed value τ to allow for the possibility that they are similar but shifted in time. The cross-correlation function $R_{x_1x_2}(\tau)$ for the variables is defined as:

$$R_{x_1x_2}(\tau) = \lim_{T\to\infty} \frac{1}{T} \int_{-\frac{T}{2}}^{\frac{T}{2}} x_1(t)\, x_2(t+\tau)\, dt. \tag{6.7}$$

Since the mean value of any variable is zero, on average each variable has equal positive and negative content. If the variables are quite similar, their cross-correlation (for some values of τ will be large; if they are not similar, the product $x_1x_2$will sometimes be negative and sometimes positive so that the integral over all time will provide a much smaller value, regardless of the choice of τ). A simple result is:

$$R_{x_1x_2}(\tau) = R_{x_2x_1}(-\tau). \tag{6.8}$$

The autocorrelation is the cross-correlation of a variable with itself. Then the autocorrelation of a variable $x(t)$ is defined as:

$$R(\tau) = \lim_{T\to\infty} \frac{1}{T} \int_{-\frac{T}{2}}^{\frac{T}{2}} x(t)x(t+\tau)\,dt. \tag{6.9}$$

Clearly, as $\tau \to o$, $R(\tau) \to \sigma_r^2$, the autocorrelation equals the variance (the mean square value).

The autocorrelation, $R(\tau)$ tells us about the nature of the random variable. If $R(\tau)$ drops off rapidly as the time shift τ moves away from $\tau = 0$, the variable has a broad frequency content; if it drops off more slowly and exhibits a cosine profile, the variable has a narrow frequency content centered around the frequency corresponding to the periodicity of $R(\tau)$.

This concept is extended to detect the frequency content of a random variable by cross-correlating the variable with a sine wave: sweeping the wave over a range of frequencies and examining the cross-correlation tells us whether the random variable is dominated by oscillation at particular frequencies. We begin to see that the nature of stationary, ergodic random processes is best understood by examining them in the frequency domain.

We can write $x(t)$, which contains many discrete frequencies in terms of a Fourier series expanded in N steps of a fundamental frequency ω_0:

$$x(t) = \sum_{n=1}^{N} [a_n \cos(in\omega_0 t) + b_n \sin(in\omega_0 t)] \tag{6.10}$$

where a_n and b_n are Fourier coefficients.

Keeping in mind that we will be interested only in the real part, the variance of $x(t)$ can be expressed as:

$$\sigma_r^2(x) = \lim_{T\to\infty} \frac{1}{T} \int_{-\frac{T}{2}}^{\frac{T}{2}} \sum_{n=-N}^{N} A_n^* \exp(-in\omega_0 t) A_n \exp(in\omega_0 t)dt \tag{6.11}$$

where A_n is the complex amplitude of the nth term, A_n^* is the complex conjugate of A_n.

Using the orthogonality of Fourier terms. Continuing:

$$\sigma_r^2(x) = \lim_{T\to\infty}\frac{1}{T}\int_{-\frac{T}{2}}^{\frac{T}{2}}\sum_{n=-N}^{N} A_n^* A_n dt.$$

$$= \sum_{n=-N}^{N} A_n^* A_n = \sum_{n=-N}^{N} |A_n|^2 = \sum_{n=-N}^{N} \sigma_r^2(x_n) \tag{6.12}$$

where x_n is the nth component of the Fourier series.

The contribution to the variance of x, $\sigma_r^2(x)$, at the frequency $f_n = n\omega_0/(2\pi)$, per unit frequency, is thus:

$$S_x(f_n) = \frac{2\pi}{\omega_0}A_n A_n^* \tag{6.13}$$

since we are stepping up the frequency range in steps of $\Delta f = \omega_0/(2\pi)$. The variance can, therefore, be written as:

$$\sigma_r^2(x) = \sum_{n=-N}^{N} S_x(f_n)\Delta f \tag{6.14}$$

As we examine x as a function of frequency, $S_x(f_n)$ tells us the amount of "power" (in the sense of mean square value) contained in x, per unit frequency, at the frequency f_n. As we consider smaller and smaller intervals, $\Delta f \to 0$, S_x is the PSD of the variable x:

$$\sigma_r^2(x) = \int_{-\infty}^{\infty} S_x(f)df \tag{6.15}$$

6.2.3 Random response analysis

For a system excited by some random loads or prescribed base motions, which are characterized in the frequency domain by a matrix of cross-spectral density functions, $S_{NM}(f)$, we can think of N and M as two of the degrees of freedom of the finite element model that are exposed to the random loads or prescribed base motions. In typical applications the range of frequencies will be limited to those to which we know the structure will respond—we do not need to consider frequencies that are higher than the modes in which we expect the structure to respond.

The values of $S_{NM}(f)$ might be provided by fast Fourier transform of the cross-correlation of time records or by the FT of the autocorrelation of a single time record, together with known geometric data. The system will respond to this excitation. We are usually interested in looking at the PSDs of the usual response variables—stress, displacement, etc. The PSD history of any particular variable will tell us the frequencies at which the system is most excited by the random loading.

An overall picture is provided by looking at the variance (the mean square value) of any variable; the RMS value is provided for this purpose. The RMS value is used instead of variance because it has the same units as the variable itself. It is computed by integrating the single-sided PSD of the variable, $\tilde{S}_x$ over the frequency range.

The transformation of the problem into the frequency domain inherently assumes that the system under study is responding linearly.

6.2.4 The frequency response function

Random response is studied in the frequency domain. Therefore, we need the transformation from load to response as a function of frequency. Since the random response is treated as the integration of a series of sinusoidal vibrations. The discrete (finite element) linear dynamic system has the equilibrium equation:

$$\delta u^N (M^{NM} \ddot{u}^M + C^{NM} \dot{u}^M + K^{NM} u^M) = \delta u^N F^N \tag{6.16}$$

where M^{NM} is the mass matrix, C^{NM} is the damping matrix, K^{NM} is the stiffness matrix, F^N are the external loads, u^N is the value of degree of freedom N of the finite element model, and δu^N is an arbitrary virtual variation.

The problem is projected onto the eigenmodes of the system, which are first extracted from the undamped system. The eigenmodes are orthogonal across the mass and stiffness matrices.

6.2.5 Von Mises stress computation

This is a quite critical issue, which has been largely investigated in the literature [5-9]. Noting that the von Mises stress is a quadratic stress function, therefore it is a non-Gaussian stress with positive mean value, even if its stress components are zero-mean Gaussian random stresses.

The first one and theoretical proper definition of the Mises stress in frequency domain was proposed by Preumont and Piefort [5]. An interesting paper about including the phase shift into von Mises PSD were written by Bonte et al. [6].

The computation of PSD and RMS of von Mises stresses in this chapter is based on work by Segalman et al. [7]. Under this approach, the PSD of von Mises stress at a node a is given by:

$$S^a_{\text{mises}}(f) = \sum_{\beta=1}^{m}\sum_{\alpha=1}^{m} S_{\alpha\beta}(f)\,T^{\alpha}_{\alpha\beta} \tag{6.17}$$

where m is the number of modes, $S_{\alpha\beta}(f)$ are the elements of the PSD matrix of ψ generalized displacements:

$$T^{\alpha}_{\alpha\beta} = \left[\psi^{\alpha}_{\beta}\right]^T [A]\left[\psi^{\alpha}_{\beta}\right]. \tag{6.18}$$

$\left[\psi^{\alpha}_{\beta}\right]$ are the modal stress components of the α th mode at node a, and the constant matrix A is given by:

$$\begin{pmatrix} 1 & -1/2 & -1/2 & 0 & 0 & 0 \\ -1/2 & 1 & -1/2 & 0 & 0 & 0 \\ -1/2 & -1/2 & 1 & 0 & 0 & 0 \\ 0 & 0 & 0 & 3 & 0 & 0 \\ 0 & 0 & 0 & 0 & 3 & 0 \\ 0 & 0 & 0 & 0 & 0 & 3 \end{pmatrix} \tag{6.19}$$

Similarly, RMS of von Mises stress at a node a is computed as:

$$\text{RMISES}^a(f) = \sqrt{\sum_{\beta=1}^{m}\sum_{\alpha=1}^{m} V_{\alpha\beta}(f)\,T^{a}_{\alpha\beta}}. \tag{6.20}$$

where $V_{\alpha\beta}(f)$ are the elements of the variance matrix of generalized displacements.

6.2.6 Frequency domain approaches of life estimation

Fatigue analysis can be conducted in both time and frequency domain. In time domain, rainflow counting algorithm is usually used to calculate the number of cycles at each stress and strain levels based on the time history. Due to random load in many cases, a description in frequency domain is more efficient and hence using a statistical method is the most appropriate approach.

All the shown methods are based on Palmgren-Miner's rule [10] of cumulative damage ratio:

$$E[D] = \sum_i \frac{n_i}{N_i} \tag{6.21}$$

where $E[D]$ is the expected damage ratio, n_i is the number of cycles at stress level s_i, and N_i is the number of cycles for failure at stress level s_i, given by material's *S–N* curve.

To get n_i from the PSD of the random stress response and further compute $E[D]$, a variety of statistical methods have been used to predict probability density function and rainflow count, namely: Dirlik, Steinberg, Narrow Band, Wirsching, Chaudhury and Dover, Tunna and Hancock method [11-17].

6.3 Dirlik method

The Dirlik method is a mathematical tool which used to performing fatigue analysis. To use this method, signals to be analyzed must first be transferred from the time domain to the frequency domain where they are presented in the form of a PSD function. Stress analysis in the time domain is ideally suited to signals representing periodic or continuous stress loading but for random stress loading data, prohibitively large time records are often required for an accurate analysis. There are distinct advantages in performing stress analysis in the frequency domain rather than the time domain where a random stress loading signal forms the basis for analysis. The main advantage is much less time intensive calculations although this comes at a cost of slight loss in accuracy.

Any fatigue analysis process begins with the response of the structure or component. This is normally expressed as a time history of stress or strain. Stress cycles in the time history result in fatigue and the most important aspects of these cycles are the stress amplitude ranges and the mean stress values. These values are normally analyzed using rainflow cycle counting. The time history data can be converted to data in the frequency domain by using a variation of the FT. In both formats, the *y*-axis displays the signal amplitude but in the frequency domain, the *x*-axis represents the signal frequency as opposed to time. The FT effectively breaks the signal down into discrete sinusoidal waves. These waves vary in frequency, phase, and amplitude and form the original time signal again when combined using an inverse FT.

The Dirlik method [17,18], devised in 1985, approximates the cycle-amplitude distribution by using a combination of one exponential and two Rayleigh probability densities. It is based on numerical simulations of the time histories for two different groups of spectra. This method has long been considered to be one of the best methods and has already been subject to modifications, for example, for the inclusion of the temperature effect, by Zalaznik and Nagode [19].

The Dirlik method consists of a series of calculations which are based on four moments of area of the PSD function and was taken from Halfpenny [20]. These moments of area are $m_0, m_1, m_2,$ and m_4. The nth moment of area is calculated as:

$$m_n = \int f^n . G(f) df \tag{6.22}$$

where $G(f)$ is the PSD function and f is the frequency in Hertz.

The rainflow cycle amplitude probability density, p_a estimate is given by:

$$p_a(S) = \frac{1}{\sqrt{m_0}} \left[\frac{G_1}{Q} e^{\frac{-Z}{Q}} + \frac{G_2 Z}{R^2} e^{\frac{-Z^2}{2R^2}} + G_3 Z e^{\frac{-Z^2}{2}} \right] \tag{6.23}$$

where Z is the normalized amplitude:

$$Z = \frac{S}{\sqrt{m_0}} \tag{6.24}$$

For a fatigue analysis, the moments up to m_4 are normally used. The even moments represent the variance σ_X^2 of the random process X and its derivatives:

$$\sigma_X^2 = m_0 \quad \sigma_{\dot{X}}^2 = m_2 \tag{6.25}$$

and the parameters G_1 to G_3, R, and Q are defined as:

$$G_1 = \frac{2(x_m - \alpha_2^2)}{1 + \alpha_2^2} \tag{6.26}$$

$$G_2 = \frac{1 - \alpha_2 - G_1 + G_1^2}{1 - R} \tag{6.27}$$

$$G_3 = 1 - G_1 - G_2 \tag{6.28}$$

$$R = \frac{\alpha_2 - x_m - G_1^2}{1 - \alpha_2 - G_1 + G_1^2} \tag{6.29}$$

$$Q = \frac{1,25\ (\alpha_2 - G_3 - G_2 R)}{G_1} \tag{6.30}$$

where x_m is the mean frequency, as defined by the author of the method [8]:

$$x_m = \frac{m_1}{m_0}\left(\frac{m_2}{m_4}\right)^{\frac{1}{2}} \tag{6.31}$$

where α_i is used for spectral width estimation and it has the general form:

$$\alpha_i = \frac{m_i}{\sqrt{m_0 m_{2i}}};\ i = 1, 2 \tag{6.32}$$

6.4 Frequency domain random vibration analysis

The random vibration capability of LS-DYNA originated from Boeing's in-house vibroacoustic code N-FEARA [21]. This feature computes the dynamic response of structures exposed to vibration or structural-acoustic coupling based on a known source. Various excitations and acoustic environments can be considered, including base excitations, correlated or noncorrelated acoustic waves such as plane wave, progressive wave, reverberant wave, turbulent boundary layer [22].

A keyword has been introduced in LS_DYNA to perform random vibration analysis. Through the keyword, user provides information about the location, direction, range of frequencies for the random excitation and response area can be given as node, set of nodes, set of segments, or part. The direction of load can be in any of the x, y, z directions or given as a vector. Load curve for the PSD loads in random computation is also specified under the keyword [22].

The feature of random vibration fatigue is implemented as an option of the keyword, as it is a natural extension of the random analysis procedure. The method for performing fatigue analysis is defined as well as the parts or elements where the fatigue analysis is needed. The material's *S–N* fatigue curve and some other options like the exposure time are also defined. Modal analysis is the first step for running the analysis and then the implicit solution started.

Ringeval and Huang [23] give in their paper several examples to demonstrate the effectiveness of the random vibration fatigue analysis feature with LS-DYNA. One of those examples is a simple cantilever aluminum beam subjected to base accelerations. The numerical values are compared with the experimental results and different fatigue failure theories are used to predict fatigue life. That example is considered as a benchmark for the verification of the results obtained with LS-DYNA as the authors compare their results with those found in the literature for the same subject. Also, the same problem has been simulated with ANSYS® and RADIOSS BULK and

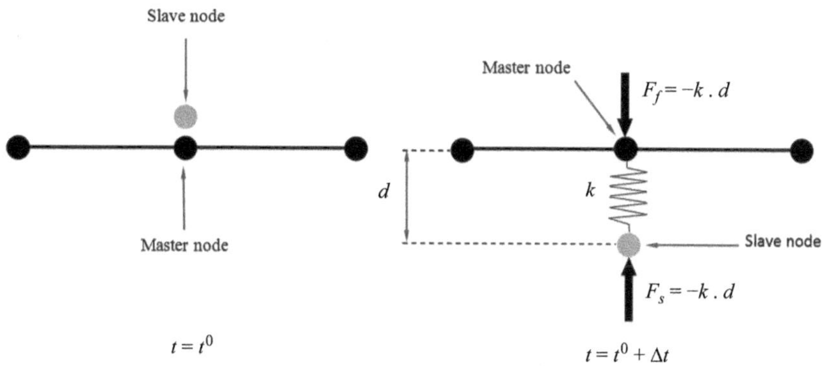

Figure 6.1 Penalty contact algorithm between master node and slave node.

the maximum RMS stress computed by LS-DYNA® is in good agreement with the results obtained by the other commercial software.

6.5 Contact algorithm

Contact algorithm has been extensively studied in computational mechanics by several authors. Details on contact algorithms can be found in Belytschko et al. [24]. Classical implicit and explicit coupling are described in detail in Longatte et al. [25] and Longatte and Kueny [26], where hydrodynamic forces from the fluid solver are passed to the structure solver for stress and displacement computation. In this paper, a coupling method based on contact algorithm is used. Since the coupling method described in this chapter is based on the penalty method for contact algorithms, the contact approach is a good introduction to this method. In contact algorithms, a contact force is computed proportional to the penetration vector, the amount the constraint is violated. In an explicit FEM method, contact algorithms compute interface forces due to impact of the structure on the fluid, these forces are applied to the fluid and structure nodes in contact in order to prevent a node from passing through contact interface. In contact algorithms, one surface is designated as a slave surface, and the second as a master surface. The nodes lying on both surfaces are also called slave and master nodes, respectively. The penalty method imposes a resisting force to the slave node, proportional to its penetration through the master segment, as shown in Fig. 6.1 describing the contact process. This force is applied to both the slave node and the nodes of the master segment in opposite directions to satisfy equilibrium.

Figure 6.2 Seat support tube (SST).

6.6 Case study of an aircraft seat support tube

There have been increasing reports of annoyance, fatigue, and even neck and back pain during prolonged operation of troop propeller aircraft, where persistent multiaxis vibration occurs at higher frequencies. A claim from failure of the aluminum tube used as a seat support tube (SST) for troop seat wall style construction. A failure occurs after about 85 flight hours. A numerical static analysis model for the tube with its holders shows that it is safe, even with impact loading.

6.6.1 Seat support tube

Aircraft industries have faced a problem with the SST currently used on the C-130 Lockheed Martin Aircraft as shown in Fig. 6.2. The SST has been continuously fracturing around the joint connection, which introduced the need to replace the SST frequently. The weight of aircraft components has become an important factor. Efforts are being made to reduce the weight of the aircraft and consequently increase the payload. The SST is made of 2024-T3 aluminum alloy which is widely used in aircraft industries due to its high strength ($\sigma_y = 341$ MPa) and low weight (density $\rho = 2700\ \mathrm{kg/m^3}$).

The *S–N* fatigue curve, Fig. 6.3, is selected according to the European Standard Eurocode 9 [27] with a reference fatigue strength $\Delta\sigma = 71$ MPa at 2×10^6 cycles and a single inverse constant slope $m = 7.0$.

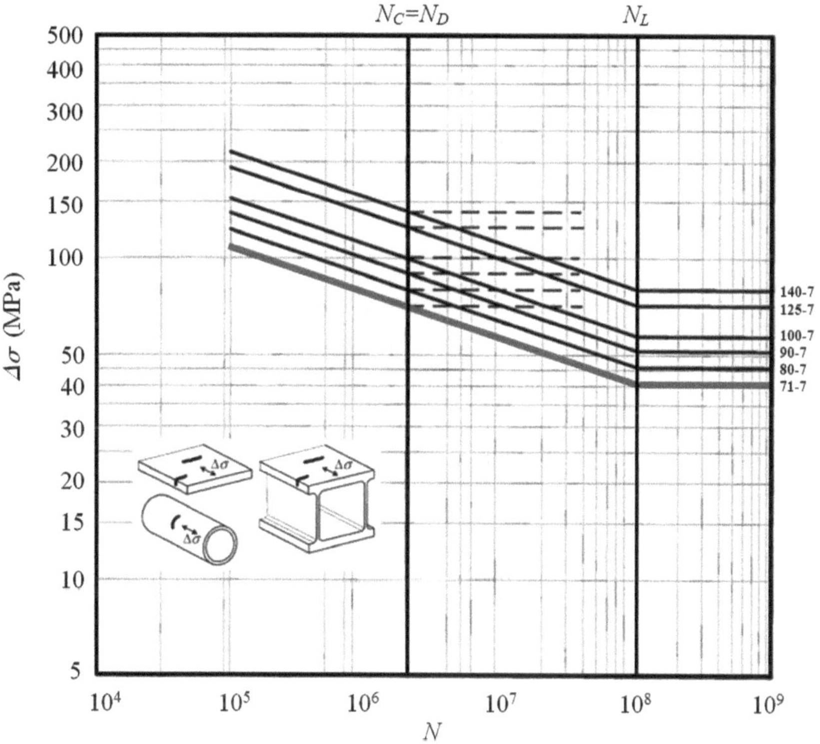

Figure 6.3 *S–N* fatigue curve used for the SST material.

6.6.2 SST model and analysis steps

The tube was meshed with 12,624 shell elements and the supporting holders and pins are meshed with 832 solid elements. Fig. 6.4 shows the finite element model of the SST.

A uniform pressure load, represents the pressure of the seat pan on the seat tube, is applied, as shown in Fig. 6.5.

The static analysis results were accessed by the random vibration analysis as a preload stresses. A suitable contact algorithm was added in order to transfer the support's movement to the tube. Tube surface considered as the slave side while the pin surface considered as the master side. Initial assumption was adopted to prevent the initial penetration.

The two fixed pins were subjected to base acceleration spectral density for the range of frequency 15–2000 Hz, as shown in Fig. 6.6. The peaks of the load curve were calculated according to the number of propeller blades. In our case, we have four blades, Fig. 6.7. The *dB Calculator* was used to calculate

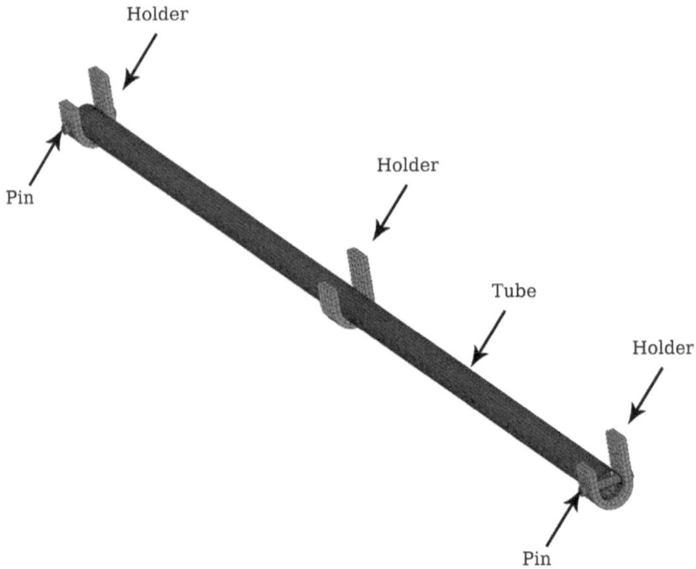

Figure 6.4 Finite element model of the SST.

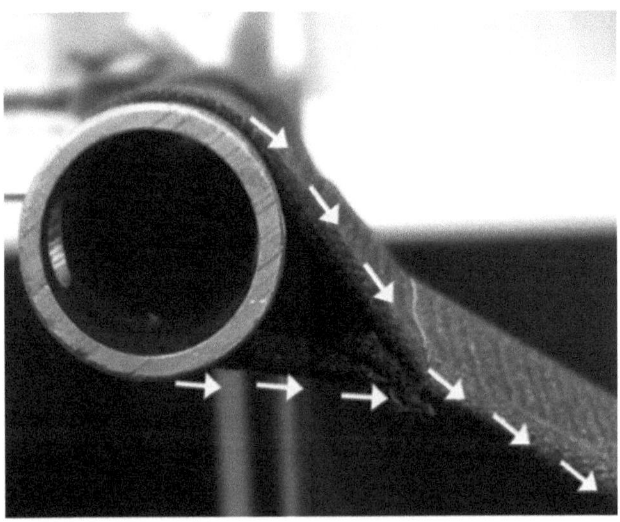

Figure 6.5 Uniform pressure load applied to the SST.

those peaks [28]. A constant damping ratio 0.02 was adopted. The SST was exposed to the random vibration load for 1 h per axis as recommended by MIL-STD-810G [29].

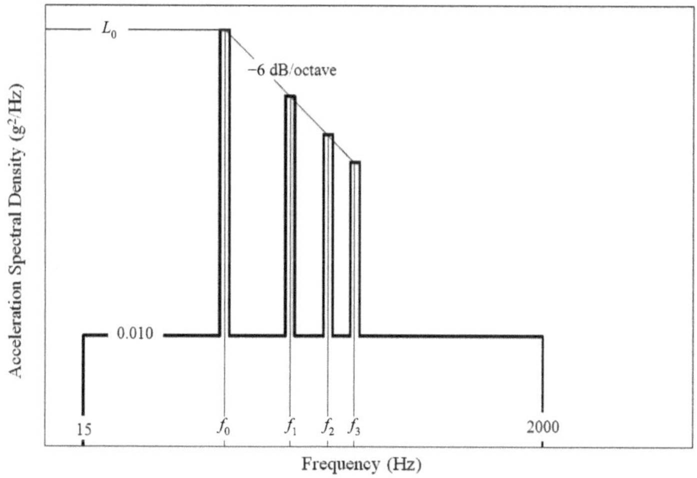

Figure 6.6 Propeller aircraft vibration exposure.

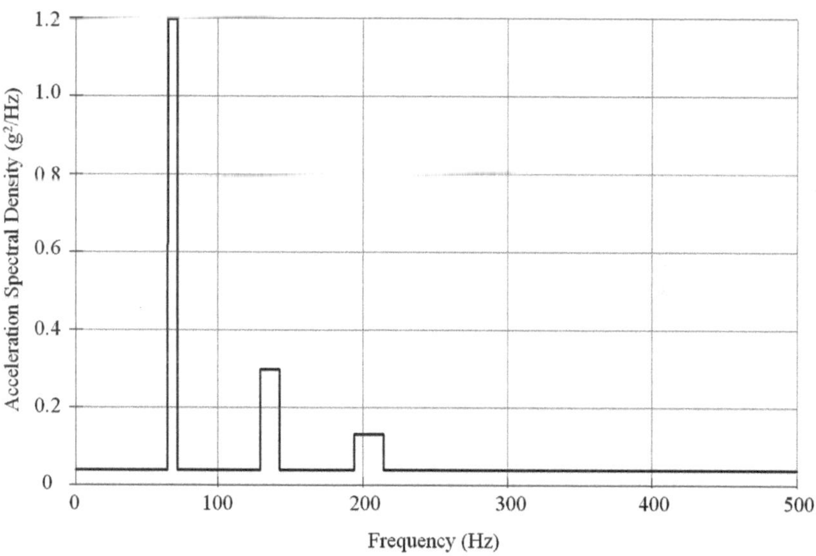

Figure 6.7 Input acceleration PSD.

The first 10 natural modes, as shown in Table 6.1, are required for the eigenvalue analysis. Lobatto's integration rule with three integration points through the thickness of the tube is applied to get the stress on the shell surface. The shear factor, which scales the transverse shear stress, is suggested to have a value equal to 5/6. For shell element with implicit solution the

Table 6.1 Mode shapes and their natural frequencies.

Mode number	Natural frequency (Hz)
1	120.3
2	354.0
3	442.7
4	445.7
5	620.4
6	1170.4
7	1214.7
8	1270.8
9	1350.9
10	1745.0

invariant node numbering is taken as 2. An infinite fatigue life for stress lower than the lowest stress on *S–N* curve (40.6 MPa) was assumed, Fig. 6.3. Due to the tube side oval holes, a further concentration factor K_t is calculated as 0.26 [30].

6.6.3 SST result analysis

The response stress PSD measured at the critical point shows that the most natural frequency that has been excited by the input loading in the range 15–2000 Hz, Fig. 6.8, equal to 1745 Hz.

Based on Dirlik method results, the service life for the SST will be equal to 83.3 h which is in a good agreement with the actual service life which is about 85 h. Fig. 6.9 provides, respectively, a photograph of the broken SST and contour plot of the cumulative damage ratio observed on the SST. Results given by the numerical analysis are in good agreement with the actual failure location.

6.7 Case study of landing gear's leg

The landing gear is a structure that supports an aircraft on ground and allows it to taxi, take off, and land [31,32]. In fact, landing gear design tends to have several interferences with the aircraft structural design. Nowadays the weight of landing gear has become an important factor. Efforts are being made to reduce the weight of the aircraft and consequently increase the payload. Design of landing gears takes the effect of static and dynamic loads as well as the impact loading. This study presents analysis for prediction of the service life for a light landing gear with a thickness optimized to only

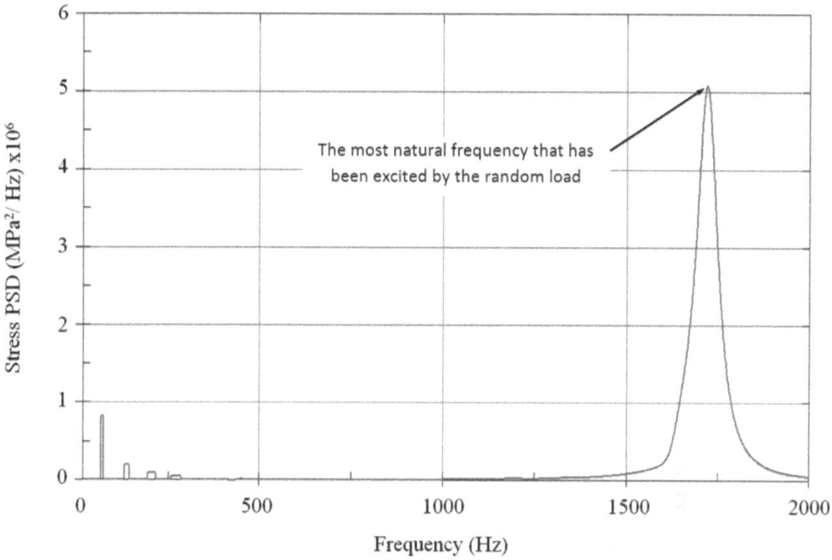

Figure 6.8 Stress PSD at critical point.

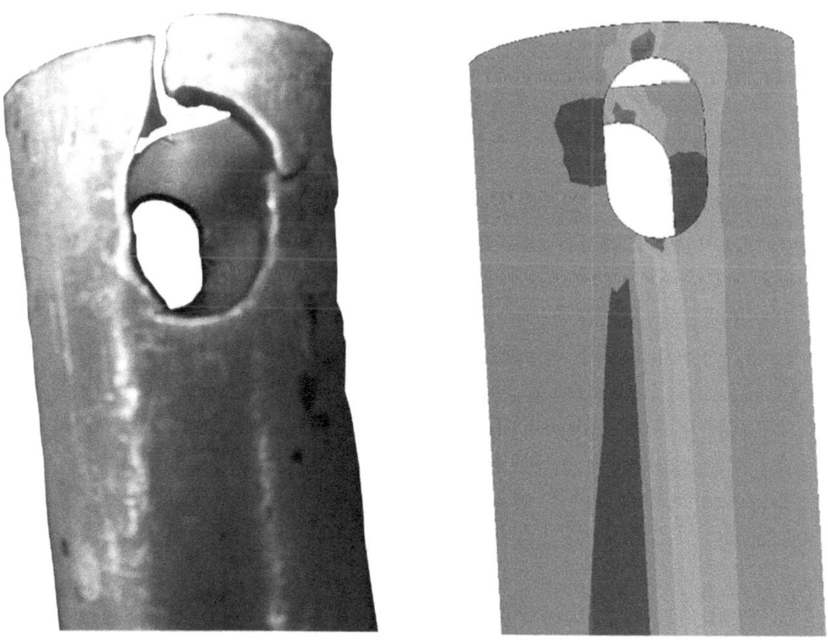

Figure 6.9 Cumulative damage ratio by Dirlik method observed on the SST.

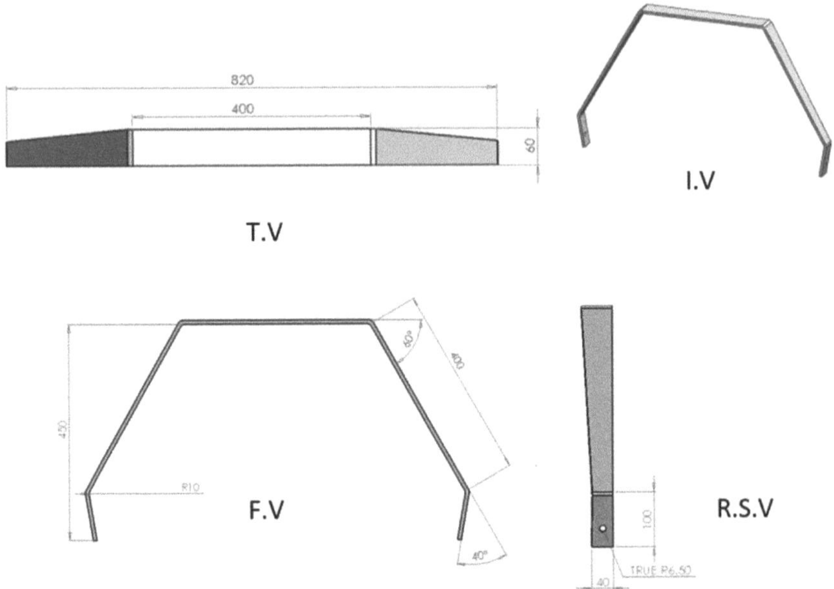

Figure 6.10 Geometry details of the landing gear (all dimensions are in millimeter).

8 mm. This design optimization where carried by Albahkali and Alqhtani [33]. Dimensions of the landing gear are shown in Fig. 6.10.

As shown in Fig. 6.11, the landing gear is clamped to the fuselage in its middle horizontal part. The two wheels attached to its ends are weighing 0.651 kg each.

Fig. 6.12 shows PSD outputs in g^2/Hz for the landing gear vibration with change in frequency. It indicates random vibration loads on the landing gear. The modal frequencies required to calculate the resultant effect of modal spectrum vibration are extracted up to the spectrum frequency.

The landing gear is made from AA7076-T6 aluminum alloy with density $\rho = 2700$ kg/m^3, Young's modulus $E = 70$ GPa and Poisson's ratio $\nu = 0.33$. The *S–N* fatigue curve, as shown in Fig. 6.3, is again selected according to the European Standard Eurocode 9 [27] with a reference fatigue strength $\Delta\sigma = 71$ MPa at 2×10^6 cycles and a single inverse constant slope $m = 7.0$.

6.7.1 Landing gear model analysis

The model is meshed with 9073 nodes and 8576 shell elements (fully integrated linear DK quadrilateral and triangular shell). The two wheels are represented by 28 mass elements with 0.0465 kg each. They distributed

Figure 6.11 Landing gear.

uniformly at the circumference of the two holes. The upper face is subjected to base acceleration spectral density for the range of frequency 0–2000 Hz. A constant damping ratio 0.035 is adopted. The landing gear is exposed to the random vibration load for 45 min (2700 s) which is equal to the flight time.

The first 10 natural modes are required for the eigenvalue analysis. Lobatto's integration rule with three integration points through the thickness of the landing gear is applied to get the stress on the shell surface. The shear factor, which scales the transverse shear stress, is suggested to have a value equal to 5/6. For shell element with implicit solution the invariant node numbering is taken as 2. An infinite fatigue life for stress lower than the lowest stress on *S–N* curve (40.6 MPa) is assumed.

6.7.2 Landing gear result analysis

The response stress PSD measured at the critical point shows that two natural frequencies are excited by the input loading in the range 10–2000 Hz (Fig. 6.13), the first natural frequency (11.711 Hz) and the fifth natural frequency (105.77 Hz). When examine the model with higher input acceleration PSD load ($g^2/Hz = 0.1$), the eighth natural frequency (287.244 Hz)

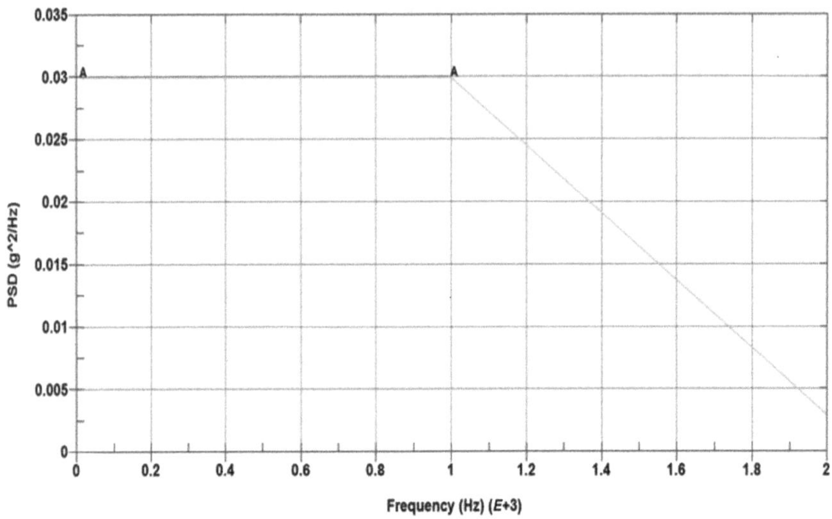

Figure 6.12 Input acceleration PSD.

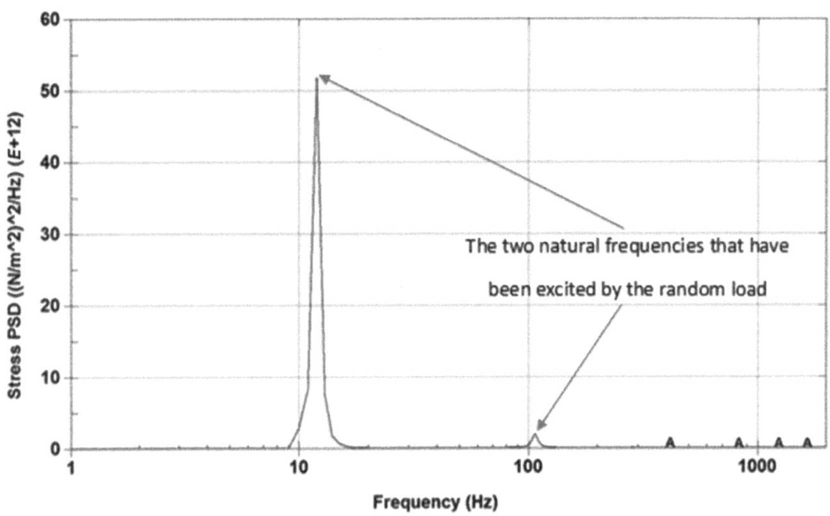

Figure 6.13 Stress PSD at critical point.

is excited also. Fig. 6.13 shows the stress PSD at critical point while Figs. 6.14 and 6.15 show contour plot for the cumulative damage ratio and the RMS of S_x stress at the critical point using Dirlik method, respectively. Based on Dirlik method results, the service life for the landing gear will be equal to 3037.6 h with a cumulative damage of 0.000246.

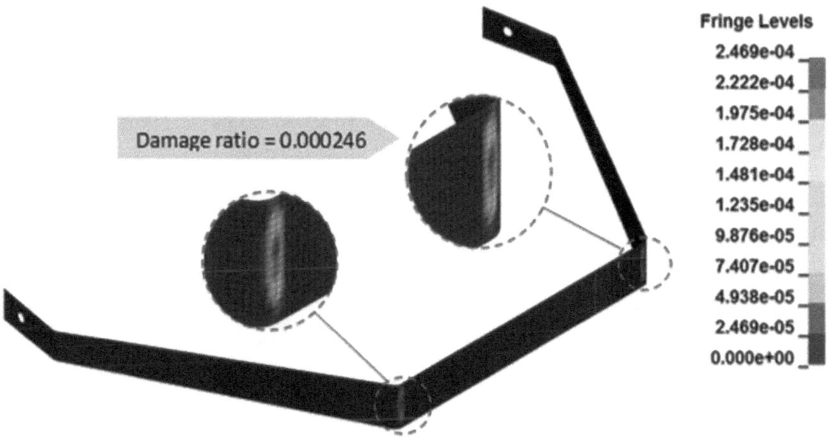

Figure 6.14 Cumulative damage ratio by Dirlik method.

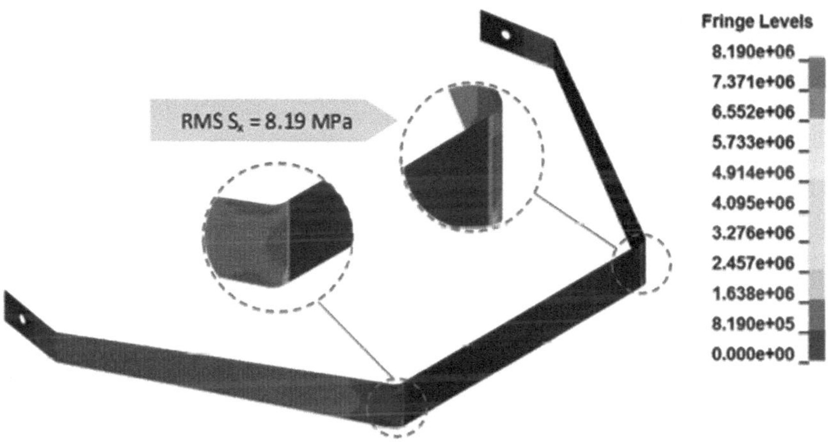

Figure 6.15 RMS of S_x stress at the critical point using Dirlik method.

6.8 Conclusion

Frequency domain random vibration analysis is applied in this chapter. The analysis provides cumulative damage ratio calculation and fatigue life prediction for SST and a unmanned aerial vehicle (UAV) landing gear subjected to random vibration excitations. Dirlik method which has long been considered to be one of the best methods is demonstrated. Experimental and numerical analysis have been performed on an aircraft SST and can be extended to more complex structures commonly used in aerospace and automotive industry. For this reason, the random vibration analysis is conducted on the assembly of the SST, where the contact algorithm was added in order to transfer the supporting pins' movement to the tube taking into consideration the prestress results in the static loading. Prestresses have two effects on the structure under vibration analysis: first, it makes it stiffer; second, it shifts the structure's natural frequencies. Due to the weight, constrain, as in the case of the SST design, the margin of the factor of safety is limited. Therefore, it is essential to conduct fatigue analysis also to predict the service life. In addition, the frequency response approach can improve understanding of the system dynamic behaviors, in terms of frequency characteristics of both structures and loads and their couplings. Using alternative mounting designs for the passenger (or troop) seats is mandatory. On the other side, it appears that the current ISO standard does not provide optimum assessment methods or guidelines for evaluating the higher frequency vibration exposures associated with troop propeller aircraft. In design of a structure in consideration of fatigue life, it is very important to evaluate not only a fatigue life under the constant loading but also a fatigue life under the service loading. Due to weight constrain, as in the case of the landing gear design, the margin of the factor of safety is limited. Therefore, it is essential to conduct fatigue analysis also to predict the service life. In addition, the frequency response approach can improve understanding of the system dynamic behaviors, in terms of frequency characteristics of both structures and loads and their couplings.

References

1. S. Rice, Mathematical Analysis of Random Noise. Selected Papers on Noise and Stochastic Process, Dover, New York, 1954.
2. J. Bendat, Probability functions for random responses: prediction of peaks, fatigue damage, and catastrophic failures, NASA Report, 1964.
3. C. Braccesi, F. Cianetti, G. Lori, D. Pioli, Fatigue behaviour analysis of mechanical components subject to random bimodal stress process: frequency domain approach, Int. J. Fatigue 27 (2005) 335–345.
4. R.W. Clough, J. Penzien, Dynamics of Structures, McGraw-Hill, New York, 1975.
5. A. Preumont, V. Piefort, Predicting random high-cycle fatigue life with finite elements, J. Vibr. Acoust. 116 (2) (1994) 245–248.
6. M.H.A. Bonte, A. De Boer, R. Liebregts, Determining the von Mises stress power spectral density for frequency domain fatigue analysis including out-of-phase stress components, J. Sound Vibr. 302 (1-2) (2007) 379–386.
7. D.J. Segalman, C.W.G. Fulcher, G.M. Reese, R.V. Field, An Efficient Method for Calculating RMS von Mises Stress in a Random Vibration Environment, Sandia Report, SAND98-0260, 1998.
8. Enrique de la Fuente, An efficient procedure to obtain exact solutions in random vibration analysis of linear structures, Eng. Struct. 30 (11) (2008) 2981–2990.
9. G. Quaranta, P. Mantegazza, Randomly excited structures reliability by means of the von Mises stress response, in: 16th National Congress (AIDAA), Palermo, 2001 Sep. 24–28.

10. M.A. Miner, Cumulative damage in fatigue, J. Appl. Mech. 67 (1945) A159–A164.
11. D.S. Steinberg, Vibration Analysis for Electronic Equipment, second ed., Wiley, New York, 1988.
12. J.S. Bendat, Probability Functions for Random Responses, NASA Report on Contact, NASA-5-4590, 1964.
13. P. Wirshing, T. Paez, K. Ortiz, Random Vibration, Wiley, New York, 1995.
14. G. Chaudhury, W. Dover, Fatigue analysis of offshore platforms subjected to sea wave loadings, Int. J. Fatigue 7 (1) (1985) 13–19.
15. J. Tunna, Fatigue life prediction for Gaussian random loads at the design stage, Fatigue Fract. Eng. Mater. Struct. 9 (3) (1986) 169–184.
16. D. Gall, J. Hancock, Fatigue Crack Growth Under Narrow and Broad Band Stationary Loading, Glasgow University, Marine Technology Centre, Glasgow, Scotland, UK, 1985.
17. T. Dirlik, Application of computers in fatigue analysis, Ph.D thesis, The University of Warwick, 1985.
18. M. Mršnik, J. Slavic, M. Boltezar, Frequency domain methods for a vibration fatigue life estimation – application to real data, Int. J. Fatigue 47 (2013) 8–17.
19. A. Zalaznik, M. Nagode, Frequency based fatigue analysis and temperature effect, Mater. Des. 32 (10) (2011) 4794–4802.
20. A. Halfpenny, A Frequency Domain Approach for Fatigue Life Estimation From Finite Element Analysis, nCode® International Ltd, Sheffield, UK.
21. LS-DYNA®Keyword User's Manual, Version 971, Livermore Software Technology Corporation, Livermore, California, 2012.
22. Y. Huang, M. Souli, C. Ashcraft, R. Grimes, J. Wang, M. Rassaian, J. Lee, Development of frequency domain dynamic and acoustic capability in LS-DYNA®, in: Proceedings of the Eighth European LS-DYNA® Users Conference, Strasbourg, 2011 May.
23. A. Ringeval, Y. Huang, Random vibration fatigue analysis with LS-DYNA®, in: Proceedings of the 12th International LS-DYNA® Users Conference, Dearborn, Michigan, USA, 2012 June 3–5.
24. T. Belytschko, W. Liu, B. Moran, K. Elkhodary, Nonlinear Finite Elements for Continue and Structures, second ed., Wiley, Hoboken, New Jersey, USA.
25. F. Longatte, Z. Bendjeddou, M. Souli, Methods for numerical study of tube bundle vibrations in cross-flows, J. Fluids Struct. 18 (2003) 513–528.
26. F. Longatte, J. Kueny, Analysis of rotor-stator-circuit interactions in a centrifugal pump, in: Proceedings of the Third ASME/JSME Joint Fluids Engineering Conference, 2009, pp. 1039–1045.
27. Eurocode 9: Design of aluminum structures – Part 1-3: Structures susceptible to fatigue, Aug. 2011.
28. Irvine T., dB Calculations for log-log plots, version 3.4, www.vibrationdata.com/tutorials_alt/dB_octave.pdf.
29. Department of Defense, Test Method Standard: Environmental Engineering Considerations and Laboratory Tests, MIL-STD-810G, Oct. 31, 2008.
30. R. Budynas, K. Nisbett, Shigley's Mechanical Engineering Design, ninth ed., McGraw Hill, New York city, New York, USA, 2011.
31. V. Nagulpalli, A. Gupta, S. Fan, Estimation of Fatigue Life of Aluminum Beams Subjected to Random Vibration, Department of Mechanical Engineering, Northen Illinois University, DeKalb, IL.
32. A. Matta, G. Kumar, R. Kumar, Design optimisation of landing gear's leg for an unmanned aerial vehicle, Int. J. Eng. Res. Applic. (IJERA) 2 (4) (2012) 2069–2075.
33. E. Albahkali, M. Alqhtani, Design of Light Landing Gear, Graduate Project, King Saud University, Riyadh, Saudi Arabia, 2011 June.

CHAPTER SEVEN

Recent developments in quasi-3D aerodynamic methods for multidisciplinary aircraft conceptual design studies

Oliviu Şugar-Gabor, Andreea Koreanschi
Aeronautical and Mechanical Engineering, University of Salford, Salford, United Kingdom

7.1 Introduction

Aircraft design involves a highly complex, multidisciplinary approach. Throughout the design process, not only the individual disciplines such as aerodynamics, structural mechanics, system engineering, etc. must be considered, but also the interactions between them. This leads to very challenging, multiphysics and multidisciplinary problems.

Early in the design process, when the focus is on generating a fit-for-purpose concept but without refining details, the amount of time spent on discipline-specific analysis must be kept as low as possible. The design of commercial aircraft, whether piston-propeller, turboprop, or turbojet benefits from the existence of a wealth of experience, translated into efficient, empirically refined analytical calculation methods. However, the designer of more unconventional vehicles such as a novel box-wing two-seater aircraft or a flapping wing micro aerial vehicle often has no alternative but to resort to discipline-specific methods and tools.

For determining the aerodynamic loads on lifting surfaces, high-fidelity computational fluid dynamics (CFD) can produce accurate results while maintaining the complexity inherent to turbulent flow, but at a significant cost in both time and resources. At the other side of the spectrum, inviscid classical methods such as the lifting-line theory (LLT) or the vortex lattice method (VLM) can generate reasonable results within seconds but are very limited in applicability due to the numerous underlying hypotheses.

In this chapter, recent developments in quasi-three-dimensional (3D) aerodynamic methods are presented. At their core, these methods are based on the LLT and VLM, but with a relaxed set of hypotheses, while also

Multiphysics Simulations in Automotive and Aerospace Applications.
DOI: 10.1016/B978-0-12-817899-7.00001-0

considering the effect of viscosity (to a certain degree) by introducing a strong nonlinear coupling with two-dimensional (2D) viscous aerofoil aerodynamics. These methods can provide more accurate results compared to their inviscid classical counterparts and have an extended range of applicability with respect to the lifting surface geometry. As expected, the necessary computational time is higher due to the relatively large nonlinear systems of equations requiring iterative solutions, but it is still at least one order of magnitude lower compared to CFD. The good accuracy achieved for relatively low computational time requirement makes quasi-3D methods a solid choice for conducting conceptual level multidisciplinary optimizations of lifting surfaces.

7.2 Nonlinear lifting line method

Owing to its relatively good accuracy in the range of linear aerodynamic behavior, the LLT has been used for various applications including the design of low-speed aircraft wings, boat sails, propellers, or wind turbine blades. However, the original theory was applicable only for straight wings with moderate to high aspect ratio. Authors have proposed enhancements to the original theory to enhance its range of applicability and the accuracy of the predicted results.

By replacing the 2D form of the Kutta–Joukowski theorem with the fully 3D cortex lifting law [1], the LLT could be easily applied to lifting surfaces with arbitrary camber, sweep angle, and dihedral [2]. The model developed in [3,4] used the effective angle of attack updated strategy of van Dam and a lifting-line model based on the works of Weissinger to include viscous flow effects into the LLT formulation. Results showed good predictions for the stall and poststall flow regime but were restricted to wings of elliptical planform.

Unsteady extensions of the theory have also been developed and applied to the analysis of flapping bird wings in forward flight [5] and the design and optimization of wind turbine blades [6,7,8].

In [44], a nonlinear LLT model has been developed to design and analyze the performance of ship propellers. The helical wake created by constant pitch propeller was modeled by splitting the typical horseshoe vortices into segments whose position is updated as the wake evolves. A morphing wing design was introduced in [46] with the aim of increased stall recovery, its performance being numerically predicted with a nonlinear LLT model. It used 2D Reynolds-Averaged Navier-Stokes (RANS) results to predict the viscous aerodynamic characteristics of the aerofoil sections and a linearization of Prandtl's original integral–differential equation into which the RANS result for the lift coefficient were directly introduced. Good agreement was obtained with low-speed wind tunnel results.

Other nonlinear lifting-line models can be found in [52,53,54], developed based on various approaches. These include modifying Weissinger's original method by including viscous aerofoil results [52], determining a circulation strength correction based on lookup tables of viscous aerofoil results and iterating until the difference between inviscid and viscous circulation values become negligible [53], or by updating the coordinates of the control point according to the viscous lift curve slope followed by an application of a Pistolesi-type boundary condition to build a linear system of equations for the circulation values [54].

A nonlinear LLT model was developed and presented in [9,10]. The model uses the fully 3D vortex lifting law, as in [1], but also reformulates the equations to allow coupling with 2D experimental aerofoil results provided via a prebuilt database file.

In the model, the continuous distributions of bound vorticity over the wing surface and of trailing vorticity in the wing wake are approximated using a finite number N of horseshoe vortices, each vortex having its own strength (circulation) Γ_i. The bound segment of the vortices is aligned with the wing quarter-chord line, while the trailing segments are aligned with the direction of the freestream. The 3D vortex lifting law [11] is applied to express the inviscid force $\mathbf{dF}_i$ acting on the bound segment of each horseshoe vortex:

$$\mathbf{dF}_i = \rho \Gamma_i \mathbf{V}_i \times \mathbf{dl}_i \tag{7.1}$$

In Eq. (7.1), ρ is the fluid density, $\mathbf{V}_i$ is the local airspeed vector, and $\mathbf{dl}_i$ is the spatial vector along the lifting line differential segment, aligned according to the local vorticity.

The local airspeed vector over one bound vortex segment is equal to the sum of the freestream velocity $\mathbf{V}_\infty$ and the velocities induced by all the other horseshoe vortices distributed over the wing surface and wake:

$$\mathbf{V}_i = \mathbf{V}_\infty + \sum_{j=1}^{N} \Gamma_j \mathbf{v}_{ij} \tag{7.2}$$

The magnitude of the force acting on a wing strip of area A_i and having a local airfoil lift coefficient C_{l_i} is given by:

$$\mathbf{F}_i = \frac{1}{2}\rho V_\infty^2 A_i C_{l_i} \tag{7.3}$$

Prebuilt databases of experimental aerofoil results (or, equivalently, numerical results provided either by RANS-based CFD or panel-boundary layer codes such as XFOIL) are used to provide the C_{l_i} values. These are precalculated at a given number of flow conditions (Reynolds number, angle of attack), while for any other flow condition not in the database, a simple linear interpolation is used between the two closest data points available.

If the wing strips are taken such that each horseshoe vortex-bound segment corresponds to one strip, then the modulus of the force given by Eq. (7.1) can be set equal to the one given by Eq. (7.3), since the bound segment is the only segment upon which the surrounding fluid exerts a force, the trailing segments being aligned with the freestream. Thus, for each vortex over the wing surface, the following equation can be written as:

$$\rho\Gamma_i\left(\mathbf{V}_\infty + \sum_{j=1}^{N}\Gamma_j\mathbf{v}_{ij}\right)\times \mathbf{dl}_i - \frac{1}{2}\rho V_\infty^2 A_i C_{l_i} = 0, \quad i = 1, 2, \ldots, N \tag{7.4}$$

The nonlinear system of equations obtained can be solved using Newton's classical method, since the Jacobian matrix elements can be analytically determined. The aerodynamic force and moment about the root chord quarter-chord point can then be determined, using Eqs. (7.5) and (7.6):

$$\mathbf{F} = \rho\sum_{i=1}^{N}\left[\left(\mathbf{V}_\infty + \sum_{j=1}^{N}\Gamma_j\mathbf{v}_{ij}\right)\Gamma_i \times \mathbf{dl}_i\right] \tag{7.5}$$

$$\mathbf{M} = \rho\sum_{i=1}^{N}\mathbf{r}_i \times \left[\left(\mathbf{V}_\infty + \sum_{j=1}^{N}\Gamma_j\mathbf{v}_{ij}\right)\Gamma_i \times \mathbf{dl}_i\right] + \mathbf{dM}_i \tag{7.6}$$

where $\mathbf{dM}_i$ is the local, 2D pitching moment of the aerofoil section (again obtained from the prebuilt database), and $\mathbf{r}_i$ is a vector from the position of the chosen moment reference point to the control point.

Typical nonlinear LLT models found in the literature (see, e.g., refs. [44,46,53]) require significant under-relaxation if the variable being updated from iteration to iteration is the circulation or the local lift coefficient value. The current model shows much better convergence properties which are attributed to the strong coupling between the circulation and the term representing the viscous correction. It has been shown by other authors [3] that such a strong coupling resulted in an increased computation effort per iteration but achieved significantly faster convergence with minimal under-relaxation for a nonlinear LLT utilizing an effective angle of attack correction approach.

It must be noted that the method does not use Trefftz plane analysis for the induced drag but calculates the overall aerodynamic force vector using on-body velocities only. The overall accuracy in estimating induced drag with the near field approach remains good. The profile drag can also be approximated by:

$$C_{D_0} = \frac{1}{S}\int_{-b/2}^{b/2} C_d(y)c(y)dy \cong \frac{1}{S}\sum_{i=1}^{N} C_{d_i}c_i\Delta y_i \tag{7.7}$$

where S is the wing area, C_{d_i} is the 2D aerofoil drag coefficient, c_i is the local chord, and Δy_i is the width of the wing strip (which coincides with the distance between the two trailing segments of a horseshoe vortex).

A first validation test is done using geometrical and experimental data from the NACA 1270 Technical Note [12]. The wing is a straight wing having an aspect ratio of 12, a taper ratio of 0.285 and a twist of 3°. The aerofoil section progressively changes from a NACA 4422 at the root section to a NACA 4412 at the wing tip. The experimental results were obtained for an airspeed of 65 m/s and a Reynolds number equal to 4×10^6, as calculated with the mean aerodynamic chord value. The numerical results are obtained with 35 strips per semispan, while the sectional aerofoil database is generated using the 2D XFOIL solver [13].

Fig. 7.1 shows a comparison between numerical and experimental results in term of lift, drag, and pitching moment coefficients. The estimation of the lift coefficient for angles of attack up to 12.5° is very accurate, as expected for a straight, high aspect ratio wing. The stalling angle and $C_{L_{max}}$ are overestimated, however this is due, at least in part, to the overestimation of the same parameters by XFOIL for the NACA 44-series aerofoils. The drag polar prediction is overall good, with underestimations at both very low and very high C_L values. Since no data are provided in [12] on the turbulence intensity levels, the drag underestimation at low lift might be due to an extended laminar boundary layer in the XFOIL analysis compared to the actual wind tunnel conditions. Accurate prediction of pitching moment coefficient values is known to be generally very difficult. The model captures

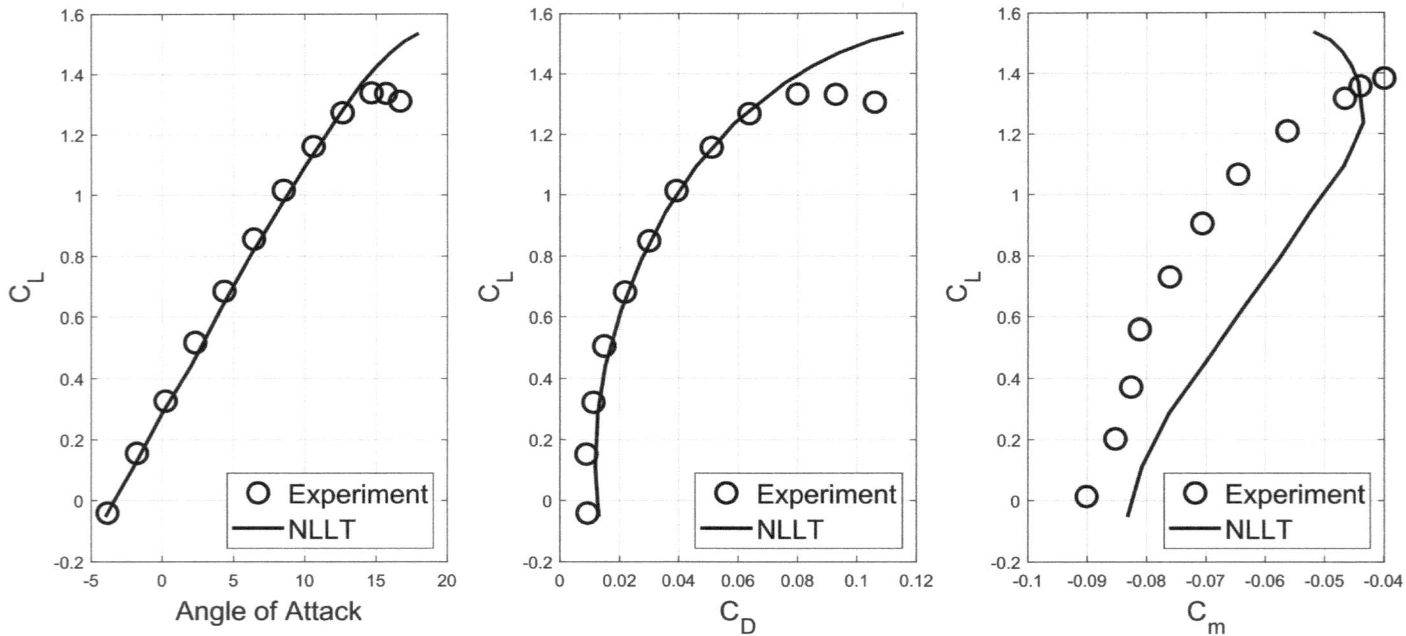

Figure 7.1 Comparison of lift, drag, and pitching moment coefficient values obtained with the nonlinear LLT method with experimental data published for the NACA TN1270 high aspect ratio straight wing geometry.

the variation trend across the lift coefficient range, but the prediction if offset by a relatively constant value.

A second validation test of is done using geometrical and experimental data from the NACA L50F16 Research Memorandum [14]. The wing geometry chosen is a moderate aspect ratio shape, with $AR=4$, having a moderate quarter-chord sweep angle of 30°, a taper ratio of 0.6 and no geometric twisting or dihedral. This wing is constructed using the very thin NACA 65A006 aerofoil. As for the previous case, all experimental results were obtained using the NACA variable density subsonic wind tunnel, but for an airspeed of 80 m/s and a Reynolds number equal to 3×10^6, as calculated with the mean aerodynamic chord value. The numerical results are obtained with 50 strips per semispan, while the sectional aerofoil database is constructed using the experimental results provided in [15].

The comparison between numerical and experimental results in term of lift, drag, and pitching moment coefficients is presented in Fig. 7.2. An excellent agreement exists for both lift and drag coefficient results for the C_L range below 0.60. In addition, the pitching moment values are accurately predicted for this range. The lifting line model underestimates the stalling angle and cannot capture the lift coefficient plateau observed in the experimental results around $C_{L_{max}}$. This behavior could be given by boundary layer separation in the region close to the tip of the swept wing, phenomenon which would be compounded with the LLT inherent limitations in capturing true 3D interactions. While it is difficult to address this drawback in the simple mathematical framework of the lifting-line, a promising alternative has been recently proposed in [47,55]. In this so-called 2.5D approach, the usual 2D viscous lift curves are corrected either by analytically applying sweep theory to the results (as in [47]) or by utilizing 3D RANS results for a swept wing with infinite span (as in [55]). While leaving the mathematics of the lifting-line model unchanged, the approach has been reported to significantly improve the 3D prediction accuracy for high angles of attack and for the wing tip region of swept wings.

7.3 Nonlinear vortex lattice method

The VLM [16,17] represents a powerful tool for preliminary wing design and optimization. It has been used in a very wide range of applications, from multiobjective optimization studies for existing commercial aircraft [19], the development of morphing wings [20], unmanned aerial vehicles (UAV) aerodynamic performance optimizations [21], for the design of

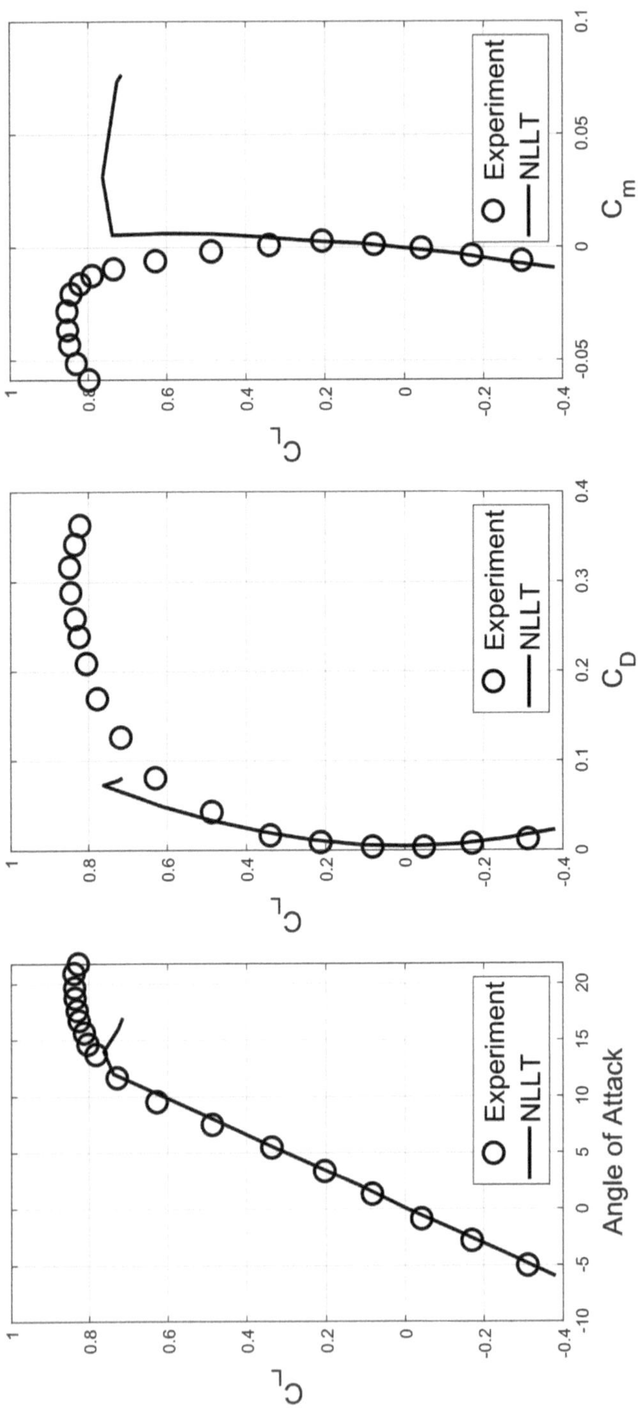

Figure 7.2 Comparison of lift, drag, and pitching moment coefficient values obtained with the nonlinear LLT method with experimental data published for the NACA L50F16 moderate aspect ratio moderate sweep wing geometry.

nonconventional blended wing body aircraft geometries [22], up to unsteady variants of the method (unsteady vortex lattice method [UVLM]) used to calculate aerodynamic loads for aeroelasticity and flight dynamics simulations [18].

Nonlinear extension to the VLM has been proposed in the literature. It must be noted that the present model is based on a nonlinear correction of the circulation to improve prediction accuracy, and not nonlinearities introduced due to wake relaxation models (such as in [50]). In [45], a model has been developed to predict the stall and poststall characteristics of aircraft wings. It used 2D RANS simulations to predict the boundary layer separation point, and the Kirchhoff flow approach to model the nonlinear lift variation of the RANS solution into the VLM. The model obtained improved results compared to other approaches such as iterative decambering. The stall behavior of a horizontal tail was investigated in [47] with a nonlinear VLM. The approach used van Dam's loosely coupled algorithm based on the effective angle of attack correction but incorporated 2.5D aerofoil characteristics based on the XFOIL solver and wing sweep theory.

Recently, it has been shown in [48] than nonlinear VLM results can be successfully used to predict the base flow field with sufficient accuracy for aero-acoustic noise calculations for micro air vehicle rotors. The method used an iterative correction of the sectional circulation values based on a lookup table of aerofoil viscous XFOIL or CFD results, the wake being modeled as vortex particles rather than rings to reduce the computational effort. An unsteady nonlinear VLM has been developed and applied to the analysis of wind turbine rotors in [49]. The method can be considered as more coupled compared to other published approaches, as the viscous corrections are determined via solving an implicit system of equations. Another recent model has been proposed in [51], based on a more loosely coupled iterative correction of the vortex strengths, using lookup table lift coefficient values. Results showed very good agreement with experimental data for the MEXICO wind turbine.

Other nonlinear VLM approaches have been developed in [55,56] based on the effective angle of attack updated strategy of van Dam and utilizing 2.5D RANS sectional data in order to increase accuracy for swept wings (as in [55]), or utilizing an iterative decambering approach implemented in a nonlinear Newton–Raphson iteration scheme so as to drive the potential flow results toward viscous data saved in lookup tables [56].

In the classical VLM approach, the unknown intensities of all the vortex rings distributed over the wing surface are determined by requiring that the zero normal flow boundary condition would be satisfied for all collocation points, leading to the following linear system of equations:

$$\sum_{j=1}^{N} \boldsymbol{v}_{ij} \cdot \boldsymbol{n}_i \Gamma_j = -\boldsymbol{V}_\infty \cdot \boldsymbol{n}_i \quad i = 1,\ 2, \ldots, N \tag{7.8}$$

In Eq. (7.8), $\mathbf{V}_\infty$ is the freestream velocity, N is the total number of vortex rings over the wing surface, $\mathbf{v}_{ij}$ is the velocity induced by the unit strength vortex ring j at the ith panel collocation point, and $\mathbf{n}_i$ is the surface normal vector calculated at the ith panel collocation point.

A novel nonlinear VLM approach was introduced in [23,24], which allows the introduction of viscous effects into the method. The intensities of the vortex rings obtained by solving the linear system presented in Eq. (7.8) are adjusted using 2D aerofoil data. For each vortex ring, a correction $\Delta\Gamma$ is defined, so that the final values of the vortex intensities become:

$$\Gamma_j \to \Gamma_j + \Delta\Gamma_j \quad j = 1, 2, \ldots, N \tag{7.9}$$

By considering that a variation in the intensity of a vortex ring determines a variation in the velocities induced by that vortex ring, the introduction of the $\Delta\Gamma_j$ corrections is followed by the introduction of a secondary induced velocity field $\mathbf{V}_i^T$ over the wing surface. Thus, for the nonlinear VLM approach, Eq. (7.8) is replaced by:

$$\left(\mathbf{V}_\infty + \sum_{j=1}^{N} (\Gamma_j + \Delta\Gamma_j)\mathbf{v}_{ij} + \mathbf{V}_i^T \right) \cdot \mathbf{n}_i = 0 \tag{7.10}$$

To obtain the second set of equations needed for the problem resolution, a nonlinear viscous pressure coefficient distribution is required. Prebuild databases of experimental or numerical aerofoil results are again used, but this time containing pressure coefficient distributions for several Reynolds number values Re_i and angles of attack α_i:

$$CP_i^{\text{visc}} = f(\text{airfoil}_i,\ Re_i, \mathbf{V}_i, \alpha_i) \tag{7.11}$$

where CP_i^{visc} represents the viscous pressure coefficient distribution, Re_i is the Reynolds number, $\mathbf{V}_i$ is the modulus of the total velocity at the control point, and α_i represents the effective angle of attack, all variables being calculated for the ith wing strip. In the case where the wing has a large sweep angle, the strip theory analysis is extended using sweep theory, and the local values of the effective angle of attack and of the strip airfoil

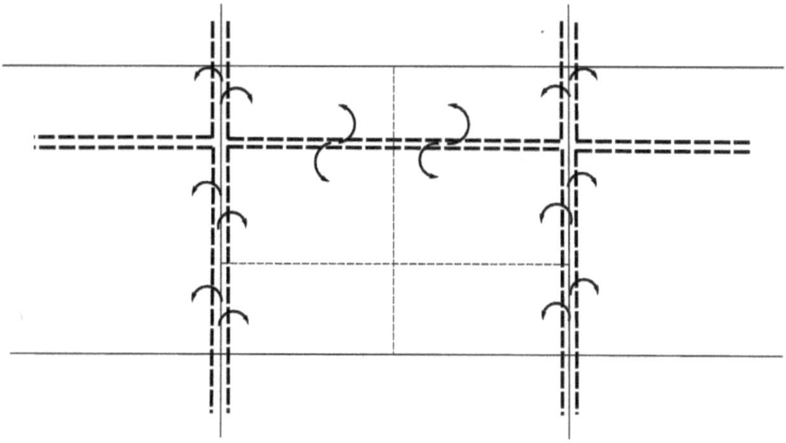

Figure 7.3 Illustration of the vortex segments on a typical panel.

aerodynamic characteristics are corrected using the iterative methodology presented in [25]. These pressure coefficient distributions are precalculated at a given number of flow conditions, while for any other flow condition not in the database, a simple linear interpolation is used between the two closest data points available.

The equations needed to calculate the vortex rings' intensity corrections are constructed from the assumption that for all N panels on the wing surface, the pressure coefficient variation ($\Delta CP = CP_{\text{upper-surface}} - CP_{\text{lower-surface}}$) obtained from the vortex rings' intensities is equal to the nonlinear viscous pressure coefficient variation ΔCP^{visc} obtained from the database. For all panels, the following equality is written:

$$-\mathbf{F}_i \cdot \mathbf{n}_i + A_i Q_\infty \Delta CP_i^{\text{visc}} = 0 \quad i = 1, 2, \ldots, N \tag{7.12}$$

In Eq. (7.12), $\mathbf{F}_i$ is the aerodynamic force generated by all the vortex lines placed on the panel, $\mathbf{n}_i$ is the surface normal vector calculated at the panel collocation point, A_i is the panel area, and Q_∞ is the freestream dynamic pressure.

Following the notations indicated in Fig. 7.3, the force can be calculated using the 3D vortex lifting law [11]:

$$\begin{aligned}\mathbf{F}_i =& \rho(\Gamma_i - \Gamma_U)\mathbf{V}_i \times \gamma_{12} + \rho(\Gamma_i - \Gamma_R)\mathbf{V}_i \times \gamma_{23} + \rho(\Gamma_i - \Gamma_L)\mathbf{V}_i \times \gamma_{61} + \\ &+ \rho(\Gamma_U - \Gamma_{UR})\mathbf{V}_i \times \gamma_{34} + \rho(\Gamma_U - \Gamma_{UL})\mathbf{V}_i \times \gamma_{56}\end{aligned} \tag{7.13}$$

where ρ is the air density, Γ is the strength of a vortex ring, $\mathbf{V}_i$ is the local velocity at the panel collocation point, and $\boldsymbol{\gamma}$ is the supporting geometric segment of a vortex line.

In addition, the local velocity at the panel collocation point will be written as:

$$\mathbf{V}_i = \mathbf{V}_\infty + \sum_{j=1}^{N} (\Gamma_j + \Delta\Gamma_j)\mathbf{v}_{ij} \tag{7.14}$$

By combining Eqs. (7.13) and (7.14), the following expression is obtained for the normal force acting on each of the wing surface panels:

$$\begin{aligned}\mathbf{F}_i \cdot \mathbf{n}_i = &\rho\left[\mathbf{n}_i \times \left(\mathbf{V}_\infty + \sum_{j=1}^{N}(\Gamma_j + \Delta\Gamma_j)\mathbf{v}_{ij}\right)\right] \\ &\cdot [(\Gamma_i - \Gamma_U)\gamma_{12} + (\Gamma_i - \Gamma_R)\gamma_{23} + (\Gamma_i - \Gamma_L)\gamma_{61} + (\Gamma_U - \Gamma_{UR})\gamma_{34} \\ &+ (\Gamma_U - \Gamma_{UL})\gamma_{56} + (\Delta\Gamma_i - \Delta\Gamma_U)\gamma_{12} + (\Delta\Gamma_i - \Delta\Gamma_R)\gamma_{23} \\ &+ (\Delta\Gamma_i - \Delta\Gamma_L)\gamma_{61} + (\Delta\Gamma_U - \Delta\Gamma_{UR})\gamma_{34} \\ &+ (\Delta\Gamma_U - \Delta\Gamma_{UL})\gamma_{56}] \quad i = 1, 2, \ldots, N\end{aligned} \tag{7.15}$$

By introducing the normal force $\mathbf{F}_i$ given by Eq. (7.15) into the equality presented in Eq. (7.12), and by coupling the resulting equations with Eq. (7.10), a nonlinear system of $2N$ equations is obtained:

$$\mathbf{R} = \left\{\begin{array}{c} \vdots \\ -\mathbf{F}_i \cdot \mathbf{n}_i + A_i Q_\infty \Delta CP_i^{\text{visc}} \\ \vdots \\ \hline \vdots \\ \sum_{j=1}^{N} \mathbf{v}_{ij} \cdot \mathbf{n}_i \Delta\Gamma_j + V_i^T \\ \vdots \end{array}\right\} = 0 \tag{7.16}$$

The unknown variables of the system are the N values of the vortex rings' intensity corrections and the N values of the surface transpiration velocities:

$$\mathbf{X} = \{\ \cdots\ \ \Delta\Gamma_i\ \ \cdots\ \ \ \cdots\ \ V_i^T\ \ \cdots\ \}^{\mathbf{T}} \tag{7.17}$$

The nonlinear system of equations obtained can be solved using Newton's classical method, since the Jacobian matrix elements can be analytically determined:

$$\begin{aligned} \mathbf{J}\left(\mathbf{X}^k\right)\Delta\mathbf{X} &= -\mathbf{R}\left(\mathbf{X}^k\right) \\ \mathbf{X}^{k+1} &= \mathbf{X}^k + \Omega\Delta\mathbf{X} \end{aligned} \tag{7.18}$$

In Eq. (7.18), $\mathbf{J}(\mathbf{X}^k)$ is the Jacobian matrix of first-order partial derivatives, calculated with the current estimate of the solution vector, -$\mathbf{R}(\mathbf{X}^k)$ is the system residual, calculated with the current estimate of the solution vector, $\Delta\mathbf{X}$ is the solution increment and Ω is an under-relaxation factor. The iterative solution procedure continues until the magnitude of the largest residual becomes smaller than the desired convergence criteria.

Once the nonlinear system (7.16) is solved and the corrected vortex ring strengths are determined from (9), the aerodynamic force and moment acting on each vortex segment on the wing can be determined using equations very similar to (7.5) and (7.6), and the profile drag coefficient can be determined using Eq. (7.7).

A verification and validation test is done using geometrical and experimental data from the NACA Technical Note 1208 [26]. The wing has a sweep angle of 45°, an aspect ratio of 8 and a taper ratio of 0.4. The experimental results were obtained for an airspeed of 65 m/s and a Reynolds number equal to 4×10^6, as calculated with the mean aerodynamic chord value. The numerical results are obtained with a mesh of 18 chordwise panels and 35 spanwise panels per wing semispan, while the sectional aerofoil database is generated by XFOIL. For reference purposes, linear VLM results obtained with the widely used XFLR5 code are also included in the comparison.

In Fig. 7.4, the results for the wing lift coefficient and quarter-chord pitching moment coefficient are compared with the experimental data. The nonlinear VLM predicts the lift curve characteristics very well, with a slight overestimation in predicted C_L values for angles of attack higher than 10°. A very good agreement exists for $C_{L_{max}}$ (1.01 for the experiment vs. 1.04 in the numerical results), but there is an underestimation of the stall angle by about 1.5°. The linear variation of the pitching moment coefficient is very well captured, but there are some differences for the nonlinear higher lift conditions, where the swept back wing experiences an early tip stall phenomenon which is difficult to capture. Fig. 7.5 shows a comparison of the spanwise wing loading for an angle of attack of 4.7°, the agreement being good.

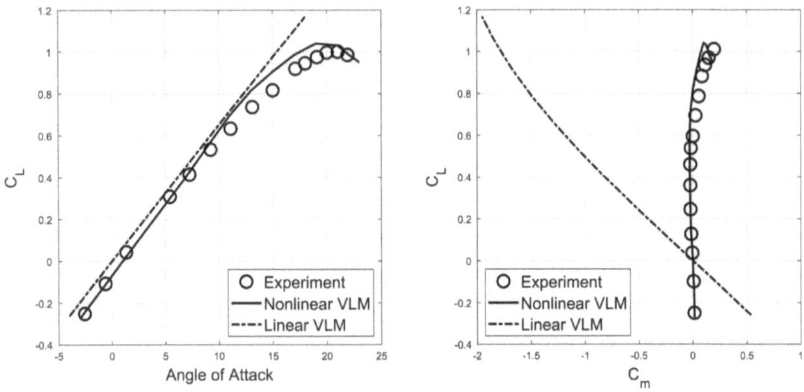

Figure 7.4 Comparison of lift and pitching moment coefficient values obtained with the nonlinear VLM method and the XFLR5 VLM code with experimental data published for the NACA TN1208 high aspect ratio high sweep wing geometry.

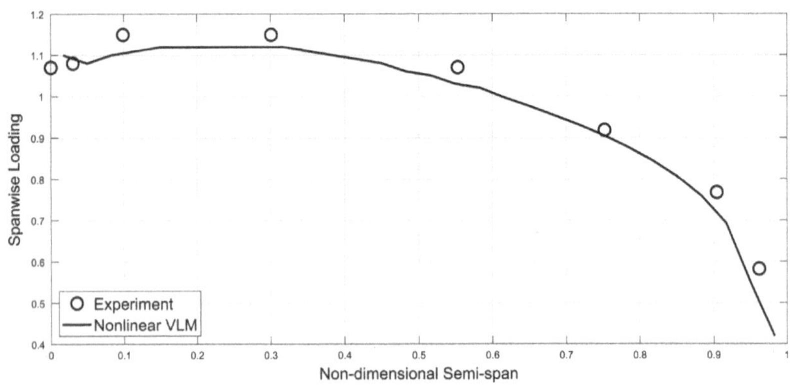

Figure 7.5 Comparison of spanwise loading obtained with the nonlinear VLM method with experimental data published for the NACA TN1208 high aspect ratio high sweep wing geometry.

7.4 Case study: aerodynamic shape optimization of a generic UAV wing

The nonlinear VLM is very well suited to be used in the early design phases of subsonic aircraft lifting surfaces, as it provides sufficiently accurate estimations of viscous aerodynamic characteristics for only a fraction of the computational requirements needed to perform a 3D CFD calculation.

The aerodynamic shape optimization of a generic UAV is performed. The wing is chosen to be representative for a tactical UAV with an all-up mass under 20 kg, having an aspect ratio of 8, a span of just over 4 m, a taper ratio of 0.5, and a NACA 2412 aerofoil section. The optimization is done using the Artificial Bee Colony algorithm, as described in [27], with the objective of improving the wing lift-to-drag ratio L/D over a rage of fixed angle of attack values, at an airspeed of 50 m/s and with a Reynolds number of 2.13×10^6, as calculated with the mean aerodynamic chord of the original wing geometry. The optimization procedure is focused on aerodynamic performance, and so no structural or weight aspects are considered.

Two optimization cases are considered. The first case uses the wingspan, taper ratio, and quarter-chord sweep angle as the optimization variables. The wingspan is constrained between 3.5 m and 5 m, the quarter-chord sweep angle between 0° and 30° and the taper ratio between 0.3 and 1. The second case changes the aerofoil shape in addition to the wing planform variables indicated for the first case. To achieve this, a Non-Uniform Rational B-Splines (NURBS) [28] parameterization of the NACA 2412 is introduced:

$$\begin{aligned}
\mathbf{C}(u) &= \sum_{i=1}^{k} \frac{N_{i,n} w_i}{\sum_{j=1}^{k} N_{j,n} w_j} \mathbf{P}_i \\
N_{i,1} &= \begin{cases} 1, & \text{if } t_i \le u \le t_{i+1} \\ 0, & \text{otherwise} \end{cases} \\
N_{i,n} &= \frac{u - t_i}{t_{i+n} - t_i} N_{i,n-1} + \frac{t_{i+n+1} - u}{t_{i+n+1} - t_i} N_{i+1,n-1}
\end{aligned} \tag{7.19}$$

In Eq. (7.19), u is the curve parameter, ranging from 0 (the start of the curve) to 1 (the end of the curve), k is the number of control points, n is the order of the curve, w_i are the weights associated with the control points, t_i are the knots, $N_{i,n}$ are the basis functions, and $\mathbf{P}_i = [x_i, y_i, z_i]$ are the control points. In the numerical optimization, the change of the aerofoil shape is achieved by changing the coordinates of the NURBS control points.

The initial coordinates of the control points $\mathbf{P}_i$ are determined through a nonlinear regression in which the NURBS aerofoil shape generated with Eq. (7.19) is iteratively driven toward the actual NACA 2412 aerofoil shape. In the numerical optimization, the change of the aerofoil shape is achieved by changing the coordinates of the NURBS control points, which act as optimization variables. Only the upper surface of the aerofoil between the leading edge and $0.5c$ can change, and the control points are constrained to move only on the y-axis and by a maximum of $0.05c$. To capture the effects of the aerofoil shape change on the wing performance, prebuild data based on 2D results are not used in this scenario. Rather, the XFOIL solver

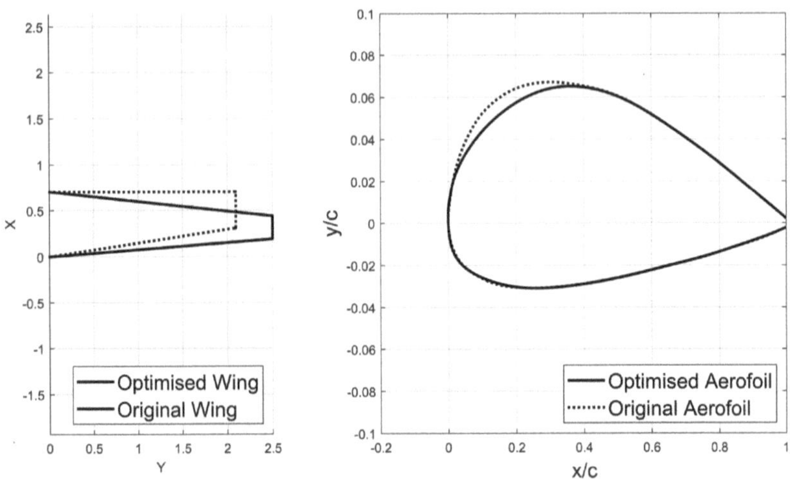

Figure 7.6 Comparison between the original and the redesigned wing and airfoil shapes.

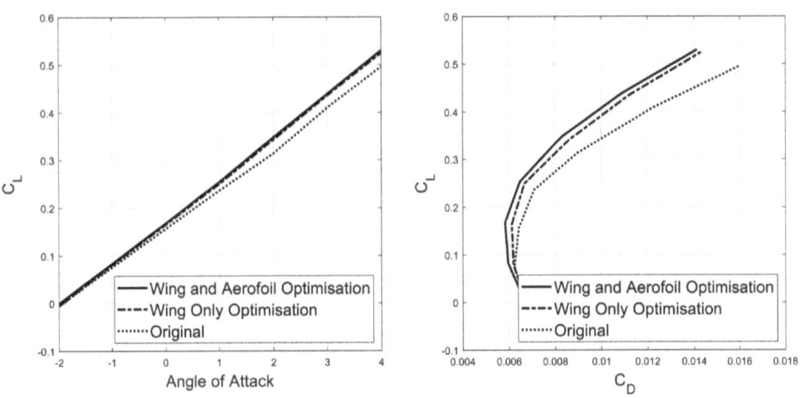

Figure 7.7 Lift curve and drag polar for the original wing, the redesigned wing using only wing planform parameters and the redesigned wing including aerofoil shape optimization.

is called while the optimization procedure runs during each iteration of the nonlinear VLM system solution and for each spanwise wing strip.

Fig. 7.6 presents a comparison between the original and redesigned wing shapes, as well as between the original and optimized (morphed) airfoil, while in Fig. 7.7, a comparison is presented between the lift curve and drag polar for the original wing, the planform-only optimized wing and the planform and aerofoil optimized wing.

It must be noted that the planform shape of the optimized wing in the two cases was approximately the same regardless of whether the aerofoil optimization variables were included or not in the process. The results hold no surprises, as the case study is meant to be more of a test for the sensitivity of the nonlinear VLM with respect to aerofoil shape changes rather than a rigorous optimization scenario. As expected, the optimal values of the planform variables provide a higher aspect ratio, lower sweep, lower taper wing, which achieves a higher L/D than the original design at any C_L value. This result is visible on both the increased lift curve slope and the reduced drag at any fixed lift coefficient value. The second optimization case, in which the parameterized aerofoil shape is added to the optimization variables, provides more significant results. As can be seen from Fig. 7.7, the nonlinear VLM is sensitive enough to capture the effects of small changes in the aerofoil thickness on the aerodynamic performance of the wing. This represents a net advantage over the classic VLM, which considers the wing a zero-thickness surface. The small decrease in aerofoil thickness between the original and optimized (morphed) aerofoil shapes over $0.3c$ amounts to a further reduction in drag over the entire range of angles of attack considered. This further reduction is only due to changes in the wing profile drag, as the planform shape obtained is essentially unchanged between the two optimization cases.

7.5 A simultaneous analysis and design methodology

Using global optimization procedures such as the Artificial Bee Colony algorithm or a Genetic Algorithm can provide the global optimum point in the design space, however doing so requires a very high number of objective function evaluations, and thus runs of the aerodynamic solver. Additionally, any increase in the number of design variables leads to further, significant increases in the number of evaluations, to a point where even using relatively simple methods such as the nonlinear LLT or VLM becomes too much time consuming for conceptual design purposes.

Gradient-based optimization algorithms can arrive at the global optimum point in a deterministic way and with a much smaller number of evaluations, provided the objective function is differentiable and a suitable method for evaluating the gradient is found. The adjoint equation approach has become particularly popular in the field of aerodynamic shape optimization because it allows the evaluation of the objective function gradient with a computational cost roughly equal to one additional aerodynamic solver run, regardless of the dimension of the design space. This approach has been extensively used over the last years, for a wide variety of problems across various industries. These include helicopter rotors [29], automobiles [30], wide-body transport aircraft [31], supersonic aircraft configurations [32], tidal turbines [33], and ship hulls [34].

The nonlinear LLT and VLM methods presented earlier are particularly suited for a gradient-based adjoint equation implementation, since all required matrices can be analytically calculated and efficiently implemented. Details of a methodology for further reducing the computational cost of a discrete adjoint method, with the construction of an adjoint-based simultaneous analysis and design (SAND) technique are presented in detail in [35] and will be summarized here.

To minimize the objective functional $J(\boldsymbol{w}, \alpha), J : \mathbb{R}^N \times \mathbb{R}^D \to \mathbb{R}$, subject to the equality and inequality constraints given by $\mathbf{G}(\boldsymbol{w}, \alpha) \leq 0$, $\mathbf{G} : \mathbb{R}^N \times \mathbb{R}^D \to \mathbb{R}^K$ and to satisfy the system of equations $\boldsymbol{R}(\boldsymbol{w}, \alpha) = 0$, $\boldsymbol{R} : \mathbb{R}^N \times \mathbb{R}^D \to \mathbb{R}^N$ that models the aerodynamic problem, a Lagrange functional can be defined [36]:

$$L(\boldsymbol{w}, \psi_1, \alpha, \psi_2) = J(\boldsymbol{w}, \alpha) + (\psi_1, \boldsymbol{R}(\boldsymbol{w}, \alpha))_N + (\psi_2, \mathbf{G}(\boldsymbol{w}, \alpha))_K \tag{7.20}$$

where $\boldsymbol{w} \in \mathbb{R}^N$ are the system-dependent variables and $\alpha \in \mathbb{R}^D$ are the system design parameters, whose values must be given in order to determine a solution for the system of equations, while $\psi_1 \in \mathbb{R}^N$ and $\psi_2 \in \mathbb{R}^K$ are two sets of Lagrange multipliers which will also play the role of the adjoint variables and $(^\circ,^\circ)_N$ signifies an appropriately defined inner product in the $\mathbb{R}^N$ space.

The critical points of the objective functional J will be among the points that cause the Gateaux variation δL of the Lagrange functional (7.20) to become zero for arbitrary variations of the independent variables, design variables and Lagrange multipliers. The Gateaux variation of (7.20) is given by:

$$\delta L = \frac{d}{d\varepsilon}\left[J + \varepsilon\delta J + (\psi_1 + \varepsilon\delta\psi_1, \boldsymbol{R} + \varepsilon\delta\boldsymbol{R})_N + (\psi_2 + \varepsilon\delta\psi_2, \mathbf{G} + \varepsilon\delta\mathbf{G})_K\right]\Bigg|_{\varepsilon=0} \tag{7.21}$$

Using the linear properties of the inner product, the Gateaux variation of the Lagrange functional can be rewritten as:

$$\delta L = \left(\frac{\partial J}{\partial \boldsymbol{w}}, \delta \boldsymbol{w}\right)_N + \left(\frac{\partial J}{\partial \alpha}, \delta \alpha\right)_K + \left(\psi_1, \frac{\partial \boldsymbol{R}}{\partial \boldsymbol{w}} \delta \boldsymbol{w}\right)_N + \left(\psi_1, \frac{\partial \boldsymbol{R}}{\partial \alpha} \delta \alpha\right)_N +$$

$$+(\boldsymbol{R}, \delta \psi_1)_N + \left(\psi_2, \frac{\partial \mathbf{G}}{\partial \boldsymbol{w}} \delta \boldsymbol{w}\right)_K + \left(\psi_2, \frac{\partial \mathbf{G}}{\partial \alpha} \delta \alpha\right)_K + (\mathbf{G}, \delta \psi_2)_K \quad (7.22)$$

In the n-dimensional real space $\mathbb{R}^n$, a linear operator L^+ is the adjoint to the operator L if the following equation can be written for any two vectors $\boldsymbol{x}, \boldsymbol{y} \in \mathbb{R}^n$, where again $(^\circ, ^\circ)$ denotes the inner product:

$$(L\boldsymbol{x}, \boldsymbol{y}) = (\boldsymbol{x}, L^+ \boldsymbol{y}) \quad (7.23)$$

Using the above defined property of adjoint operators, the Jacobian matrices appearing in Eq. (7.22) can be transferred to the Lagrange multipliers and replaced with their corresponding adjoint, which eventually allows writing:

$$\delta L = \left(\frac{\partial J}{\partial \boldsymbol{w}} + \frac{\partial \boldsymbol{R}}{\partial \boldsymbol{w}}^+ \psi_1 + \frac{\partial \boldsymbol{G}}{\partial \boldsymbol{w}}^+ \psi_2, \delta \boldsymbol{w}\right)_N + (\boldsymbol{R}, \delta \psi_1)_N +$$
$$+ \left(\frac{\partial J}{\partial \alpha} + \frac{\partial \boldsymbol{R}}{\partial \alpha}^+ \psi_1 + \frac{\partial \boldsymbol{G}}{\partial \alpha}^+ \psi_2, \delta \alpha\right)_K + (\mathbf{G}, \delta \psi_2)_K \quad (7.24)$$

Requiring that the Gateaux variation δL of the Lagrange functional becomes zero for arbitrary variations of the independent variables, design variables, and Lagrange multipliers leads to the following set of equations:

$$\frac{\partial J}{\partial \boldsymbol{w}} + \frac{\partial \boldsymbol{R}}{\partial \boldsymbol{w}}^+ \psi_1 + \frac{\partial \boldsymbol{G}}{\partial \boldsymbol{w}}^+ \psi_2 = 0 \quad (7.25)$$

$$\boldsymbol{R} = 0 \quad (7.26)$$

$$\frac{\partial J}{\partial \alpha} + \frac{\partial \boldsymbol{R}}{\partial \alpha}^+ \psi_1 + \frac{\partial \boldsymbol{G}}{\partial \alpha}^+ \psi_2 = 0 \quad (7.27)$$

$$(\mathbf{G}, \delta \psi_2)_K = 0 \quad (7.28)$$

The solution of the above system, $(\boldsymbol{w}, \boldsymbol{\psi}_1 \boldsymbol{\alpha}, \boldsymbol{\psi}_2)^T$, includes the optimal values of the design parameters as required to minimize the objective functional J, the solution of the system of equations $\boldsymbol{R}$ modeling the aerodynamic problem as obtained with the optimal design parameters, and the values for the two vectors of adjoint variables. To obtain the solution of the nonlinear system of equations presented in (7.25)–(7.28), the trust-region method is

Table 7.1 Wing and winglets geometry parameters.

Parameter	Value
Wing root chord (m)	0.10
Wing taper ratio	0.70
Wing span (m)	1
Wing aspect ratio	11.75
Wing sweep (°)	0
Wing aerofoil section	NACA 4412
Winglet root chord (m)	0.07
Winglet taper ratio	0.70
Winglet span (m)	0.10
Winglet sweep (°)	0
Winglet cant angle (°)	0
Winglet initial toe angle (°)	0
Winglet aerofoil section	NACA 4409

used [37]. The choice is justified by the fact that numerical tests have shown that the Jacobian matrix of the system can become ill-conditioned or even singular at some points.

The nonlinear LLT has been integrated in the SAND approach and applied to the problem of optimizing the geometry of a winglet fitted to a generic UAV wing geometry whose details are presented in Table 7.1.

Morphing wing tip devices [38,39] have attracted much research, as being a promising solution that could efficiently increase a wing's lift-to-drag ratio. The optimization variable is the winglet toe angle, and the objective function is improving the wing lift-to-drag ratio C_L/C_D for three angle of attack values, namely $-3°$, $0°$, and $3°$, at an airspeed of 10 m/s. The toe angle is constrained between a lower limit of $\tau_{\min} = -10°$ and an upper limit of $\tau_{\max} = 10°$. The wing is modeled using 60 horseshoe vortices, clustered toward the wing tips, while 20 horseshoe vortices are used for each winglet, clustered toward both the winglet tip and the junction with the wing tip. The variation of the 2D lift coefficient is limited to the linear region, with the lift curve slope and the zero-lift angle of attack for both NACA 4412 and NACA 4409 aerofoils being estimated based experimental data. The drag computations are limited to the induced drag component only, since 2D profile drag data is not considered.

Table 7.2 presents a comparison between the lift-to-drag ratio for the original design and for the optimized design, together with the toe angles values the morphing winglets take at each different angle of attack to achieve the indicated performance increase.

Table 7.2 Performance improvements obtained using morphing winglet.

Parameter Angle of attack	Original $\frac{C_L}{C_D}$	Optimized $\frac{C_L}{C_D}$	Winglet 1 toe (°)	Winglet 2 toe (°)
−3	19.07	21.10	−6.3208	6.3208
0	45.49	48.94	−5.9850	5.9850
3	42.31	44.55	−6.1178	6.1178

Table 7.3 Comparison of algorithm performance defined in terms of the required number of linear system solutions to obtain optimal design.

Algorithm	Adjoint simultaneous analysis and design		Interior point with adjoint gradient	
Angle of attack (°)	Iterations	System solutions	Iterations	System solutions
−3	7	7	16	264
0	15	15	18	366
3	10	10	14	275

The optimization was also done with a gradient-based optimization algorithm (the interior-point method) and using an adjoint approach to calculate the objective function gradient. It must be noted that both optimization approaches converge toward the same solution in terms of the design parameters (the two toe angles values), to an order of accuracy of two significant digits.

Before discussing the relative performance of the algorithms, it must be kept in mind that the linearized system obtained for the Newton step of the trust-region method is a system of $(N+N+D+K)$ equations, while for the interior-point optimization method, the aerodynamic model equations or the adjoint equations are only of size N. Considering that the solution process of a linear system of equations requires $O(N^3)$ operations (achievable when using high-performance linear solvers), the computational effort per linear system solution is higher with the proposed method.

It can be observed that the number of iterations required by the adjoint SAND method is slightly lower than the interior-point optimization method. The linear system obtained by applying the simultaneous method for this problem is a system of $(N+N+2+4)=(2N+6)$ equations, with $N=100$. Thus, one solution of the $(2N+6)$ linear system requires approximately $O((2N+6)^3) \approx 8 \cdot O(N^3)$, thus eight times more operations than one solution a N size linear system. Looking over the results presented in Table 7.3, it is immediately seen that even with the significantly higher computational effort per linear system solution (due to the larger matrices),

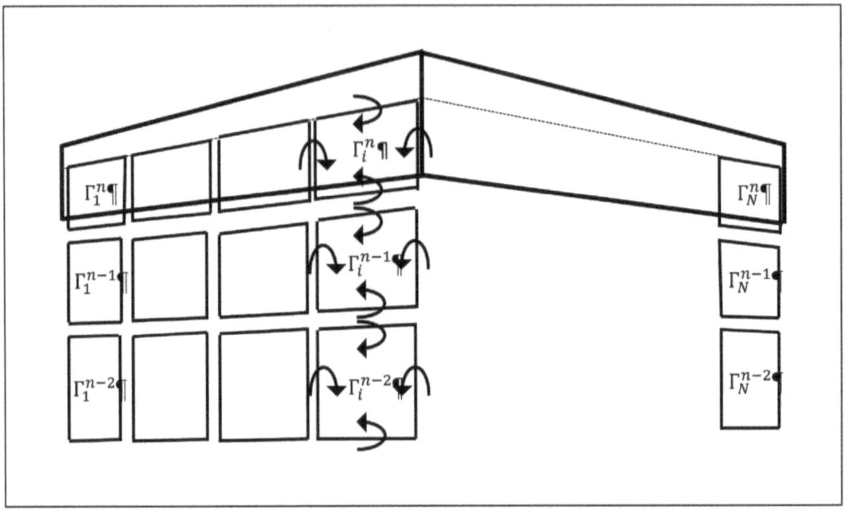

Figure 7.8 Sketch of the unsteady trailing vortex system.

the overall number of operations required by the simultaneous method is much smaller compared to the gradient-based optimization algorithm using adjoint-determined gradient. Thus, the adjoint-based simultaneous approach is more computationally efficient than the gradient-based optimization approach for nonlinear aerodynamic problems of interest having a high number of dependent variables and a relatively low number of design parameters. This is the case for many problems of engineering interest, not only conceptual aerodynamic design using potential-flow methods.

7.6 Extensions to unsteady flows

The nonlinear lifting line model summarized earlier and presented in detail in [9,10] has been extended to unsteady flows in [40]. In the context of unsteady flows, the continuous distribution of bound vorticity over the lifting surface and of trailing vorticity in the wake are approximated using a finite number of N ring vortices bound to the geometry and at each time step, a new row of N vortex rings are shed into the wake, as illustrated in Fig. 7.8.

To determine the force acting on a bound vortex segment, the following unsteady form of the vector Kutta–Joukowski theorem is introduced (a full derivation can be found in [41]):

$$\mathbf{dF} = \rho\Gamma(\mathbf{V} \times \mathbf{dl}) + \rho c \frac{\partial \Gamma}{\partial t}\mathbf{n} + \rho c \Gamma \frac{d\mathbf{n}}{dt} \tag{7.29}$$

Eq. (7.29) is written for the quarter-chord vortex segment of all vortex rings placed over the lifting surface. In addition, from classical lifting surface theory, the magnitude of the aerodynamic force acting on a spanwise strip was already presented in Eq. (7.3). Following the same idea as for the steady-state case (obtaining the 2D aerofoil aerodynamic characteristics using either experimental data or high-fidelity numerical solvers), the modulus of (7.29) can be set equal to the modulus of (7.3), and after some algebraic manipulation, the following equation is obtained:

$$\begin{aligned} &\rho\Gamma_i\sqrt{[(\mathbf{V}_i \times \mathbf{dl}_i)\cdot\mathbf{n}_i]^2 + [(\mathbf{V}_i \times \mathbf{dl}_i)\cdot\mathbf{c}_i]^2} + \rho c_i\left(\tfrac{\partial\Gamma}{\partial t}\right)_i + \rho c_i\Gamma_i\sqrt{\left[\tfrac{d\mathbf{n}_i}{dt}\cdot\mathbf{n}_i\right]^2 + \left[\tfrac{d\mathbf{n}_i}{dt}\cdot\mathbf{c}_i\right]^2} = \\ &= \sqrt{\left(\tfrac{1}{2}\rho\left[(\mathbf{V}_i\cdot\mathbf{n}_i)^2 + (\mathbf{V}_i\cdot\mathbf{c}_i)^2\right]dA_iC_{li}\right)^2 + \left(\tfrac{1}{2}\rho\left[(\mathbf{V}_i\cdot\mathbf{n}_i)^2 + (\mathbf{V}_i\cdot\mathbf{c}_i)^2\right]dA_iC_{di}\right)^2},\ i = 1,\ldots,N \end{aligned} \tag{7.30}$$

The local airspeed vector calculated at the aerodynamically bound vortex segment (the lifting surface quarter chord) is equal to the sum of the local kinematic velocity and the velocities induced by all the other vortex segments distributed in vortex rings over the lifting surface and wake. Letting M be the number of time steps performed (and thus giving the number of vortex rings rows that was shed into the wake over the time history of the unsteady analysis), and (for the purpose of simplifying the equations) letting the velocities induced by the four segments of each ring vortex be added together and treated as one velocity vector, the local airspeed vector is determined as:

$$\mathbf{V}_i = -(\mathbf{V}_0 + \mathbf{v}_{rel\,i} + \boldsymbol{\Omega} \times \mathbf{r}_i) + \sum_{j=1}^{N}\Gamma_j^n\mathbf{w}_{ij} + \sum_{k=2}^{M}\sum_{j=1}^{N}\Gamma_j^{n-k+1}\mathbf{w}_{ikj} \tag{7.31}$$

where $\mathbf{v}_{\text{rel}\,i}$ is the local relative velocity of a point on the wing with respect to the wing-fixed reference system (nonzero in the case of flapping wings, for example), $\boldsymbol{\Omega}$ is the angular velocity of the wing-fixed reference system with respect to the ground-fixed reference system, $\mathbf{r}_i$ is the position vector of a point on the wing, and M is the number of rows of vortex rings shed in the wake.

By inserting Eq. (7.31) in Eq. (7.30) and estimating the time derivative using a second-order backward difference (although other time-stepping schemes can also be used), the following time-dependent nonlinear system of equations is determined:

$$R_i(\Gamma^n) = \left(E_i(\Gamma^n) + \frac{3G_i}{2\Delta t}\right)\Gamma_i^n - \frac{4c_i}{2\Delta t}\Gamma_i^{n-1} + \frac{c_i}{2\Delta t}\Gamma_i^{n-2} - F_i(\Gamma^n) = 0, \ i = 1, \ldots, N \tag{7.32}$$

The various terms appearing in the above equation are lengthy and will not be presented here but can be found full in [40].

The nonlinear system of equations presented in Eq. (7.32) is solved at each time step to obtain updated values of the vortex ring strengths over the lifting surface. Since the Jacobian matrix can be obtained analytically (although the derivation is very lengthy), the solution is obtained using Newton's classical method for nonlinear systems:

$$\begin{aligned} &\Gamma^0 = \Gamma^{n-1} \\ &\left[\frac{\partial R_i}{\partial \Gamma_j}\right]\Bigg|_{(\Gamma^k)} \Delta\Gamma = -\left[R_i\left(\Gamma^k\right)\right] \\ &\Gamma^{k+1} = \Gamma^k + \Delta\Gamma \\ &\text{when } \mathbf{R}\left(\Gamma^{k+1}\right) < \varepsilon, \ \text{then } \Gamma^n = \Gamma^{k+1} \end{aligned} \tag{7.33}$$

The iterative solution process described in (7.33) is made assuming the lifting surface position and wake shape are frozen in time. Once the new vortex strength values Γ^n are converged to a desired precision, the position of the wake must be updated accordingly. This step represents the relaxation of the wake, and it is necessary for obtaining a physically representative force-free wake surface, but the process can introduce numerical difficulties. Because the current position $\mathbf{X}^n$ of each wake point depends on the current position of all other points, and the induced velocities $\mathbf{W}_{kj}$ depend on the current position of the vortex ring corners, the inherent nonlinearity of the wake relaxation process is handled using the following proposed fictitious time-marching scheme:

$$\begin{aligned} &\mathbf{X}^0 = \mathbf{X}^{n-1} \\ &\mathbf{X}^{t+1} = \mathbf{X}^t + \left[\frac{\mathbf{X}^t - \mathbf{X}^{n-1}}{\Delta t} - \left(\sum_{j=1}^{N}\Gamma_j^n \mathbf{W}_j(\mathbf{X}^t) + \sum_{k=2}^{M}\sum_{j=1}^{N}\Gamma_j^{n-k+1}\mathbf{W}_{kj}(\mathbf{X}^t)\right)\right]\Delta\tau \\ &\text{when } \mathbf{X}^{t+1} - \mathbf{X}^t < \varepsilon, \ \text{then } \mathbf{X}^n = \mathbf{X}^{t+1} \end{aligned} \tag{7.34}$$

Here, $\Delta\tau$ represents a fictitious time step and ε is a desired convergence criterion. The scheme in Eq. (7.25) is inspired by the dual time-stepping approach [57] used in many unsteady CFD solvers. The explicit marching in the fictitious time is setup to guarantee an implicit nonlinear approximation at the current physical time. A small number of only three iterations in the fictitious time were found to be sufficient to ensure reasonable convergence of the wake relaxation process to $\varepsilon/_b = 10^{-2}$, b being the wingspan.

The numerical simulation of flapping flight, and specifically the high-frequency, insect-type flapping motion, represents an extremely challenging problem, due to the very complex flow behavior and the development of highly nonlinear lift and thrust generation mechanism such as the clap and fling mechanism, rotational lift, wake capture (especially low advance ratio, hovering flight), laminar boundary layer separation, unsteady leading edge vortex formation, and strong aeroelastic coupling. Ho et al. [42] present a thorough review of the challenges associated with flapping flight, its numerical prediction, and associated control techniques. The analysis of avian flight (with the notable exception of the hummingbird) is somewhat less demanding, and over the last two decades has been successfully investigated using inviscid methods such as the UVLM or CFD based on the Euler equations. It has been repeatedly proven (see, e.g., ref. [43]) that the UVLM can provide unsteady lift and thrust predictions with relatively high accuracy and at low computational cost. The results obtained using the unsteady lifting line model will be verified against those determined using the UVLM for both low- and high-frequency flapping motion.

The geometry chosen has an aspect ratio of 8 and is generated using a highly cambered aerofoil from the NACA 83-series [43]. UVLM results are available for two reduced frequencies, $k_w = 0.08$ and $k_w = 1$, with $k_w = (4l\beta_0 n)/V_\infty$, β_0 being the maximum amplitude of the flapping motion, l is the wing half-span and n the flapping frequency. The lower frequency flapping case is representative of a pigeon, having $2l = 0.89$ m, $\beta_0 = 30°$, and $V_\infty = 11$ m/s. The high-frequency scenario is more representative of insect flight, and thus the parameters change accordingly, with $2l = 0.032$ m, $\beta_0 = 45°$, and $V_\infty = 1$ m/s. For the aerofoil section, only inviscid results obtained with XFOIL are used, to keep the setup as close as possible to the inviscid UVLM.

It is thus very important to note that the focal point of the comparison is not to show accurate physical modeling, but to demonstrate the ability of the unsteady lifting line model in predicting the same aerodynamic behavior as the vortex lattice in a field where it has been only very rarely used. Figs. 7.9 and 7.10 present the variation of the steady and unsteady lift components during the flapping motion as calculated by the unsteady

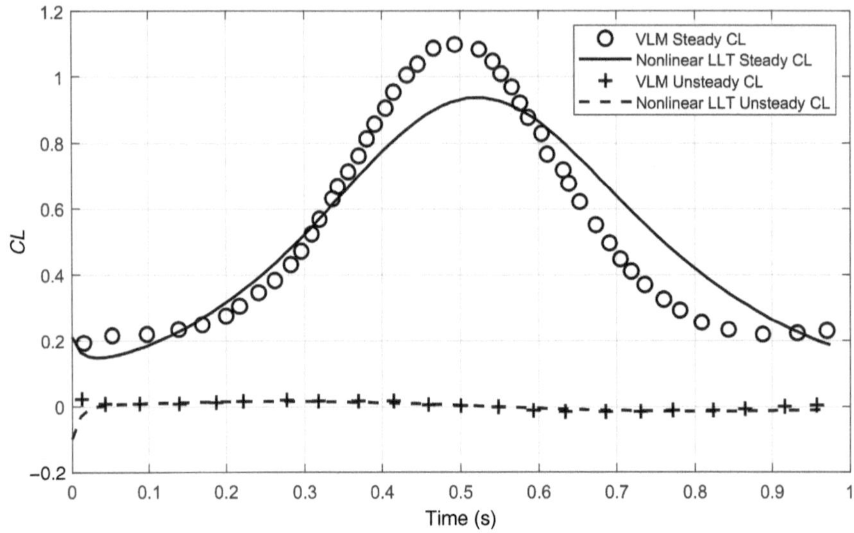

Figure 7.9 Comparison of steady and unsteady lift contributions for the flapping wing case having a reduced frequency of 0.08.

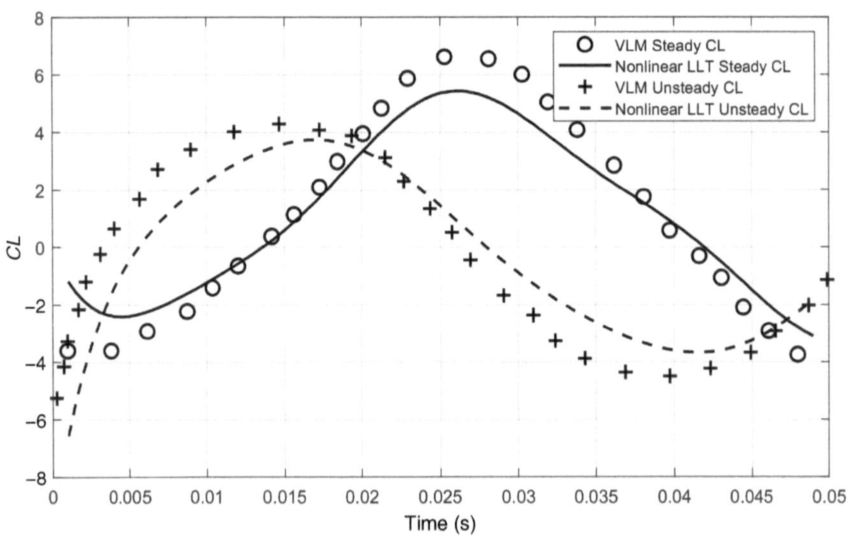

Figure 7.10 Comparison of steady and unsteady lift contributions for the flapping wing case having a reduced frequency of 1.00.

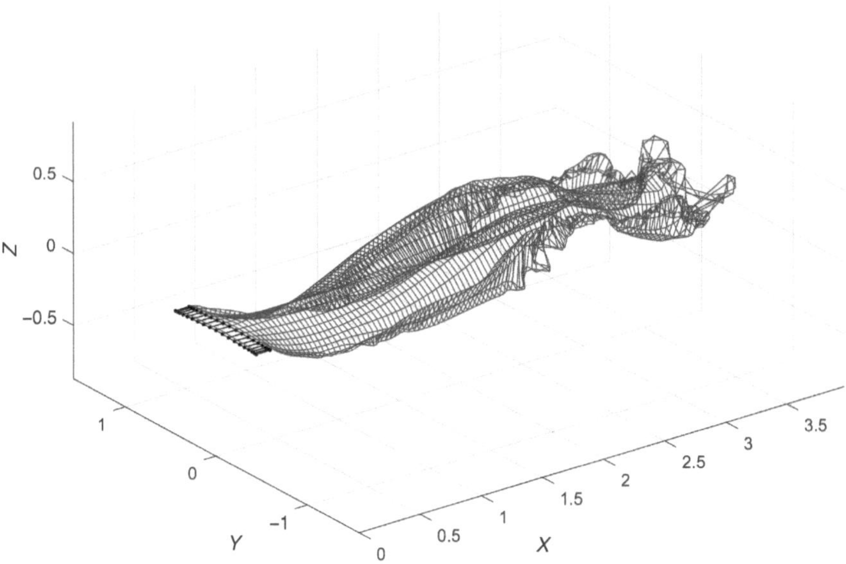

Figure 7.11 Wake development for the flapping wing case having a reduced frequency of 0.08.

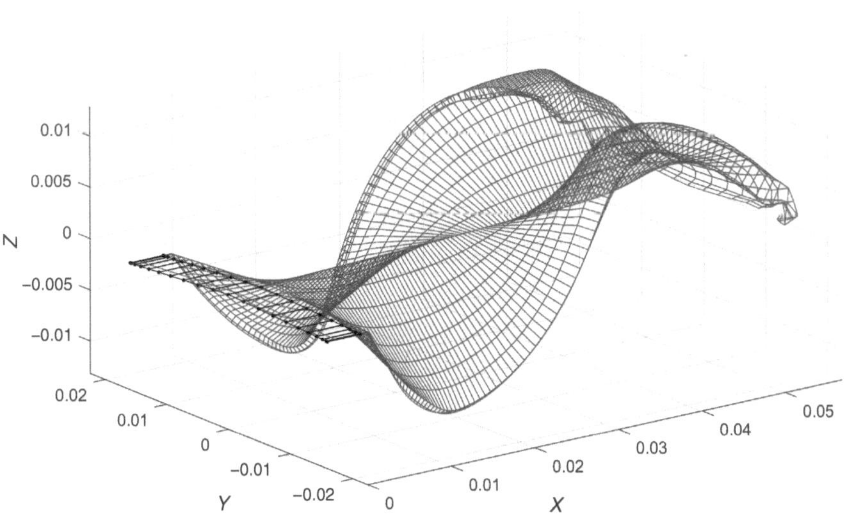

Figure 7.12 Wake development for the flapping wing case having a reduced frequency of 1.00.

lifting line and by the UVLM. The low-frequency case is dominated by the steady C_L component, while in the high-frequency case both components are significant, and out of phase. The results agree with the observation that unsteady flapping effects contribute to lift generation only if $k_w \geq 0.66$ [42]. Figs. 7.11 and 7.12 indicate how the wake development differs qualitatively between the two cases.

7.7 Conclusions

This chapter provides an overview of recent developments in quasi-3D aerodynamic models aimed at providing accurate results for multidisciplinary studies where the usage of discipline-specific tools and methods is a necessity. The nonlinear versions of the LLT and VLMs were mathematically derived by relaxing some of the underlying hypothesis present in the original formulations and by introducing the effects of viscosity through a nonlinear coupling with 2D aerofoil aerodynamics. While for the LLT this could be done relatively straightforward by means of the sectional lift coefficient, extending the VLM required using the complete viscous pressure distribution for each wing spanwise section. Comparisons done with experimental data for several different wing geometries show very good results, the novel quasi-3D methods being able to predict aerodynamic behavior in the high-lift region and accurately predicting drag for the low- to moderate-lift conditions.

Numerically, the models are implemented as nonlinear systems of equations. As such, they are easy to work with and can easily be implemented within optimization frameworks. Results show the nonlinear VLM can capture the influence of changing the aerofoil shape on the aerodynamic characteristics of a wing, where the original VLM is insensitive to aerofoil geometry. To further reduce computational time and provide conceptual designers with a fast and accurate tool, a simultaneous design and analysis method is proposed, where the optimal values of a set of design parameters is obtained simultaneously with the solution of the mathematical system describing a certain problem of interest, as the solution of a nonlinear system of equations. This completely circumvents the need of using an optimization algorithm, although the solution process itself remains iterative. For the problem of optimizing the winglet toe angle to improve a wing's lift-to-drag ratio the SAND approach determines the optimal point at a much lower overall computational effort compared to the traditional approach using a gradient-based optimization algorithm and an adjoint equation to determine the objective function gradient.

The nonlinear LLT method has also been extended to unsteady flows. Verification cases for high and low reduced frequency flapping wing problems show an accuracy comparable to the widely used UVLM. The advantage of the proposed method over the UVLM rests in the sensitivity of the results to the wing's aerofoil shape and the ability to introduce the effects of viscous effect via the coupling with the database of 2D aerofoil results. Future work includes revisiting the fictitious time-marching scheme for the wake relaxation, which is numerically stable regardless of the wake shape or complexity but very computationally expensive.

References

1. W.F. Phillips, D.O. Snyder, Modern adaptation of Prandtl's classic lifting-line theory, J. Aircraft 37 (4) (2000) 662–670.
2. R.E. Spall, W.F. Phillips, B.B. Pincock, Numerical analysis of multiple, thin-sail geometries based on Prandtl's lifting-line theory, Comput. Fluids 82 (2013) 29–37.
3. S. Gallay, E. Laurendeau, Nonlinear generalized lifting-line coupling algorithms for pre/post-stall flows, AIAA J. 53 (7) (2015) 1784–1792.
4. S. Gallay, E. Laurendeau, Preliminary-design aerodynamic model for complex configurations using lifting-line coupling algorithm, J. Aircraft 53 (4) (2016) 1145–1159.
5. P.J. Phlips, R.A. East, N.H. Pratt, An unsteady lifting line theory of flapping wings with application to the forward flight of birds, J. Fluid Mech. 112 (1981) 97–125.
6. Cline, S. and Crawford, C. Comparison of potential flow wake models for horizontal-axis wind turbine rotors. In: *48th AIAA Aerospace Sciences Meeting Including the New Horizons Forum and Aerospace Exposition, 4-7 January 2010, Orlando, Florida, USA.*
7. M.K. McWilliam, C. Crawford, Finite element based Lagrangian vortex dynamics model for wind turbine aerodynamics, J. Phys.: Conf. Ser. *524* (2014) paper 012127.
8. M. Fluck, Accelerating Unsteady Wind Turbine Aerodynamics: A Stochastic Lagrangian Vortex Model. Ph.D. Thesis, University of Victoria, 2014.
9. O. Sugar Gabor, A. Koreanschi, R.M. Botez, Analysis of UAS-S4 Éhecatl aerodynamic performance improvement using several configurations of a morphing wing technology, Aeronaut. J. 120 (1231) (2016) 1337–1364.
10. O. Sugar Gabor, A. Koreanschi, R.M. Botez, Numerical study of UAS-S4 Éhecatl aerodynamic performance improvement obtained with the use of a morphing wing approach, *American Institute of Aeronautics and Astronautics AIAA 33rd Applied Aerodynamics Conference*, 22-26 June 2015, Houston, Texas, USA.
11. P. Saffman, Vortex Dynamics, Cambridge University Press, Cambridge, UK, 1992.
12. R.H. Neely, T.V. Bollech, G.C. Westrick, R.R. Graham, Experimental and Calculated Characteristics of Several NACA 44-Series Wings with Aspect Ratios of 8, 10 and 12 and Taper Ratios of 2.5 and 3.5. NACA Technical Note No. 1270, Langley Memorial Aeronautical Laboratory, Langley Field, Va., United States, 1947.
13. M. Drela, XFOIL: *an analysis and design system for low reynolds number airfoils*, Thomas J. Mueller, Low Reynolds Number Aerodynamics, Springer, New York, NY., United States, 1989, pp. 1–12.
14. J.F. Cahill, S.M. Gottlieb, Low-Speed Aerodynamic Characteristics of a Series of Swept Wings Having NACA 65A006 Airfoil Sections, Langley Memorial Aeronautical Laboratory, Langley Field, Va., United States, 1950 NACA Research Memorandum L50F16.

15. L.K. Loftin, Theoretical and Experimental Data for a Number of NACA 6A-Series Airfoil Sections, Langley Memorial Aeronautical Laboratory, Langley Field, Va., United States, 1947 NACA Technical Report No. 903.
16. S.G. Hedman, Vortex Lattice Method for Calculation of Quasi Steady State Loadings on Thin Elastic Wings in Subsonic Flow, Aeronautical Research Institute of Sweden, Stockholm, Sweden, 1966.
17. J. Katz, A. Plotkin, Low Speed Aerodynamics: From Wing Theory to Panel Methods, McGraw-Hill, New York, 1991.
18. J. Murua, R. Palacios, J.M.R. Graham, Applications of the unsteady vortex-lattice method in aircraft elasticity and flight dynamics, Prog. Aerospace Sci. 55 (2012) 46–72.
19. L. Leifsson, K. Slawomir, A. Bekasiewicz, Fast low-fidelity wing aerodynamics model for surrogate-based shape optimization, Proc. Comput. Sci. 29 (2014) 811–820.
20. D.D. Smith, R.M. Ajaj, A.T. Isikveren, M.I. Friswell, Multi-objective optimization for the multiphase design of active polymorphic wings, J. Aircraft 49 (5) (2012) 1153–1160.
21. H. Tianyuan, Y. Xiongqing, Aerodynamic/stealthy/structural multidisciplinary design optimization of unmanned combat air vehicle, Chin. J. Aeronaut. 22 (2009) 380–386.
22. L. Peifeng, B. Zhang, C. Yingchun, Y. Changsheng, L. Yu, Aerodynamic design methodology for blended wing body transport, Chin. J. Aeronaut. 24 (2012) 508–516.
23. O. Sugar Gabor, A. Koreanschi, R.M. Botez, A new non-linear vortex lattice method: applications to wing aerodynamic optimizations, Chin. J. Aeronaut. 29 (5) (2016) 1178–1195.
24. O. Sugar Gabor, A. Koreanschi, R.M. Botez, M. Mamou, Y. Mebarki, Analysis of the aerodynamic performance of a morphing wing-tip demonstrator using a novel nonlinear vortex lattice method. American Institute of Aeronautics and Astronautics AIAA 34th Applied Aerodynamics Conference, 13–17 June 2016, Washington, DC, USA.
25. J. Mariens, A. Elham, M.J.L. van Tooren, Quasi-three-dimensional aerodynamic solver for multidisciplinary design optimization of lifting surfaces, J. Aircraft 51 (2) (2014) 547–558.
26. W.C. Schneider, A Comparison of the Spanwise Loading Calculated by Various Methods with Experimental Loadings Obtained on a Sweptback Wing of Aspect Ration 8, at a Reynolds Number of 4 Million, NACA Report No. 1208, Langley Memorial Aeronautical Laboratory, 1951.
27. O. Sugar Gabor, A. Simon, A. Koreanschi, R.M. Botez, Numerical optimization of the S4 Éhecatl UAS airfoil using a morphing wing approach. American Institute of Aeronautics and Astronautics AIAA 32nd Applied Aerodynamics Conference, 16–20 June 2014. Atlanta, Georgia, USA.
28. L. Piegl, W. Tiller, The NURBS Book, second ed., Springer-Verlag, Berlin, Germany, 1997.
29. E. Fabiano, A. Mishra, D. Mavriplis, K. Mani, Time-dependent aero-acoustic adjoint-based shape optimization of helicopter rotors in forward flight, in: 57th AIAA/ASCE/AHS/ASC Structures, Structural Dynamics, and Materials Conference, AIAA Paper 2016–1910, 2016.
30. T. Blacha, M.M. Gregersen, M. Islam, H. Bensler, Application of the adjoint method for vehicle aerodynamic optimization, in: SAE 2016 World Congress and Exhibition, 2016 SAE Technical Paper 2016-01-1615.
31. G.K.W. Kenway, J.R.R.A. Martins, Multipoint high-fidelity aerostructural optimization of a transport aircraft configuration, J. Aircraft 51 (1) (2014) 144–160.
32. J. Reuther, J.J. Alonso, M.J. Rimlinger, A. Jameson, Aerodynamic shape optimization of supersonic aircraft configurations via an adjoint formulation on distributed memory parallel computers, Comput. Fluids 28 (4) (1999) 675–700.
33. S.W. Funke, P.E. Farrell, M.D. Piggott, Tidal turbine array optimisation using the adjoint approach, Renew. Energy 63 (2015) 658–673.

34. S.A. Ragab, Shape optimization of surface ships in potential flow using an adjoint formulation, AIAA J. 42 (2) (2004) 296–304.
35. O. Sugar Gabor, Discrete adjoint-based simultaneous analysis and design method for conceptual aerodynamic applications, INCAS Bull. 9 (3) (2017) 133–147.
36. DG. Cacuci, Sensitivity and Uncertainty Analysis, Volume I: Theory, CRC Press, Boca Raton, Florida, USA, 2003.
37. R.H. Byrd, J.C. Gilbert, J.A. Nocedal, Trust region method based on interior point techniques for nonlinear programming, Math. Program. 89 (1) (2000) 149–185.
38. L. Falcão, A.A. Gomes, A. Suleman, Aero-structural design optimization of a morphing wingtip, J. Intell. Mater. Syst. Struct. 22 (10) (2011) 1113–1124.
39. D.D. Smith, R.M. Ajaj, A.T. Isikveren, M.I. Friswell, Multi-objective optimization for the multiphase design of active polymorphing wings, J. Aircraft 49 (4) (2012) 1153–1160.
40. O. Sugar Gabor, A general numerical unsteady nonlinear lifting line model for engineering aerodynamics studies, Aeronaut. J. 122 (1254) (2018) 1199–1228.
41. O. Sugar Gabor, Nonlinear lifting-line model using a vector formulation of the unsteady Kutta-Joukowski theorem, INCAS Bull. 11 (1) (2019) 189–203.
42. S. Ho, H. Nassef, N. Pornsinsirirak, Y.C. Tai, C.M. Ho, Unsteady aerodynamics and flow control for flapping wing flyers, Prog. Aerospace Sci. 39 (2003) 635–681.
43. T.E. Fritz, L.N. Long, Object-oriented unsteady vortex lattice method for flapping flight, J. Aircraft 41 (6) (2004) 1275–1290.
44. J.R. Chreim, F.R. Esteves, M.M. Pimenta, G.R. Assi, J.L.D. Dantas, A.M. Kogishi, Validation of a novel lifting-line method for propeller design and analysis, in: Proceedings of the 14th Practical Design of Ships and Other Floating Structures-PRADS, 2019.
45. C.R. Dos Santos, F.D. Marques, Lift prediction including stall, using vortex lattice method with Kirchhoff-based correction, J. Aircraft 55 (2) (2018).
46. L.L. Gamble, A.M. Pankonien, D.J. Inman, Stall recovery of a morphing wing via extended nonlinear lifting-line theory, AIAA J. 55 (9) (2017) 2956–2963.
47. H. Goitia, R. Llamas, Nonlinear vortex lattice method for stall prediction, MATEC Web Conf. 304 (2019) 1–8.
48. Y. Jo, T. Jardin, R. Gojon, M.C. Jacob, J.M. Moschetta, Prediction of noise from low Reynolds number rotors with different number of blades using a non-linear vortex lattice method, in: 25th AIAA/CEAS Aeroacoustics Conference (Aeroacoustics 2019), 20–23 May 2019, Delft, Netherlands.
49. H. Kim, S. Lee, S. Lee, Numerical analysis on the aerodynamics of HAWTs using nonlinear vortex strength correction, Curr. Appl. Phys. 10 (2010) 311–315.
50. T. Lee, S.O. Park, Improved iteration algorithm for nonlinear vortex lattice method, J. Aircraft 46 (6) (2009) 2174–2176.
51. H. Lee, D.J. Lee, Numerical investigation of the aerodynamics and wake structures of horizontal axis wind turbines by using nonlinear vortex lattice method, Renew. Energy 132 (2019) 1121–1133.
52. D. Owens, Weissinger's model of the nonlinear lifting-line method for aircraft design, in: 36th AIAA Aerospace Sciences Meeting and Exhibit, 1998 AIAA Paper 98-0597.
53. A. van Garrel, Development of a Wind Turbine Aerodynamics Simulation Module. ECN-C-03-079 Report, 2003.
54. G. Vernengo, L. Bonfiglio, S. Brizzolara, Supercavitating three-dimensional hydrofoil analysis by viscous lifting-line approach, AIAA J. 55 (12) (2017) 4127–4141.
55. M. Parenteau, K. Sermeus, E. Laurendeau, VLM coupled with 2.5 D RANS sectional data for high-lift design, in: Proceedings of the 2018 AIAA Aerospace Sciences Meeting, 2018.

56. R.C. Paul, A. Gopalarathnam, Iteration schemes for rapid post-stall aerodynamic prediction of wings using a decambering approach, Int. J. Numer. Methods Fluids 76 (4) (2014) 199–222.
57. A. Jameson, Time dependent calculations using multigrid, with application to unsteady flows past airfoils and wings, in: 10th AIAA Computational Fluid Dynamics Conference, June 1991, AIAA Paper 91-1596.

CHAPTER EIGHT

Design optimization and dynamic testing of CFRP for helicopter loading hanger

Zahra Andleeb[a], Sohail Malik[a], Ghulam Hussain[a], Hassan Abbas Khawaja[b,c], Mojtaba Moatamedi[c,d]
[a] Ghulam Ishaq Khan Institute of Engineering Sciences and Technology, Topi, Swabi, Pakistan
[b] UiT – The Arctic University of Norway, Tromsø, Norway
[c] Al Ghurair University, Dubai, United Arab Emirates
[d] Oslo Metropolitan University, Oslo, Norway

8.1 Introduction

Aerial cranes are being used in the wide variety of applications such as in construction, transport, emergency, military, etc. [1–3]. Fig. 8.1(A) shows an aerial crane operation. Aerial crane operation involves helicopter, loading hanger, and the lifted weight. Loading hanger is comprised of parts, such as loading line, attachment plate, slings, etc. Fig. 8.1(B) shows a close-up view of the loading hanger [4].

In this work, focus is on the loading hanger. The loading hanger is connected between helicopter and the load. This connection is activated using electromagnetic switch, which can be detached on unloading or emergency in case of unstable helicopter flight. The objective here is to design this part, select appropriate material, use Multiphysics tools for optimization, test material against dynamic loading in shock tube experimental setup and after shock impact analyze chosen material specimens for any changes in tensile and flexural properties in material testing system (MTS). This work presents five phases, which includes (1) geometric modeling, (2) material selection, (3) design optimization, (4) shock tube experiment, and (5) MTS.

The geometric modeling phase begins with a sketch of the attachment plate. Fig. 8.2(A) and (B) shows the top and side views of the attachment plate based on preset design requirements. A set of requirements is provided to enclose a physical problem [5,6]. There are four sling holes surrounded by an attachment area. Each sling hole has a diameter of 3 cm. In addition to that, each hole needs to be 30 cm from the center of the attachment plate and equidistant from each other. The center area is for joining the attachment

Multiphysics Simulations in Automotive and Aerospace Applications.
DOI: 10.1016/B978-0-12-817899-7.00005-8

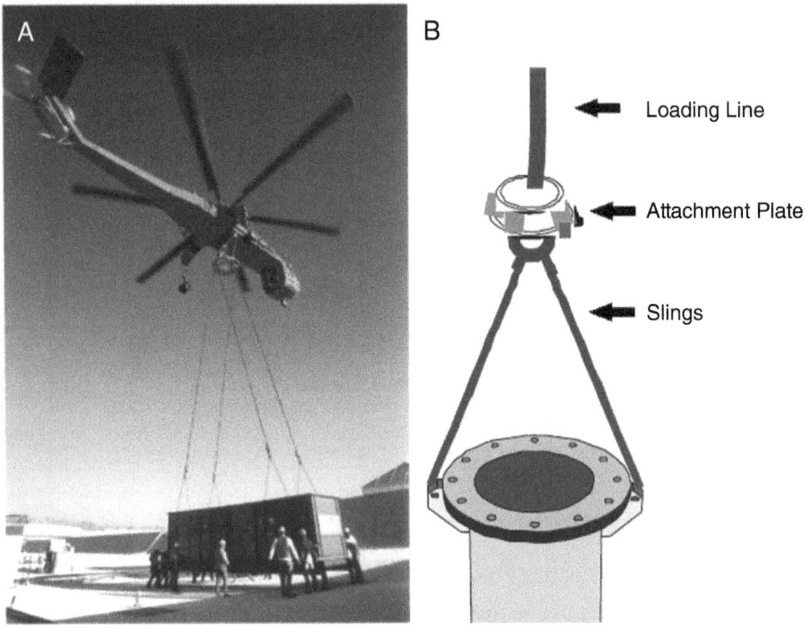

Figure 8.1 (A) Aerial crane operation is shown. (B) A zoom in view of loading hanger is shown comprising of three main segments, loading line from the helicopter, the attachment plate, and slings to attach the weight [8].

plate with the helicopter. It is also preset to 100 cm^2. Only requirement here is that this area should be axisymmetric to the center of the attachment plate. The helicopter maximum loading capacity is up to 1 metric ton (1000 kg).

Material selection phase is the choosing of an appropriate material for the attachment plate. It is desired that the designed part is efficient and safe. In this case, strength and stiffness are both important. In this work, appropriate material is chosen from a list of available materials. Another important selection criterion is strength to weight ratio. The factor of safety (FOS) is considered to be three in the selection process [7,8].

The optimization phase includes modifications in the geometry of the attachment plate while fulfilling all of the above given requirements. The focus is on the structural stresses and displacements. The aim is to reduce the weight of the attachment plate.

The composite materials are increasingly used in a number of applications, requiring high strength and light-weight properties. With the increase in usage of composites, it is also becoming essential to study them [9]. Researchers have tested various materials subjected to shock wave [10].

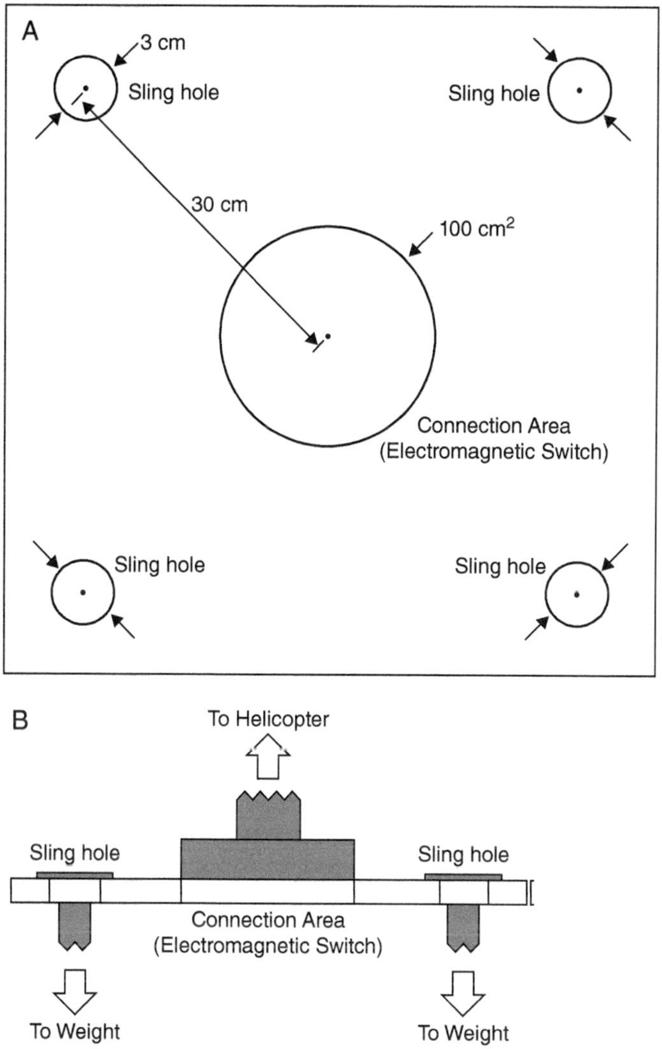

Figure 8.2 The attachment plate with preset dimensions' requirements.

Usually, such experiments are carried out using explosives submerged in water [11,12]. The test specimens are placed at different distances to control the intensity of shock waves [13,14]. This method is both expensive and potentially dangerous and requires careful planning [11,15]. In addition, complicated calculations and estimations are required to analyze the results. Another method is based on creating shock waves in a shock tube [16]. Such method has considerable advantages in comparison to earlier methods

[17,18]. In this method, either a solid projectile or a high pressure fluid is injected into a tube containing liquid instantaneously. This generates a shock wave which propagates along the tube. A test specimen subjected to the shock wave can be placed along the tube. The intensity of shock wave is controlled using velocity of entering projectile or the pressure of fluid [46,47]. Further control can be added using a T-junction in the tube [17]. Pressure transducers and strain gauges can be placed along the tube for the collection of data. The shock tube experimental setup used in this study is given in Section 2.5.

8.2 Methodology

The methodology followed in this study involves development of computer aided design (CAD) models using SoildWorks® [19] and Multiphysics analysis using ANSYS® software [20]. These are followed by evaluation of the obtained results against requirements of an aerial crane operation and repeating these steps until most optimized part is obtained. Fig. 8.3 shows the road map of the methodology.

8.2.1 CAD modeling

The development of CAD model includes building of 3D models using SolidWorks® CAD package. Various parts are built as shown in Fig. 8.4. All of these parts meet the dimensions stated earlier. Multiphysics analysis is conducted using ANSYS® Multiphysics package [21]. Two types of boundary conditions are applied, which are displacement constraint and distributed forces (pressure). The forces are equally distributed in each sling line and applied on the edge of each sling hole. The boundary conditions are illustrated in Fig. 8.5.

8.2.2 Multiphysics analysis

The Multiphysics analysis is performed by applying linear, elastic, isotropic, and homogeneous materials. The list of materials used is given in Table 8.1. The von Mises yield criterion [22,23] is employed for detecting the failure. The use of isotropic model for composite materials is inappropriate; however, one of the requirement is to have effective stresses as low as one-third of the yield strength (FOS equal to three). At low values of effective stresses, von Mises criterion is close to Tsai–Wu criterion [24], which is valid over anisotropic materials such as carbon fiber reinforced polymer (CFRP)[41] .

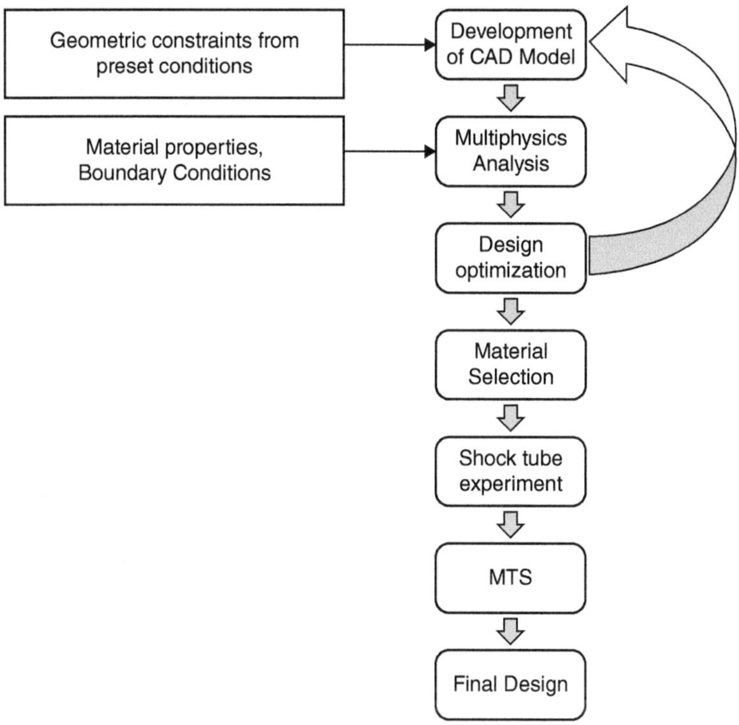

Figure 8.3 Roadmap of methodology.

Table 8.1 Properties of materials; yield strengths, densities, and specific strengths.

Materials	Yield strength (MPa)	Density (kg/m^3)	Specific strength (yield strength/density)
Aluminum alloys	120	2700	0.04
Stainless steel alloys	500	8000	0.0625
Titanium alloys	600	4500	0.13
Nickel alloys	900	7800	0.115
Carbon fiber reinforced polymer (CFRP)	320	1800	0.17

8.2.3 Design optimization

The designs shown in Fig. 8.4 show high stresses in the sharp corners [25,26]. This clearly indicates that sharp corner produces stress concentration points which must be avoided [27]. It is also realized that the problem is analogous to a simple bending beam considering that the attachment plate is quad

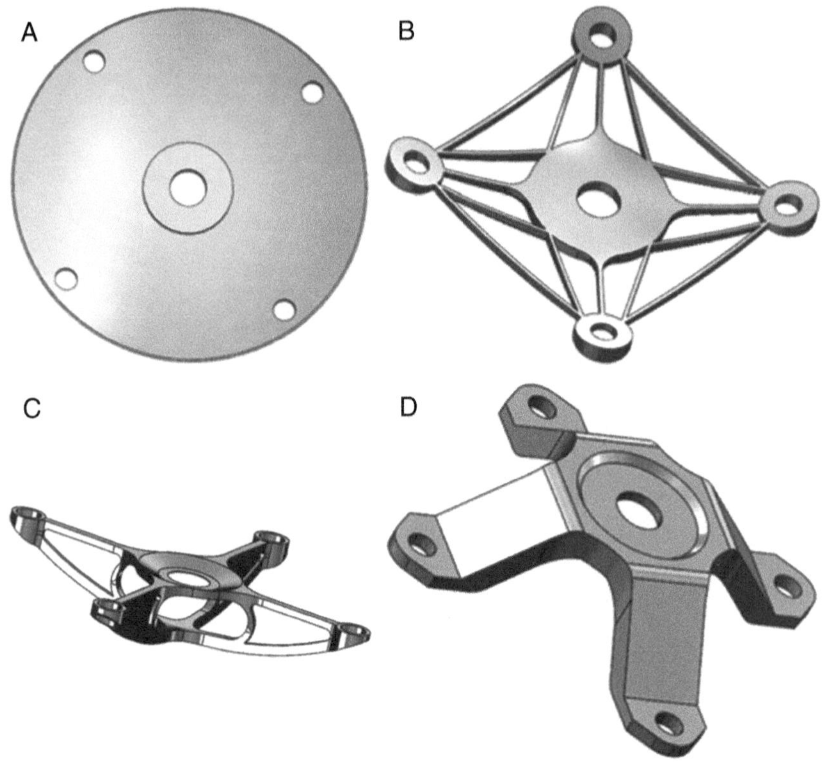

Figure 8.4 Various 3D parts built in SolidWorks® [9] for the attachment plate.

symmetric. Solution to such a problem can be found by applying Euler–Bernoulli bending theory [28,29,44,45]. The Euler–Bernoulli equation for the quasi-static bending of slender, isotropic, and homogeneous beams under a transverse load is given in Eq. (8.1).

$$\frac{d^2}{dx^2}\left(EI\frac{d^4w(x)}{dx^4}\right) = EI\frac{d^4w(x)}{dx^4} = q(x) \tag{8.1}$$

where x is a unit dimension in longitudinal direction, $q(x)$ is a distributed load, E is the Youngs modulus, I is the area moment of inertia, and $w(x)$ is the deflection from the neutral axis of the beam. Similarly, bending moment $M(x)$ and shear force $Q(x)$ can also be expressed in terms of Youngs modulus E, the area moment of inertia I, and deflection $w(x)$ as given in Eqs. (8.2) and (8.3) [14,28]. Similar design optimization studies have been performed

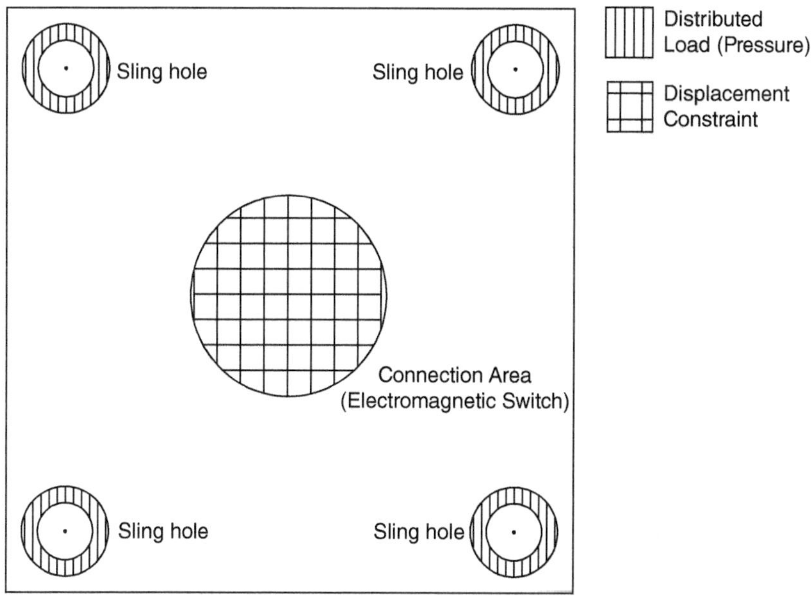

Figure 8.5 Boundary conditions applied to the model. (A) Displacement constraint applied at the connection between helicopter and the attachment plate. (B) Distributed load (pressure) is applied at the sling holes.

by [30].

$$M(x) = -EI\left(\frac{d^2w(x)}{dx^2}\right) \tag{8.2}$$

$$Q(x) == -EI\left(\frac{d^3w(x)}{dx^3}\right) = \frac{dM(x)}{d(x)} \tag{8.3}$$

8.2.4 CFRP samples

The material of choice is CFRP, since it has the highest strength to weight ratio [31,32]. For this study, quasi-isotropic shell structures are made by strengthening fibers in all respective 45° angles. A quasi-isotropic shell structure consists of a lay-up of laminates at 0°, +45°, +90°, −45° (135°), 0°, and so on as shown in Fig. 8.6. This makes the composite shell structure equally strong in two-dimensional plane. Different strengths and mechanical properties can be obtained by laying fabrics in different angular positions [33]. This is useful if required in particular application [34][48].

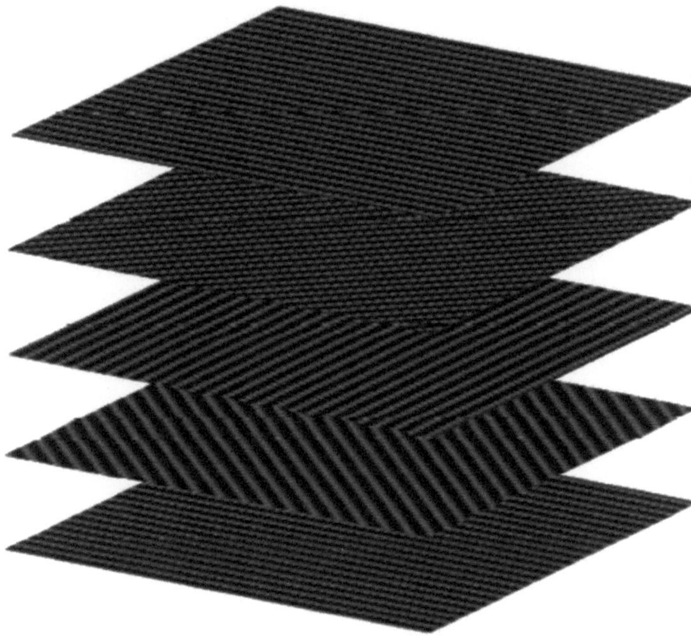

Figure 8.6 A 3D model showing a quasi-isotropic lay-up of carbon fiber fabrics.

In this work, carbon fiber reinforced polymers (CFRPs) shell specimens are fabricated using Multipreg E720 laminates [42]. These laminates consist of two main components [42]. The matrix material (also known as resin) is an adhesive material, which helps to glue carbon fiber fabrics together. The matrix material also aids in increasing the stiffness and strength of the composite material. A generic tensile stress/strain behavior of fibers, resin, and composite is given in Fig. 8.7 [35]. The failure mechanism of composite material is given by [24].

8.2.5 Shock tube experiment setup

The shock tube consists of two main parts: the driver section and the driven section (Fig. 8.8) [36]. The driver section contains compressed air, which is released into the driven section in order to create a shock wave [37]. This is generally triggered with a burst plate or a high speed valve. In given experiments, shock wave is generated in water by releasing high pressure air through high speed valve [5]. The shock wave propagates through water in the driven section [17]. At the end of the driven section, a T-section is fitted, enabling a secure mount of a test specimen (Fig. 8.9). In these experiments,

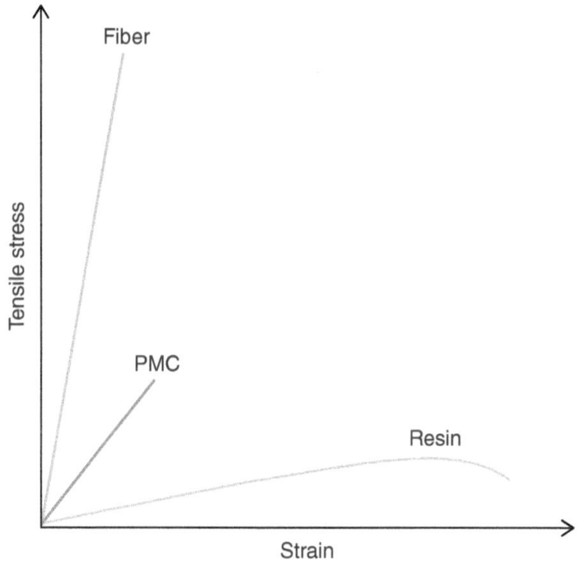

Figure 8.7 A stress/strain diagram, showing the fiber sheets in green, the resin (matrix) material in yellow, and the combined composite material (PMC) in the middle [16].

Figure 8.8 The shock tube. Driver section to the right (containing compressed air) and driven section to the left containing uncompressed water and a T-section for mounting the test specimen.

three pressure sensors (one static and two piezoelectric sensors) are placed at three different places in the shock tube to capture the propagation of the shock wave. The test specimen is mounted with a rosette strain gauge in order to capture the small strains created by the shock wave impact (Figs. 8.9 and 8.10) [43].

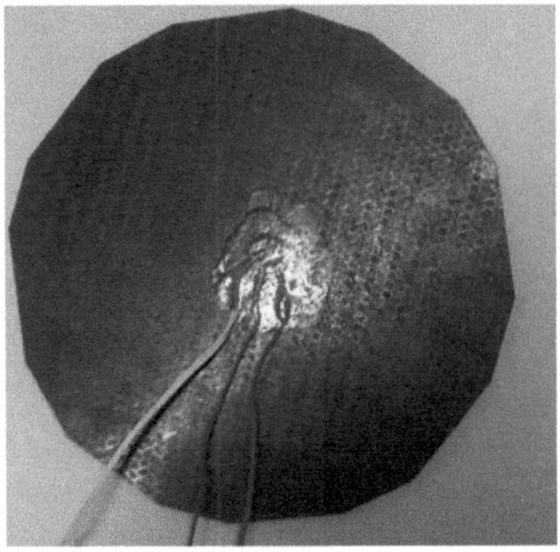

Figure 8.9 Rosette strain gauge attached to a CFRP specimen using cyanide-based glue. The three-colored wires are connections to three strain gauges at 45° angles. The strain gauge is placed in the geometric center of the plate.

Figure 8.10 The T-section. The CFRP specimen is mounted on the T-section where the three-colored wires are the connection of the rosette strain gauge.

8.2.6 Material test system

The MTS is employed to examine the change in properties of test specimens [38,39]. Given setup (Fig. 8.11) can apply static and dynamic loads up to 100 metric tons. These tests were performed on specimens with and without being subjected to shock impact (Fig. 8.12).

Two kinds of tests were performed using MTS system: tensile (Fig. 8.13) and flexural tests (Fig. 8.14). Rosette strain gauges were attached to the middle of each test piece for local strain. An average displacement gauge was also added to measure the total displacement of the test piece. The MTS was preprogrammed to apply tensile load, thus stretching the test piece until it ruptured. Tensile tests proved to be useful; however, clamps may create stress

Figure 8.11 The MTS during tensile testing of a steel sample.

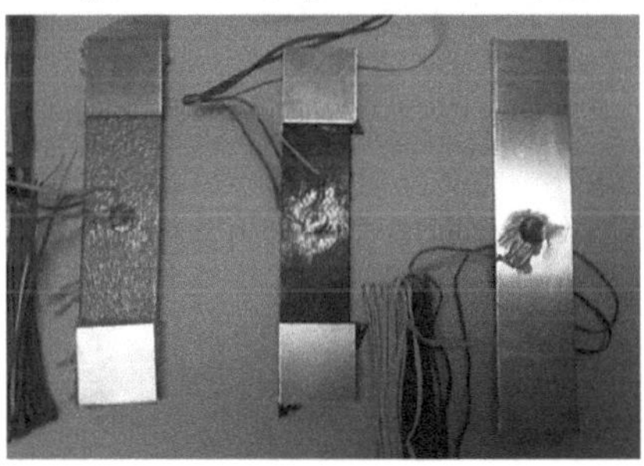

Figure 8.12 Kevlar, CFRP, and aluminum test samples. The ends are reinforced using aluminum plates for holding into the clamps. Rosette strain gauge is mounted for measuring the local strains.

Figure 8.13 Tensile loads applied to a test piece. The material is stretched.

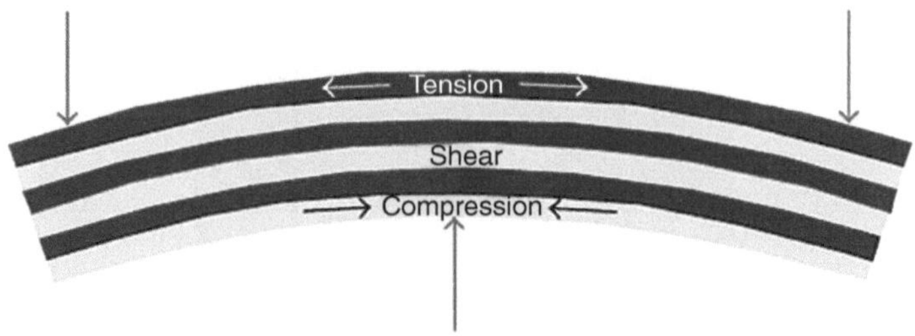

Figure 8.14 Flexural loads applied to a material. Such load will give a tensile stress in the top surface, a shear stress in the middle, and compressive stresses in the bottom layer.

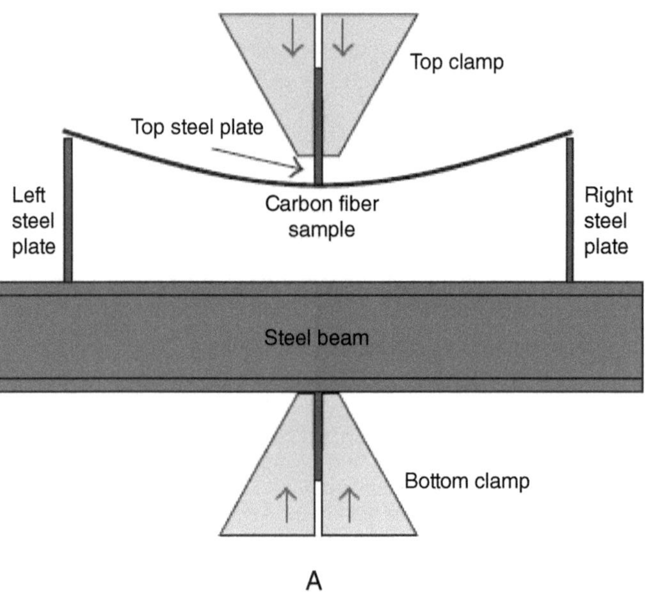

Figure 8.15 (A) A sketch of the flexural tests.

concentration points and hence lead to early failure. This can be avoided by performing flexural tests [44]. Flexural testing is more accurate in order to find the mechanical properties of test specimens.

The experiment has been performed by bending the material using a pressure line in the middle of the test pieces (applied top down), while being placed on two supports (one on the left hand bottom side and one on the right hand bottom side) as shown in Fig. 8.15(A) and 8.15(B). The middle

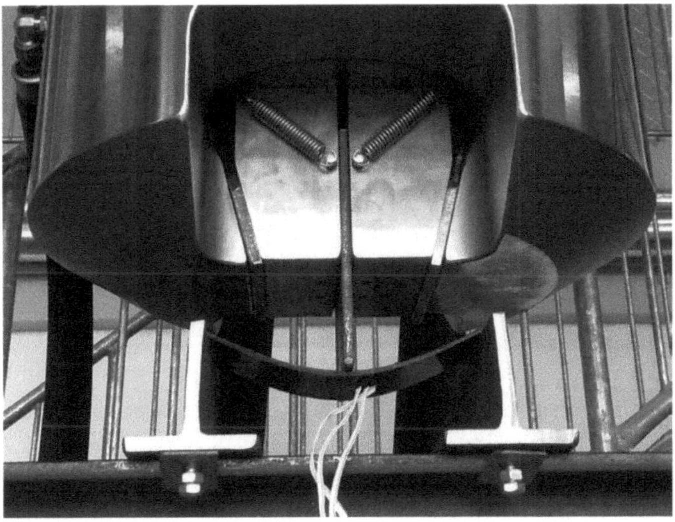

B

Figure 8.15 (B) The actual flexural test performed in MTS.

(top down) line is applying pressure from the top onto the middle of the plane, bending the material until it suffers from total mechanical failure. The test piece used in flexural tests was prepared in the same way as for the tensile tests. In both tests, rosette strain gauge was placed in the geometric center of the test piece. It is expected that the strain in longitudinal direction (from support to support) is going to be the highest with almost no strain in lateral direction.

8.3 Results and discussion

8.3.1 Design optimization results

For the beam cross sections that are symmetrical about the plane parallel to the transverse direction, it can be shown that the bending tensile stress is as given in Eq. (8.4) [18,19].

$$\sigma(x) = -Ey\frac{d^2w(x)}{dx^2} = \frac{M(x)y}{I} \tag{8.4}$$

where $\sigma(x)$ is the bending tensile stress and y is a unit dimension in the transverse direction. It is shown from Eq. (8.4) that the bending tensile stress is directly proportional to the moment. The value of moment is proportional to applied force and distance between the constraint and the loading point. Hence, by reducing the distance between constraint and loading point, the stresses can be reduced. Keeping this in consideration, the models are rebuilt and tested for stresses. The final optimized three-dimensional CAD model of an attachment plate is shown in Fig. 8.16(A). Fig. 8.16(B) and (C) shows the von Mises stress and displacement contours. The material of choice is CFRP, since it has the highest strength to weight ratio. The net weight of the optimized model is 0.25 kg. The maximum stress is about 48 MPa, which is less than a half to the requirement (one-third of yield strength of CFRP = 107 MPa; Table 8.1). This gives FOS of six. Higher FOS is also an advantage against aging failures such as fatigue and creep [26]. Maximum displacement is about 0.034 mm, which is very small in comparison to the part dimensions.

8.3.2 Shock tube experiment results

It is found that strain profile is linear mapping of driven pressure profile which can be seen from Fig. 8.17(A) and (B). The experiments are repeated on the same test piece to confirm the validity of the results. Repeated experiments proved that CFRP test pieces can withstand shock pressure (Fig. 8.17(A)); however, it is worth noting that applied shock pressure in these experiments did not result in strains for more than 25% of strain limit for rupture. It can be deduced that material behaves linearly under shock pressure loading which is further confirmed using transient finite element method (FEM) simulation by applying driven pressure loading data from shock tube experiments.

8.3.3 Material test system results

At this stage, it is important to see the response of CFRP test samples under slow loading; therefore, MTS tests are employed. At first, tensile tests are performed on samples as shown in Fig. 8.12 where stress and strain data are recorded as shown in Fig. 8.18. Tensile test is run until rupture of the test piece. As shown in Fig. 8.18, that CFRP test sample behaved linearly until rupture. The test specimen ruptured after approximately 37.4 s, gaining 10.8 millistrains. The overall Youngs modulus is found to be 28.897 GPa, while the modulus according to standard is provided as 26.751 GPa. A slight

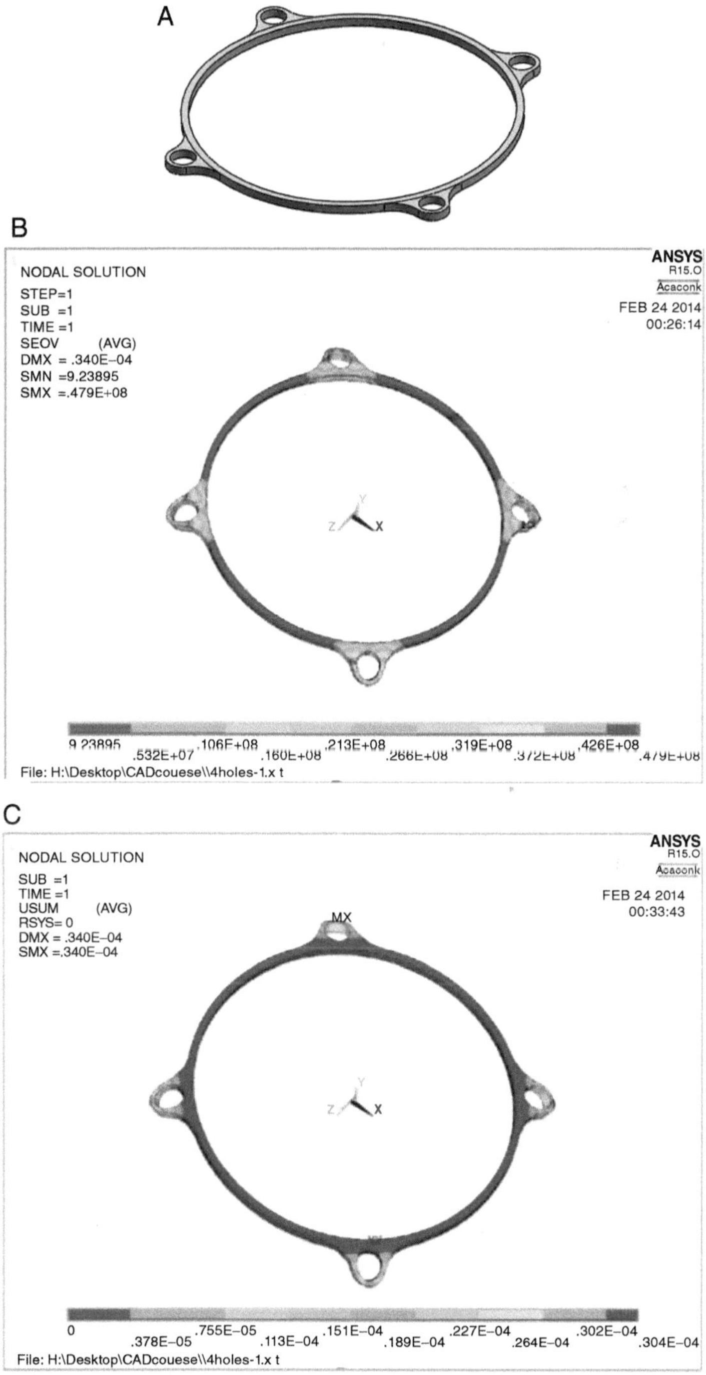

Figure 8.16 (A) Optimized CAD model. (B) The von Mises stress ($N/m^2 = 10^{-6}$ MPa) contours. (C) The resultant displacement (m) contours.

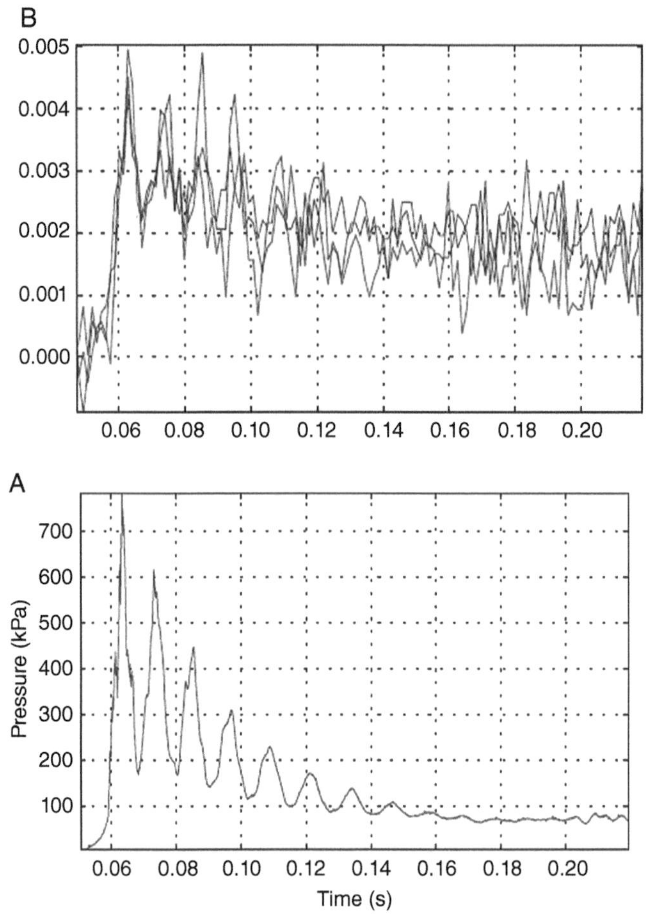

Figure 8.17 (A) Driven pressure profile in shock tube. (B) Rosette strain gauges' response from four layers of CFRP specimen. Noise in strains after 0.16 s is due to experimental setup.

deviation is observed in results with repetition in the tests for which a likely cause is variation in the curing process in preparation of the samples. Further, to confirm the linearity of CFRP response, flexural tests (Fig. 8.15(A) and 8.15(B)) are performed in MTS. The results show the strains in 0° (*X*), 45° (*XY*), and 90° (*Y*) directions. The *X*-strain is largest (red), followed by the *XY* strain (green), and the smallest is the *Y*-strain (blue) as shown in Fig. 8.19. The *X*-strain shows a maximum of 0.013 before the material ruptured after 298 s. This shows that the quasi-isotropic CFRP test specimen behaves linearly during low velocity, dynamic loads.

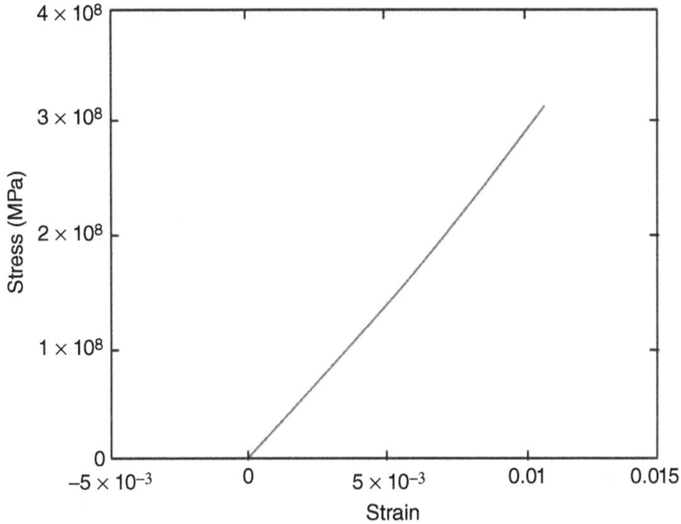

Figure 8.18 Stress/strain curve from the initial tensile tests on CFRP test specimens. The test piece ruptured after 37.4 s, gaining a strain of 10.8 millistrains in total.

8.3.4 Finite element method simulations results

The FEM simulations are performed using experimental data gathered from shock tube experiments and the MTS, similar study is performed by [6,40][49]. The 8-node 281-shell element provided in ANSYS 14.0 is used [21]. Mesh sensitivity analysis is carried out to ensure the quality of simulation results. In the analysis, the maximum strain is monitored with the mesh density. The value of maximum strain stabilizes with the increase in mesh density as shown in Fig. 8.20. The FEM mesh used in simulation is shown in Fig. 8.21. This represents the FEM model of 1200 elements (mesh density of "4*x*" from Fig. 8.20). The simulated shell structure is 2.145 mm thick and 3.70 cm in diameter. The outer edges of the circular mesh are constrained in all degrees of freedom. Transient pressure data from shock wave experiments are applied to the surface area. Obtained results are shown in Fig. 8.22. The strains obtained from simulations are one-to-one mapping to the pressure data and strain, confirming the results earlier seen in experiments. This demonstrates that there is a coherent response in the structure with the application of dynamic loading. Thus, the simulations also confirm that the response is linear during transient impulsive loading for quasi-isotropic CFRP.

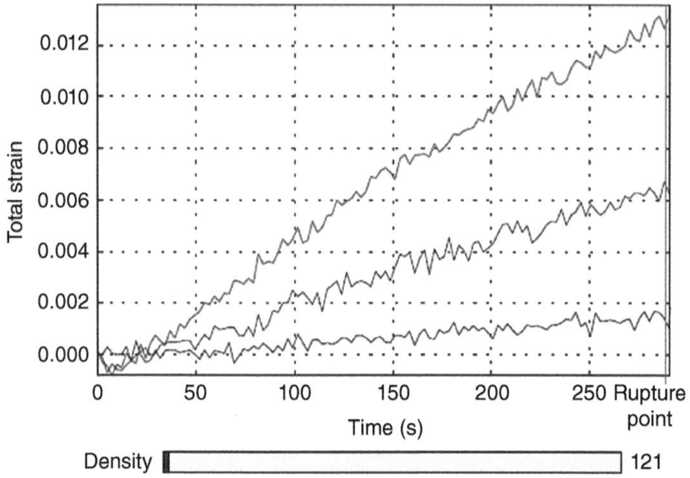

Figure 8.19 Strains over time gathered during flexural testing. X-strain in red, XY-strain in green, and Y-strain in blue. X-strain is the largest strain, XY-strain is about half of X-strain and Y-strain is the least.

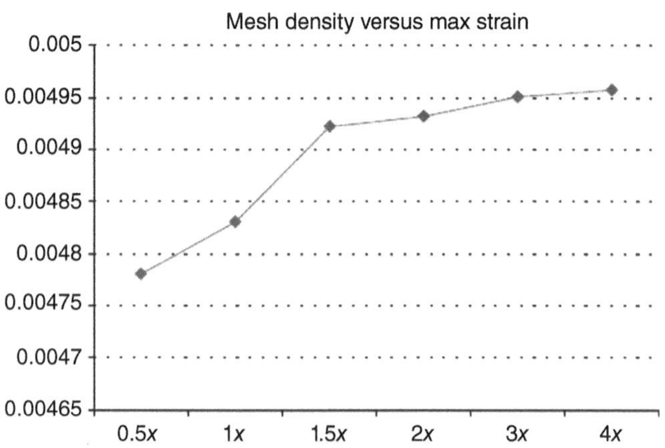

Figure 8.20 Results from the mesh sensitivity analysis ($x = 300$ elements).

8.4 Conclusion

Given chapter presents an optimization of an attachment component of a loading hanger with set of requirements. The optimization methodology is a cyclic iterative process. CAD model and Multiphysics analysis (e.g., FEMs) are the key inputs for the optimization process. CFRPs are the material of choice based on higher strength and light-weight properties.

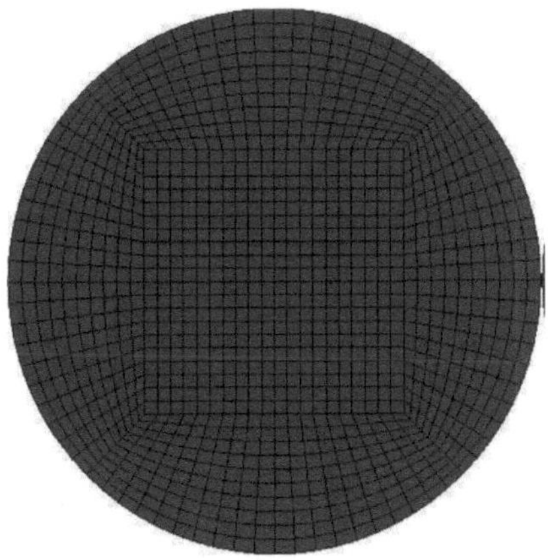

Figure 8.21 Mesh used in finite element analysis; shown is "4x" mesh density containing 1200 elements.

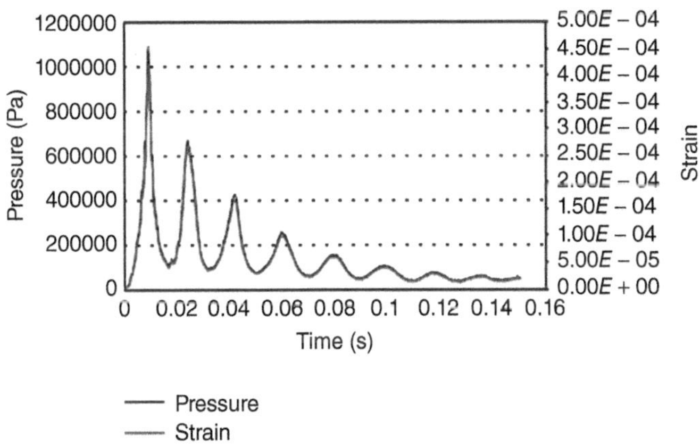

Figure 8.22 Pressure from experiment compared to equivalent strains from numerical simulation for an eight-layer CFRP.

Results from the shock tube experimental setup, MTS (with both tensile and flexural testing), and the FEM simulations using ANSYS software proved that, under dynamic loading, CFRPs E720 composite behaves linearly. A good agreement between simulations and experimental results confirms that FEM can be used for modeling of deforming structures under dynamic loading especially in the case as that discussed. The results from the given case prove that the mechanical behavior for quasi-isotropic CFRPs E720 laminates is as similar during high speed, short time dynamic shock wave loads as it is for long time static, low velocity loads.

References

1. Keshava Arora, Vikas Shinde, Aspects of Materials Handling, USP/Laxmi Publications (P) Ltd, New Delhi-India, 2007.
2. T. Lancashire, et al., Investigation of the Mechanics of Cargo Handling by Aerial Crane-type Aircraft, Defense Technical Information Center, Virginia-United States, 1966.
3. L.F. Webster, The Wiley Dictionary of Civil Engineering and Construction, Wiley, New Jersey-United States, 1997.
4. I.H.S.AConstruction Health and Safety Manual, Infrastructure Health & Safety Association, Ontario-Canada, 2007.
5. H.A. Khawaja, R. Messahel, B. Ewan, S. Mhamed, M. Moatamedi, Experimental and numerical study of pressure in a shock tube, J. Press. Vessel Technol. Trans. ASME 138 (4) (2016) 041301-1–041301-6, doi:10.1115/1.4031591.
6. H. Xue, H. Khawaja, M. Moatamedi, Conceptual design of high speed supersonic aircraft: a brief review on SR-71 (Blackbird) aircraft, AIP Conf. Proc. 1637 (1) (2014) 1202–1210, doi:10.1063/1.4904694.
7. H. Khawaja, M. Moatamedi, Selection of high performance alloy for gas turbine blade using multiphysics analysis, Int. J. Multiphys. 8 (1) (2014) 91–100, doi:10.1260/1750-9548.8.1.91.
8. H. Xue, H. Khawaja, M. Moatamedi, Multiphysics design optimization for aerospace applications: case study on helicopter loading hanger, AIP Conf. Proc. 1637 (2014) 1211–1217, doi:10.1063/1.4904695.
9. D. Hull, T. W. Clyne, An Introduction to CompositeMaterials, An Introduction to Composite Materials, Cambridge University Press, Cambridge, UK, 1996.
10. Marc Meyers, Shock Wave and High-Strain-Rate Phenomena in Materials, CRC Press, Boca Raton-United States, 1992.
11. R. H. Cole, Underwater Explosions, Underwater Explosions, Princeton University Press, New Jersey-United States, 1948.
12. C. Sulfredge, R. Morris, R. Sanders, R. S.-P. of the American, and U. 2005, Calculating the Effect of Surface or Underwater Explosions on Submerged Equipment and Structures, Calculating the Effect of Surface or Underwater Explosions on Submerged Equipment and Structures, Pennsylvania State University, Pennsylvania-United States, 2005.
13. Y. Kim, S. Itoh, A study on the behavior of underwater shock waves generated in a water container and its application to magnetic refrigeration material, Int. J. Multiphys. 1 (3) (2007) 291–302, doi:10.1260/175095407782219265.
14. Y. Nishimura, N. Kawaji, S. Itoh, Numerical analysis of the behavior of shock wave in spheroid vessel, Int. J. Multiphys. 4 (4) (2010) 317–328, doi:10.1260/1750-9548.4.4.317.
15. S.Y. Knyazeva, V.K. Kedrinskiy, Hydrodynamics of Explosion: Experiments and Models, Springer, New York, NY, USA, 2006.
16. A.C. Courtney, L.P. Andrusiv, M.W. Courtney, Oxy-acetylene driven laboratory scale shock tubes for studying blast wave effects, Rev. Sci. Instrum. 83 (4) (2012), doi:10.1063/1.3702803.
17. H. Ji, M. Mustafa, H. Khawaja, M. Moatamedi, Design of water shock tube for testing shell materials, World J. Eng. 11 (1) (2014) 55–60, doi:10.1260/1708-5284.11.1.55.
18. R. Hampson, M. Moatamedi, A review of composite structures subjected to dynamic loading, Int. J. Crashworthiness 12 (4) (2007) 411–428, doi:10.1080/13588260701483334.
19. Dassault Systèmes, SolidWorks®, Dassault Group, France, 2012.
20. ANSYS Inc., ANSYS®, Academic Research, Release 14.0, ANSYS Software Company, United States, 2014.

21. ANSYS Inc., ANSYS® Academic Research, Theory Reference: Structures, Static Analysis, Release 14.0, ANSYS Software Company, United States, 2014.
22. R.V. Mises, Mechanik der festen Körper im plastisch- deformablen Zustand, Nachr. von der Gesellschaft Wissenschaften Göttingen Math. Klasse 1913 (1913) 582–592.
23. Robert Cook, Warren Young, Advanced Mechanics of Materials, Advanced Mechanics of Materials, Prentice Hall, New Jersey-United States, 1999.
24. S.W. Tsai, E.M. Wu, A general theory of strength for anisotropic materials, J. Compos. Mater. 5 (1) (1971) 58–80, doi:10.1177/002199837100500106.
25. H. Khawaja, Application of a 2-D approximation technique for solving stress analyses problem in FEM, Int. J. Multiphys. 9 (4) (2015) 317–324, doi:10.1260/1750-9548.9.4.317.
26. H. Khawaja, Applicability extent of 2-D heat equation for numerical analysis of a multiphysics problem, AIP Conf. Proc. 1798 (2017) 020075-1–020075-9, doi:10.1063/1.4972667.
27. Barry Goodno, James Gere, Mechanics of Materials, Mechanics of Materials, SI Edition, Cengage Learning, Boston-United States, 2017 SI Edition.
28. James Gere, Stephen Timoshenko, Mechanics of Materials (Pws-Kent Series in Engineering), PWS-KENT Pub. Co, United States, 1990.
29. Stephen Timoshenko, History of Strength of Materials: With a Brief Account of the History of Theory of Elasticity and Theory of Structures, History of Strength of Materials, Dover Publications, New York-United States, 1953.
30. H. Khawaja, et al., Optimization of elstomeric micro-fluidic valve dimensions using non-linear finite element methods, Int. J. Multiphys. 3 (2) (2009) 187–200, doi:10.1260/175095409788837847.
31. C.H. Strand, Z. Andleeb, H.A. Khawaja, Multiphysics impact analysis of carbon fiber reinforced polymer (CFRP) shell, Explos. Shock Waves High Strain Rate Phenom. 13 (2019) 115–120, doi:10.21741/9781644900338-20.
32. Z. Andleeb, C. Strand, S. Malik, G. Hussain, Multiphysics analysis of CFRP Charpy tests by varying temperatures, Int. J. Multiphys. 14 (2) (2020) 143–160, doi:10.21152/1750-9548.14.2.143.
33. D.C. Barnes, H. Stephen, M. Kusek, O. Cross, R. Dennis, A. Lascola, "Quasi-Isotropic Composite Isogrid Structure and Method of Making Same," Google Patent, 1999.
34. P.D. Soden, M.J. Hinton, A.S. Kaddour, "Lamina Properties, Lay-up Configurations and loading Conditions for a Range of Fibre-Reinforced Composites Laminates," Google Patent, 1998.
35. A. B. Strong, Fundamentals of Composites Manufacturing, Materials, Methods and Applications, second ed., Society of Manufacturing Engineers, Dearborn, MI, USA, 2008.
36. H. Khawaja, J. Kapaya, M. Moatamedi, Shock Tube; Detail Overview of Equipment and Instruments in the Shock Tube Experimental Setup, Lambert Academic Publishing, Germany, 2015.
37. H.A. Khawaja, M. Moatamedi, Multiphysics investigation of composite shell structures subjected to water shock wave impact in petroleum industry, Mater. Sci. Forum 767 (2013) 60–67, doi:10.4028/www.scientific.net/MSF.767.60.
38. E. Stange, Z. Andleeb, H.A. Khawaja, Qualitative visualization of the development of stresses through infrared thermography, Vestn. MGTU 22 (4) (2019) 503–507, doi:10.21443/1560-9278-2019-22-4-503-507.
39. E. Stange, Z. Andleeb, H. Khawaja, M. Moatamedi, Multiphysics study of tensile testing using infrared thermography, Int. J. Multiphys. 13 (2) (2019) 191–202, doi:10.21152/1750-9548.13.2.191.
40. U.N. Mughal, H.A. Khawaja, M. Moatamedi, Finite element analysis of human femur bone, Int. J. Multiphys. 9 (2) (2015) 101–108, doi:10.1260/1750-9548.9.2.101.
41. Sergio Oller, Eduardo Car, Jacob Lubliner, Definition of a general implicit orthotropic yield criterion, Computer Methods in Applied Mechanics and Engineering 192 (7–8) (2003) 895–912, doi:10.1016/S0045-7825(02)00605-9.

42. H.A. Khawaja, T.A. Bertelsen, R. Andreassen, M. Moatamedi, Study of CRFP Shell Structures under Dynamic Loading in Shock Tube Setup, Journal of Structures (2014), doi:10.1155/2014/487809.
43. Zahra Andleeb, et al., Thermoelastic Investigation of Carbon-Fiber-Reinforced Composites using Drop Weight Impact Test, Applied Sciences 11 (1) (2021) 207, doi:10.3390/app11010207.
44. Hui Xue, Hassan Khawaja, Investigation of Ice-PVC separation under Flexural Loading using FEM Analysis, The International Journal of Multiphysics 10 (3) (2016) 247–264, doi:10.21152/1750-9548.10.3.247.
45. Hui Xue, Hassan Khawaja, Analytical and Case Studies of a Sandwich Structure using Euler-Bernoulli Beam Equation, Mathematics in Engineering, Science and Aerospace (MESA) 7 (4) (2016) 599–612.
46. H.A. Khawaja, et al., Fluid solid interaction simulation of CFRP shell structure, Mathematics in Engineering, Science and Aerospace (MESA) 8 (3) (2017) 311–324.
47. Odd Myrli, Hassan Khawaja, Fluid-Structure Interaction (FSI) Modelling of Aquaculture Net Cage, The International Journal of Multiphysics 13 (1) (2019) 97–111, doi:10.21152/1750-9548.13.1.97.
48. Z. Andleeb, et al., Multiphysics Study of Infrared Thermography (IRT) Applications, The International Journal of Multiphysics 14 (3) (2020) 249–271, doi:10.21152/1750-9548.14.3.249.
49. Mojtaba Moatamedi, Hassan Khawaja, Finite Element Analysis, CRC Press, Boca Raton-United States, 2018.

CHAPTER NINE

Numerical investigation on the use of wood–plastic composites for the thermoforming of NACA profile for UAVs

Fouad Erchiqui[a], Mhamed Souli[b]

[a] Université du Québec en Abitibi-Témiscamingue, École de génie,445, boul. de l'Université, Rouyn-Noranda (Québec) J9X 5E4, Canada

[b] Unité de Mécanique de Lille (UML), Université de Lille Sciences et Technologies, France

9.1 Introduction

Conditioned by the environmental and economic requirements, the market trends show that biocomposite materials will have a significant influence on the manufacturing industry [1]. In fact a manufacture with biocomposite materials would take advantage of both the intrinsic characteristics of the fibers and the plastic matrix, and a low production cost. In comparison to man-made reinforcements, natural fibers possess many qualities such as their low density, their low cost and their nonabrasive nature. Moreover, they are easily biodegradable and recyclable due to their high cellulose content. The mechanical performance of cellulose fibers has encouraged researchers to develop techniques for their extraction and modification from bioresources [2, 3]. In addition, studies have shown that chemical and physiochemical interactions between the polymer matrix and the fibers are determining factors in obtaining materials with good mechanical [4] and dielectric [5] properties.

Many applications of the biocomposites are found in different industries; however, their potential for the thermoforming of unmanned aerial vehicle (UAV) blades has not yet been fully elucidated in the literature. This chapter is an important contribution to the sector. It aims at studying the potential use of biocomposite materials for the manufacture of small UAV blades based on wood fibers in the form of sawdust particles and high-density polyethylene (HDPE).

Multiphysics Simulations in Automotive and Aerospace Applications.
DOI: 10.1016/B978-0-12-817899-7.00008-3

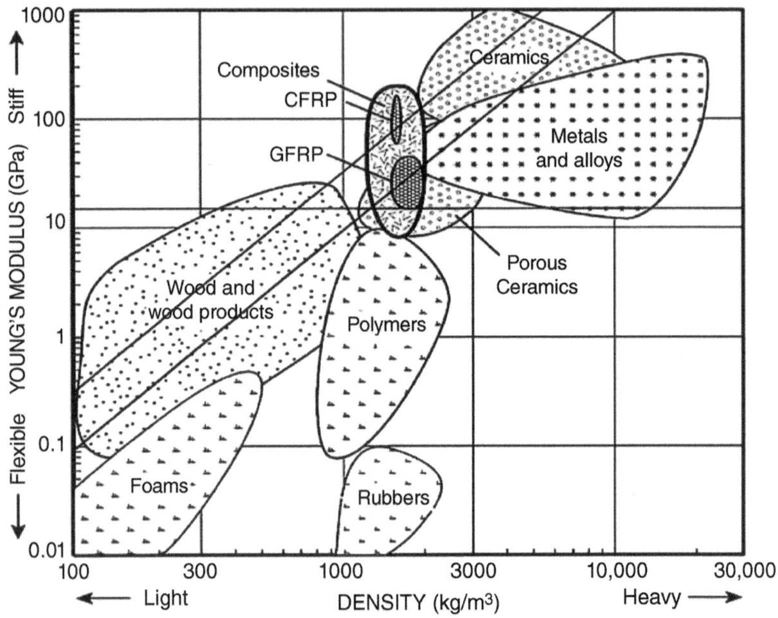

Figure 9.1 Blade materials and their mechanical properties [7].

9.1.1 Materials for the manufacture of UAV blades

The main materials currently used in the manufacture of UAV blades include wood, plastic, aluminum alloys, and composites [6, 7]. Fig. 9.1 illustrates the materials used and their characteristics [7]. The proper use of a UAV blade and its lifetime generally depend on the following three considerations:

1. The conceptual aspect of the blade including its aerodynamic profile (extrados).
2. The properties of the material constituting the blade including the lightness, nondeformability, homogeneity resistance to fatigue, vibration, and erosion.
3. The process including the total cost of manufacture. It must be simple with a profitable cost of implementation.

From the material point of view [7], treated wood shows very good mechanical characteristics such as good fatigue resistance, flexibility, and lightness which can contribute to the production of small blades. In the same order, glued laminated timber, which is a composite consisting of a stack of lamellas of one or different species of wood glued together are also good candidates for the production of blades with good fatigue resistance. However, there exist several disadvantages of using these wooden materials. They include their lack of homogeneity, their sensitivity to erosion, their low resistance to deformation, and [7] their mass variation over time. These disadvantages can then induce the destructive vibrations of the material structure. The metallic materials such as aluminum and its alloys are also good candidate for the production of blades of different lengths due to their resistance to corrosion and fatigue. In the case of composite materials [7], which are considered to be excellent candidates and the most widely used for blades production, they offer the possibility of producing various profiles and dimensions while varying the amount of material along the blade, and ensuring very good corrosion and fatigue resistance.

9.1.2 Behavior and design of drone blades

The mechanical behavior of a biocomposite blade when subjected to the various functional stresses such as wind pressure, external load, and temperature help define its performance. However, this performance must pass through the design studies, which in turn require both the knowledge of the behavior of the materials used and the modeling and numerical simulation tools. In order to determine the mechanical behavior of these materials, there is need to apply a range of experimental techniques and identification methods such as the least squares [8] and the neural [9] methods. Other identification techniques and applications found in the literature include the extension technique [10], the simultaneous thin cylinder blow-molding extension [11], the free blowing [8], the technique based on reported rheological tests [12], and the one-dimensional mechanical stretching tests [13].

There is a lack of scientific information about the characterization of biocomposite materials made of thermoplastics reinforced with plant fibers for aerospace blade production. In this regard, no work about different aspects of their properties such as behavior, damage, service life, and shaping have been reported so far in scientific journals. Some interesting works on the mechanical and rheological properties of plant-based biocomposite materials have been reported by authors like Erchiqui et al. [14], Tazi et al. [12], and Soroush Fini et al. [15] concerning, respectively, polyethylene terephthalate reinforced with hemp fibers, HDPE reinforced with wood particles, and the impact of northern conditions (temperature ranging from −50°C to 0°C) on the mechanical properties of HDPE reinforced with wood particles.

9.1.3 UAV blades' manufacturing techniques

The various blade manufacturing processes are closely related to several factors highlighted in the literature [6, 7]. They include the structural design, the material used, as well as the profitability and reliability of the blade.

In the case of aluminum and its alloys, the processes are similar to those used in aeronautic industry. In the case of composite materials, infusion and preimpregnation are the two techniques often used. The principle of the first method consists in "sucking" under vacuum a resin into either the reinforcing fibers or the reinforcing woven fabrics which are deposited on the surface of a mold. The vacuum reduces the pressure at one end of the stack, allowing atmospheric pressure to force the resin into the stack. Preimpregnation technology is a process in which a layer of fiber or woven fabric is impregnated with a high-viscosity resin to form a homogeneous composite precursor. The latter is treated in a mold between 80°C and 120°C for molding parts that are applicable in various industries. For all practical purposes, a good pale must be light, nondeformable, homogeneous, resistant to fatigue, vibration, corrosion, and erosion. In addition, it must not deteriorate or undergo premature ageing; moreover, it must be resistant to environmental conditions such as UV rays, humidity, heat, and lightning. Therefore, innovating in the structural design of a biocomposite UAV blade also means respecting these considerations. In order to achieve this goal, the dynamic behavior of its materials with regard to temperature and pressure have to be investigated, the mechanical tests performed about its reaction to static and fatigue stresses as well as the chemical tests on erosive wear, corrosion, and water ageing (service life). Furthermore, some studies have to be carried on the dimensioning and optimization of the aerodynamic profile in order to increase lift, reduce the effect of turbulence at the leaking edge, while taking into consideration the twisting and chord length variations. Finally, the choice of the manufacturing process must be made based on its rentability.

9.1.4 Thermoformability of the biocomposite materials for the manufacture of UAV blades

The thermoformability of biocomposite materials made of thermoplastic matrices reinforced with wood particles for the manufacture of UAV parts have not yet been sufficiently elucidated in the literature. The thermoforming process is applicable for the manufacturing of thin parts through the following stages and transformations. The material sheet is heated up, allowed to softened, blown or pressurized to fit the mold, then the formed part is cooled down and ejected. These, respectively, correspond to the phenomena of heat transfer, thermomechanics, large deformation mechanics, crystallization, and contact mechanics. In addition to these combination of physical phenomena, the part must also meet certain mechanical criteria. All these aspects can be influenced by the rheological parameters of the composite materials, the operating conditions, and the distribution of the molecular orientation in the manufactured part. An optimal application of those thermoformed parts in the industry requires a proper identification of the parameters influencing its shaping capacity. Such parameters include the pressures, temperatures, and reinforcement concentrations. A successful design with an improved quality of the intended thermoformed part necessitates the use of/or the development of modeling, optimization, and characterization tools. This is the context in which this introductory chapter on the potential use of wood–plastic composites for the manufacture of UAV blades through the thermoforming process is developed, according to the methodology formulated in the next section.

- *Wood–plastic composites (WPC) elaboration and characterization*: This section gives a summary of the methods of the elaboration of a family of biocomposites, based on HDPE and sawdust particles, as well as both their thermophysical and thermorheological properties [12].
- *Numerical modeling of the thermoforming process*: In this section, the experimental thermorheological data are applied toward the identification of the laws governing the viscoelastic behavior of the biocomposites (integral Lodge viscoelastic law) [16]. This is followed by the presentation of the nonlinear explicit finite element method (formulation, hypotheses, discretization, and convergence criterion) [17]. Finally, an airflow load and pressure derived from the Van der Waals' equation are considered for the modeling process.
- *The thermoforming of a thin NACA (National Advisory Committee for Aeronautics) for UAVs*: This section deals with the thermoforming of the upper surface of an UAV blade using the viscoelastic laws obtained previously. Some simplifying hypotheses are considered for clarification purposes. The simulation operation was able to predict on the one hand the thicknesses and stresses distribution in the UAV blade and, on the other hand the work and power required for the molding process.

9.2 Wood–plastic composite and characterization

9.2.1 Wood–plastic composites elaboration

9.2.1.1 Materials

The sawdust used in this study was supplied by Tembec sawmill located in Béarn (QC, Canada). It consists of a blend of 65% white spruce (*Picea glauca*),

Table 9.1 Average geometrical dimensions of the sawdust particles [12].

Area (mm^2)	Long axis (mm)	Short axis (mm)	Shape ratio (*L/D*)
0.1154	0.574	0.302	1.97

20% of black spruce (*Picea marica*), and 15% balsam fir (*Abies balsamea*). The particle size analysis of the sawdust is performed on more than 5000 particles using a Fiber Quality Analyser from Optest Equipment (Hawkesbury, Ontario). Table 9.1 shows the average geometry of the sawdust particles. The wood particles were sieved to a diameter below 700 microns (þ < 0.7 mm). The polymer matrix was a HDPE (SclairA59), with a melting temperature of 138°C, and a density of 980 kg/m^3. The coupling agent was Fusabond 226 from DuPont, which is a polyethylene grafted with maleic anhydride. It represented 3% (w/w) of the composite formulation; such amount is well known to improve the homogeneity of the mixture while ensuring a better fibers distribution in the polymer matrix [18, 19]. The sieved sawdust particles were predried for 24 h at 105°C. Consequently, the residual moisture content in the sawdust particles used for samples preparation is estimated to be less than 3% (w/w).

9.2.1.2 Method

The composite samples were prepared using a twin-screw counter-mix rheometer (Haake Rheometrix with roller rotors) with an angular speed of 90 rpm at the temperature of 170°C. In fact, this temperature was chosen to be above the melt point of the matrix and lower than the onset of the wood fibers degradation temperature reported around 200°C [20]. During the preparation of the composite material, HDPE of brand Sclair A59 and Fusabond the coupling agent were first mixed for 2 min in order to ensure complete melting of the materials. Then, the fiber particles were added to the mixture and blended for 5 min. Four different biocomposite formulations reinforced with 0%, 20%, 40%, and 50% (w/w) sawdust particles were investigated. In the next step, some flat disk samples of 20 mm diameter and 4 mm thickness were prepared by compression molding with a hot press heated at 150°C using aluminum mold plates. This temperature was selected above the melting point of the polymer which is 138°C. In order to ensure that the mixture was properly softened, a preload was first applied for 2 min. Such preload was also beneficial for getting rid of the probable accumulated air in the material. Then, a load of 8 MPa was applied on the material for 4 min to complete its shape.

Table 9.2 Mechanical properties of HDPE reinforced with 0%, 20%, and 40% (w/w) sawdust particles [21].

% Sawdust particles	% HDPE	*E* (MPa)	R_m (MPa)	ε (%)
0	100	420 ± 26	23.0 ± 0.77	108.3 ± 11.2
20	80	1210 ± 33	26.0 ± 0.26	6.86 ± 0.26
40	60	4331 ± 300	35.4 ± 0.24	3.36 ± 0.19

Table 9.2A Temperature variations of the linear regression of the specific capacity of HDPE reinforced with 20%, 40%, and 50% (w/w) sawdust particles [12].

$-C_{p_0} = a.T$				$C_p - C_{p_1} = \frac{A}{\omega\sqrt{\frac{\pi}{2}}}.exp^{-2\frac{(T-Tm)^2}{(\omega)^2}}$				
	$0 < T < 80°C$			$81°C < T < 200°C$				
Composites	$C_{p_{10}}$	$a.^{*}10^4$	R^2	C_{p_1}	tm	A	ω	R^2
HDPE	2.73	89.	98	4.02	138.80	283.5	13.53	0.99
HDPE-20% wood	2.45	116	98	3.81	140.98	315.5	16.79	0.98
HDPE-40% wood	2.16	93	99	3.64	138.64	161.9	12.40	0.98
HDPE-50% wood	0.75	62	99	1.64	139.06	83.76	11.58	0.97

9.2.2 Mechanical properties of WPC

The tensile tests were carried on in order to determine the values of the modulus of elasticity (E), maximum mechanical strength (R_m), and elongation at break (ε) of each investigated biocomposite formulation. The results are summarized in Tables 9.2 and 9.2A.

It can be noticed that the modulus of elasticity increases with the sawdust particles content. These results were predictable due to the marked difference between the values of the moduli of elasticity of the wood particles and HDPE. However, at room temperature, the reinforcement of the matrix resulted in a decrease in the mechanical strength of the biocomposites as a function of the sawdust particles content. The chemical incompatibility between the wood particles and the polymer matrix results in weak interactions between the two elements, leading to a high stress concentration at the edges of the sawdust particles. Consequently, the biocomposites become more brittle with the failures occurring at lower stress levels than the virgin polymer [21].

9.2.3 Thermophysical properties of the WPC

A numerical simulation of the composites behavior as a function of temperature requires a precise description of its thermophysical properties as a

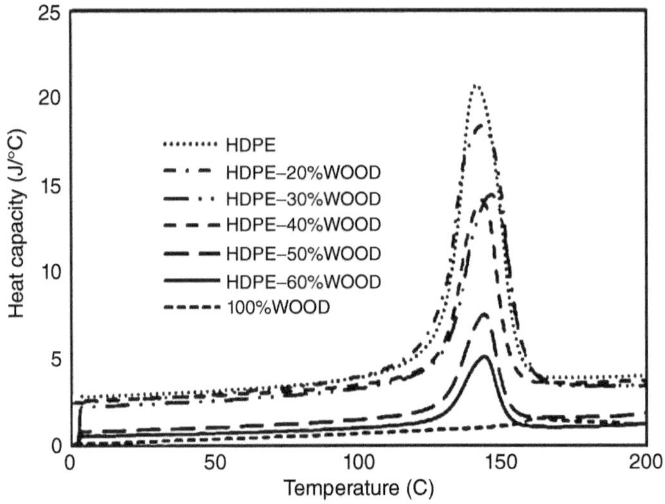

Figure 9.2 Temperature–variations of the heat capacities of HDPE reinforced with 20%, 30%, 40%, 50%, and 60% (w/w) sawdust particles [12].

function of both the temperature and its degree of reinforcement with the wood particles.

9.2.3.1 Heat capacity of the WPC

The DSC (Differential Scanning Calorimetry) thermograms of the investigated biocomposite formulations were used to evaluate the variations of their heat capacities. In Fig. 9.2, the variations of the specific heat capacities (ΔC_p) of the biocomposites are shown with both the degree of reinforcement and the temperature. It shows that the heat capacities of HDPE–wood composites are lower than that of the matrix during the melting phase. This can be attributed to the fact that wood particles hinder HDPE's molecular chains mobility around its melting temperature. These results are similar to those reported in many studies about biocomposites reinforced with plant fibers. Bendahou et al. have investigated the effect of the reinforcement's percentage and the coupling agent on the physical properties of the biocomposites made of polyolefin reinforced by short palm tree fibers [22]. They reported a slight increase of the biocomposites' melting temperature independently of the percentage of the natural fibers, and also a considerable improvement in the structural arrangement of the polymers with an increase of the reinforcement's concentration. According to the

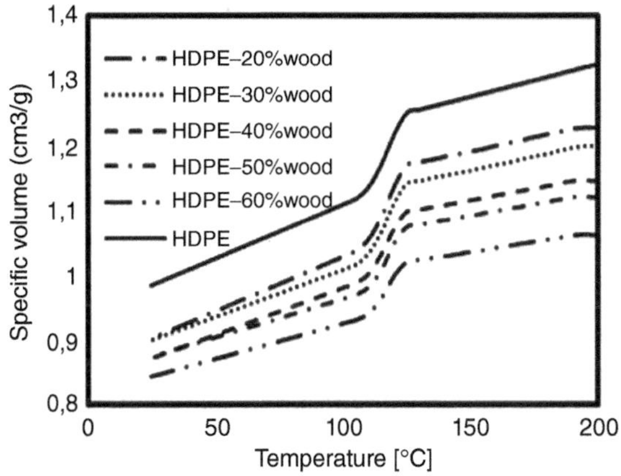

Figure 9.3 Temperature-variations of the specific volume of HDPE reinforced with 20%, 30%, 40%, 50%, and 60% (w/w) sawdust particles [12].

authors, the natural fibers in the polymers act as nucleation site promoting a high degree of crystallinity of the polymer chains around the fibers.

Table 9.2 gives the mathematical expressions of the heat capacity of HDPE reinforced with 20%, 40%, and 50% (w/w) sawdust particles with respect to the temperature, based on regression analysis.

9.2.3.2 Specific volume

The numerical simulation of the biocomposites behavior as a function of temperature also requires a precise description of the variations of their specific volume with the temperature. In this regard, Fig. 9.3 shows the experimental results obtained on the variations of the specific volume with both the temperature and the reinforcement concentration in the biocomposites. It shows that the specific volume increases with temperature and decreases with an increase in the sawdust particles concentration in the biocomposites. The maximum change starts at the transition temperature and continues through the melting phase. In the case of the biocomposites with a high reinforcement concentration, such observation can be explained by the fact that as the density of wood particles per unit volume increases, the level of interaction between them also becomes important.

Table 9.3 gives the mathematical expressions of the heat specific of HDPE reinforced with 20%, 40%, and 50% (w/w) sawdust particles based on regression analysis on the temperature–variation curves.

Table 9.3 Temperature-variations of the linear regression of the specific volume of HDPE reinforced with 20%, 30%, 40%, 50%, and 60% (w/w) sawdust particles [12].

$\vartheta = A_2 + \frac{(A_1-A_2)}{1+\exp\frac{(T-T_0)}{dT}}$						$\vartheta - \vartheta_0 = a.T$		
20 < *T* < 130						**131 < *T* < 200**		
Composites	A_1	A_2	T_0	dT	R^2	ϑ_0	$a.10^4$	R^2
HDPE	0.97	80	302	31	0.98	1.12	10.10	0.99
HDPE-20% wood	0.89	79	302	31	0.98	1.06	8.05	0.99
HDPE-40% wood	0.85	62	303	32	0.98	1.01	6.92	0.98
HDPE-50% wood	0.86	79	292	28	0.98	0.99	6.53	0.98

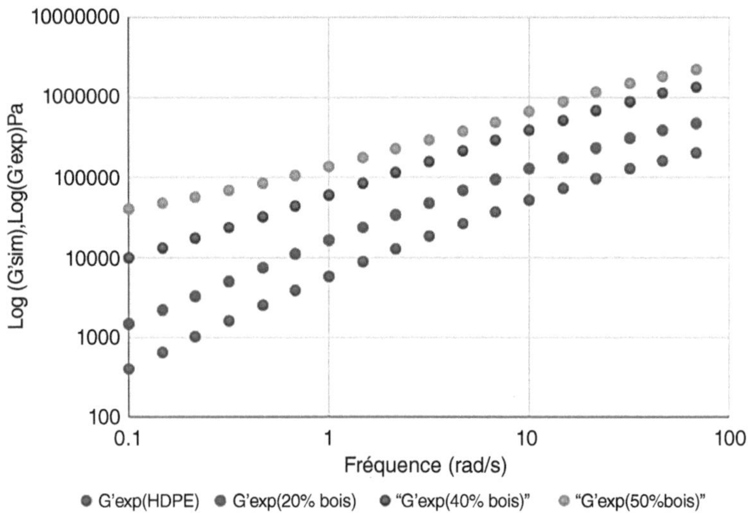

Figure 9.4 Experimental results of the storage moduli (G^1) as a function of the sawdust particles content.

9.2.4 Rheological properties of the WPC

The numerical simulation of thermoforming of thin parts such as drone blades requires an adequate description of the viscoelastic behavior of the biocomposite materials as a function of both the temperature and the reinforcement concentration. In order to achieve this purpose, small amplitude oscillatory shear tests [23] were considered to determine the storage and loss moduli of the formulations as a function of frequency. The results are shown in Figs. 9.4 and 9.5 for HDPE reinforced with 0%, 20%, 40%, and 50% (w/w) sawdust particles at 170°C.

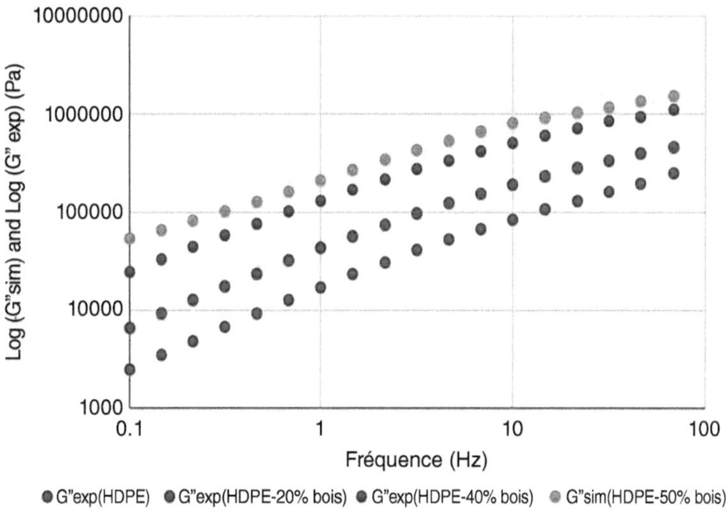

Figure 9.5 Experimental results of the loss moduli (G'') as a function of the sawdust particles content.

9.3 Numerical modeling of the blade's thermoforming process

9.3.1 Characterization of the WPC's viscoelastic behavior

This section is based on the data previously acquired on the samples of HDPE reinforced with sawdust particles in the Laboratory of Bioplastics of the University of Québec in Abitibi-Témiscamingue. These biocomposite materials were assumed to be isotropic, homogeneous, and isothermal for modeling purposes. In regard to many literature references, the Lodge's integral viscoelastic model which is suitable for representing the behavior of HDPE and HDPE-based biocomposites [24] were applied to capture the viscoelastic response of the WPC.

The Lodge constitutive model forms the basis of most integral viscoelastic models [16]. It is often used to describe the incompressible behavior of thermoplastic polymers in their liquid state. However, Erchiqui et al. [17] have shown that this model can be used for semisolid materials as well. According to this model, the knowledge of the material's strain history is essential to determine the Cauchy stress tensor described by Eq. (9.1).

$$\sigma(t) = -p(t)I + \int_{-\infty}^{t} \sum_{k=1}^{n} \frac{g_k}{\tau_k} e^{\left(-\frac{t-\tau}{\tau_k}\right)} B(t,\tau)\, d\tau \tag{9.1}$$

where p, I, B, t, and τ are the hydrostatic pressure, identity tensor, finger deformation tensor, time, and relaxation time, respectively. The parameters g_k and τ_k are, respectively, the stiffness modulus and the relaxation time associated with the mode. The tensor is related to the deformation gradient tensor as shown in Eq. (9.2):

$$B(t,\tau) = \left[F^T(t,\tau) \cdot F(t,\tau)\right]^{-1} \tag{9.2}$$

There are several approaches, which combine experimental and numerical data to identify the rheological properties of the Lodge model. In this work, the experimental data from previous studies done using low amplitude oscillatory shear tests on the WPC, which respectively provide the storage and loss moduli G' and G'' are considered. Moreover, the least squares method is used to minimize the discrepancies between the experimental and theoretical values during the identification of the relaxation spectrum for each type of WPC. This method is described by reducing the objective function defined by Eq. (3), where N is the number of experimental data points:

$$Z = \sum_{i=1}^{N} \left[\frac{G'_{i,\text{exp}} - G'_{i,\text{th}}}{G'_{i,\text{exp}}}\right]^2 + \left[\frac{G''_{i,\text{exp}} - G''_{i,\text{th}}}{G''_{i,\text{exp}}}\right]^2 \tag{9.3}$$

The parameters $G'_{i,\text{ exp}}$ and $G''_{i,\text{ exp}}$ represent the dynamic moduli from the experimental data while $G'_{i,\text{ th}}$ and $G''_{i,\text{th}}$ represent the theoretical values given by Eq. (9.4) [23].

$$G'(\omega) = \sum_{i=1}^{N} \frac{g_i \tau_i^2 \omega^2}{1 + \tau_i^2 \omega^2} \text{ and } G''(\omega) = \sum_{i=1}^{N} \frac{g_i \tau_i \omega}{1 + \tau_i^2 \omega^2} \tag{9.4}$$

The parameter g_i is the stiffness constant and τ_i is the relaxation time associated with the mode i, while ω is the frequency. The results obtained are given in Table 9.4. Figs. 9.6 and 9.7 show the results of the optimization represented by the dotted lines in comparison with those of the experiment represented by the points. These figures show an excellent correlation with an average error of less than 1%.

9.3.2 Finite element formulations

In this section, the explicit dynamic finite element method with both space and time discretization is used to simulate the forming of the biocomposites

Table 9.4 Stiffness modulus and relaxation time for the biocomposites at $T = 170°C$.

HDPE		WPC20		WPC40		WPC50	
g_k (MPa)	τ_k (s)	g_k (MPa)	τ_k (s)	g_k (MPa)	τ_k (s)	g_k (MPa)	τ_k (s)
0.181	0.0006148	0.3314	0.0006218	0.6811	0.000734	1.1246	0.0006693
0.0635	0.006148	0.1219	0.006218	0.2605	0.00734	0.4343	0.006693
0.0298	0.06148	0.0655	0.06218	0.1696	0.0734	0.2596	0.06693
0.011	0.6148	0.0302	0.6218	0.1029	0.734	0.1249	0.6693
0.0033	6.148	0.015	6.218	0.061	7.34	0.1166	6.693
0.0031	61.48	0.0109	62.18	0.0486	73.4	0.1011	66.93

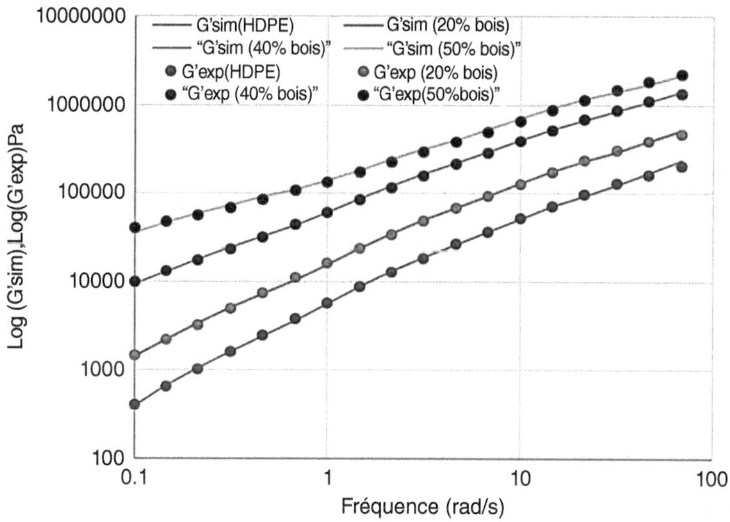

Figure 9.6 Results of the optimization with the experimental storage moduli G^1 as a function of the sawdust particles content.

part made of different concentrations of sawdust particles and a thermoplastic matrix. The principle of virtual work is expressed on the undeformed configuration for the inertial effects and internal work [23].

9.3.2.1 Finite element discretization

Both spatial and temporal discretizations are necessary for the virtual work due to the presence of the force of inertia. In the case of spatial discretization, the finite element method approach is considered [23]. However, for temporal discretization, the centered finite difference method, which is conditionally stable, is used. After summing up the contributions of all the elements, the inflation problem is then reduced to the following discrete

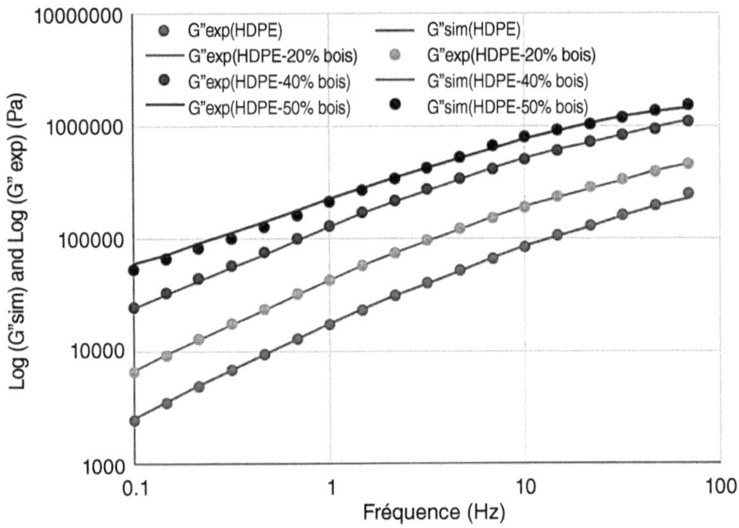

Figure 9.7 Results of the optimization with the experimental loss moduli G'' as a function of the sawdust particles content.

system of equations [23]:

$$M \times \ddot{u}(t) = F_{ext} + F_{grav} - F_{int} \tag{9.5}$$

where F_{ext}, F_{grav}, and F_{int} are the global nodal external, body, and internal force vectors applied to the thermoplastic membrane. **M** is the mass matrix. Using the diagonalization method of Eq. (9.5), the matrix **M** is transformed into a diagonal matrix and each degree of freedom can be handled independently. Finally, using a centered finite difference scheme, Eq. (9.5) can be rewritten as Eq. (9.6)

$$u_i(t + \Delta t) = \frac{\Delta t^2}{M_{ii}^d}\left(F_i^{\text{ext.}}(t) + F_i^{\text{grav.}}(t) - F_i^{\text{int.}}(t)\right) + 2u_i(t) - u_i(t - \Delta t) \tag{9.6}$$

where $M_{ii}^{\text{d}}$are the diagonal components of the M^{d} matrix.

9.3.2.2 Convergence stability criteria

The convergence behavior of the explicit dynamic finite element method for nonlinear problems is controlled by the Courant–Friedrichs–Lewy criterion which stipulates that the time step must be smaller than the critical time steps

$\Delta t \leq t_{\text{crit}}()$:

$$\Delta t_{\text{crit}} = \varepsilon \frac{l}{c} \tag{9.7}$$

Here, c is the wave speed in the medium and l is the element size. The quantity l/c is the time required for a wave to propagate across an element of size l. The proportionality constant ε depends on the integration scheme used. The initial conditions are further required in order to solve the problem at hand. The displacement and velocity vectors at the initial time are assumed zero for the material forming process.

9.3.2.3 Plane-stress assumption

In order to successfully model and simulate the inflation phase of the polymeric membrane which is first heated and then blown, an accurate evaluation of the internal force is required. The relationship between deformations and stresses must be computed for each element. Using the plane-stress assumption along the incompressibility conditions of the thermoplastic material, it follows that the components of the Cauchy stress tensor satisfy the conditions, listed in Eq. (9.8).

$$\sigma_{13} = \sigma_{23} = \sigma_{31} = \sigma_{32} = \sigma_{33} = 0 \tag{9.8}$$

9.3.2.4 Constitutive equation

In order to address the behavior of the isotropic thermoplastic composite, the Lodge constitutive model [16] is applied. In this model, the Cauchy stress tensor $\sigma(t)$ is related, at time t, to the history of the Finger deformation tensor $B(t, \tau)$ by Eq. (9.9):

$$\sigma(t) = -p(t)I + \int_{-¥}^{t} \sum_{k} \frac{g_k}{\tau_k} e^{-(t-\tau)/\tau_k} B(\tau, t)\, d\tau \tag{9.9}$$

where the finger tensor **B** is related to the right Cauchy Green deformation tensor **C** by Eq. (9.10).

$$B = C^{-1} = \left(F^T F\right)^{-1} \tag{9.10}$$

F is the deformation gradient tensor, p is the hydrostatic pressure, while g_k and τ_k represent the stiffness moduli and relaxation time, respectively. The temperature dependence of these models is accounted for using the WLF (Williams-Landel-Ferry equation) function [25].

9.3.2.5 Van der Waals equation of state and pressure loading

In this study, it is considered that the external pressure induced by the airflow load is responsible for the blowing of the membrane. The pressure inside the thermoplastic-based biocomposite membrane is closely related to the internal volume of the parison, according to the Redlich–Kwong gas equation of state given in Eq. (9.11) [26].

$$P(t) = \frac{n(t)RT_{\text{gas}}}{V(t) - nb(t)} - \frac{n^2(t)a}{V(t)(V(t) + n(t)b)\sqrt{T_{\text{gas}}}} \tag{9.11}$$

where $n(t)$ is the number of gas moles introduced to inflate the thermoplastic-based biocomposite membrane, $P(t)$ is the internal pressure, $V(t)$ is the volume occupied by the membrane at time t, T_{gas} is the absolute gas temperature, and R is the universal gas constant (=8.3145 J/mol/K). a and b are constants evaluated from the critical state of the gas [26]:

$$a = 0.42748\frac{\bar{R}^2 T_c^{2.5}}{p_c} \quad \text{and} \quad b = 0.08664\frac{\bar{R}T_c}{p_c} \tag{9.12}$$

where T_c and p_c are the critical temperature and pressure of the gas, respectively. The critical temperature of a material is the temperature above which distinct liquid and gas phases do not exist. As the critical temperature is approached, the properties of the gas and liquid phases become the same, resulting in only one phase known as the supercritical fluid. Above the critical temperature, a liquid cannot be formed by an increase in pressure, but with enough pressure, a solid may be formed. The critical pressure is the vapor pressure at the critical temperature.

In this study, the assumptions used for the calculation of the dynamic pressure are:

1. Gas temperature is assumed constant (T_{gas}).
2. The biocomposite sheet temperature is assumed constant ($T_{\text{sheet}} = T_{\text{gas}}$).
3. At every moment, the pressure between the sheet and the mold is assumed constant (ΔP).
4. The contact between the biocomposite sheet and the mold is assumed to be a sticky contact as the polymer cools and stiffens rapidly during the sheet/mold contact.

If V_0 is used to represent an initial volume enclosing the membrane at the initial time t_0 and containing a number n_0 of gas moles while assuming that the forming process temperature is constant, the Redlich–Kwong equation of state is reduced to Eq. (9.13)

$$P_0 = \frac{n_0 R T_{gas}}{V_0 - n_0 b} - \frac{n_0^2 a}{V_0 (V_0 + n_0 b)\sqrt{T_{gas}}} \tag{9.13}$$

When $n(t)$ moles of gas are introduced to inflate the thermoplastic-based biocomposite membrane, $P(t)$ is the internal pressure, and $V_0 + V(t)$ is the volume occupied by the membrane at time t, Eq. (9.14) can be rewritten.

$$\begin{aligned}\Delta P(t) &= P(t) - P_0 \\ &= \left[\frac{(n(t)+n_0)RT_{gas}}{(V(t)+V_0)-(n(t)+n_0)b} - \frac{(n(t)+n_0)^2 a}{(V(t)+V_0)((V(t)+V_0)+(n(t)+n_0)b)\sqrt{T_{gas}}}\right] \\ &\quad - \left[\frac{n_0 R T_{gas}}{V_0 - n_0 b} - \frac{n_0^2 a}{V_0(V_0 + n_0 b)\sqrt{T_{gas}}}\right]\end{aligned} \tag{9.14}$$

Eq. (9.15) represents the time evolution of pressure inside the thermoplastic-based biocomposite membrane. It is closely related to the evolution of the internal volume of the membrane under the thermodynamic state equation. In this case, the external virtual work is expressed in terms of a closed volume becomes Eq. (9.15) that follows [27].

$$\delta W^{ext} = \Delta P(t)\, \delta V \tag{9.15}$$

The advantage of using a load expressed in terms of gas flow instead of pressure load is that it allows a natural exploration of the load–deformation curve without dealing with the instability associated with the classical pressure loading. The introduction of the constant pressure as a load force instead of the gas flow velocity in the finite element formulations leads to a divergence in computations for pressure values exceeding the critical point (beginning of the unstable segment of the load–deformation curve) [26].

9.3.2.6 Work and power for the thermoforming of the UAV blade

The amount of energy required for the forming process is equivalent to the external mechanical work transferred by the external forces to the biocomposite structure during the forming stage. This work is given in compact form as Eq. (9.16).

$$W^{ext} = \{u^n\}^T \times \{F_{ext}\} \tag{9.16}$$

where $\{F_{ext}\}$ is the global nodal external force vector applied to the thermoplastic-based biocomposite membrane and $\{u^n\}^T$ is the associated global nodal displacement vector. The power associated with the global nodal external force vector is the amount of external work divided by the time interval required to form the polymeric part (energy per unit time transferred to the sheet during the forming stage):

$$P^{\text{ext}} = \frac{W^{\text{ext}}}{t_{\text{forming}}} \tag{9.17}$$

The amount of energy and the power associated with the forming phase of a part determine the cost and feasibility of its manufacturing for industrial purposes.

9.4 Thermoforming of a thin NACA profile for UAVs

The dynamic finite element method outlined in the previous section was implemented in ThermoForm, the general purpose finite element code [17]. This code was developed by Erchiqui to study stresses and deformations occurring in thermoforming sheets as well as stretch blow-molding problems.

The WPC studied was made of HDPE reinforced with 20%, 40%, and 50% (w/w) sawdust particles labeled by WPC20, WPC40, and WPC50, respectively.

The application considered was made by the thermoforming of a 25 cm by 10 cm rectangular membrane made of HDPE reinforced with sawdust particles, with a load expressed in terms of nonlinear airflow rate as shown in Fig. 9.8. Its initial thickness h_0 was 2.0 mm. The geometries of the mold and the biocomposite sheet discretized by triangular membrane elements as shown in Fig. 9.9. The edges of the biocomposite sheets were fixed for the modeling purpose. This means that the displacement of the nodes of the quarter edges of the sheet is zero. The material temperature was assumed constant at 170°C. The rheological parameters of the Lodge behavior law are given in Table 9.4.

Fig. 9.10 illustrates the evolution of the nodes, which are in contact with the mold during forming process, represented by the black dots. It appears from the simulation that the forming time increases with the sawdust particles content. In this regard, the forming time was 0.036 s for virgin HDPE and 0.064 s for WPC50. This agrees with the previous findings where wood particles made WPC to be more resistant to deformation (Fig. 9.10.1).

Fig. 9.11 shows the variations of the pressure as a function of the forming volume for HDPE reinforced with respectively 0% and 50% (w/w) sawdust

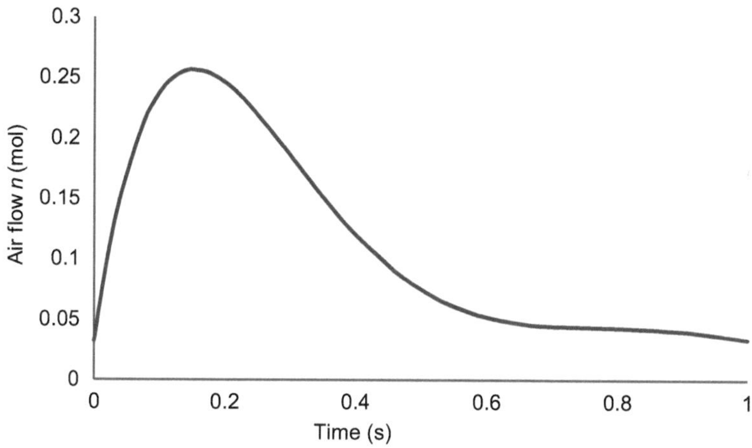

Figure 9.8 Airflow versus time.

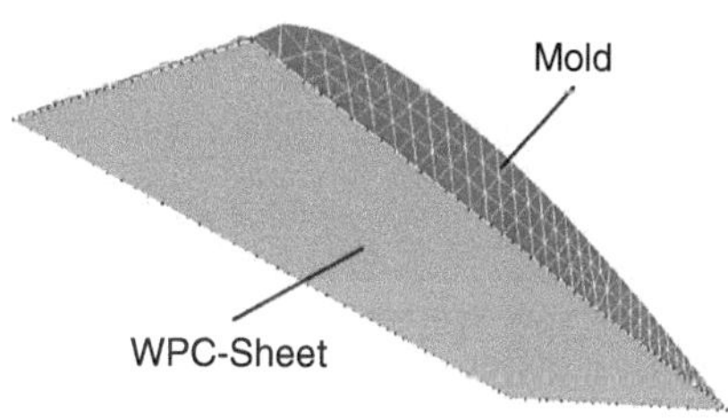

Figure 9.9 Geometries of the mold and the WPC-sheet.

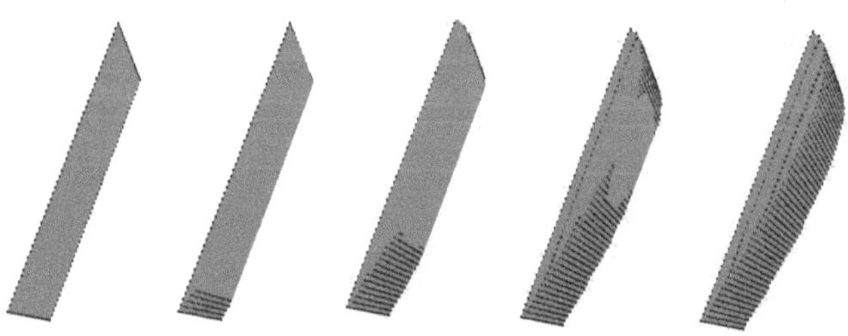

Figure 9.10 Evolution of the distribution of the contact nodes with the mold at the beginning, middle, and the end of the forming cycle.

particles and Fig. 9.12 is an histograms showing the maximum internal pressures induced at the end of forming of the biocomposite blade as a function of the sawdust particles content. According to Fig. 9.12, the maximum internal pressure induced at the end of the shaping of the biocomposite blade depends on the proportion of sawdust particles in the

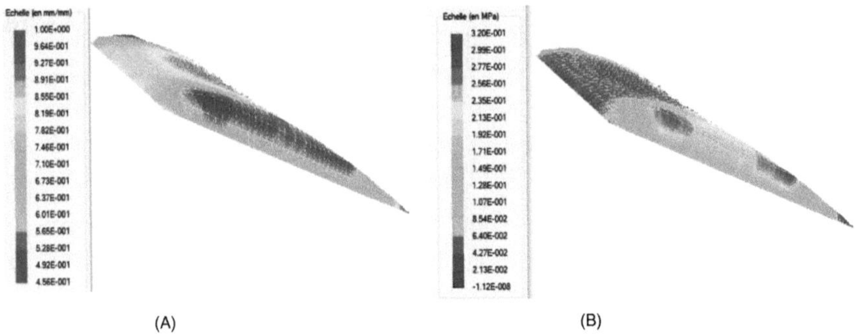

Figure 9.10.1 Distributions of the final extension (λ_3) and the von Mises stresses in the thermoformed extrados blade. (A) Extension λ_3 distribution. (B) Von Mises stresses distribution.

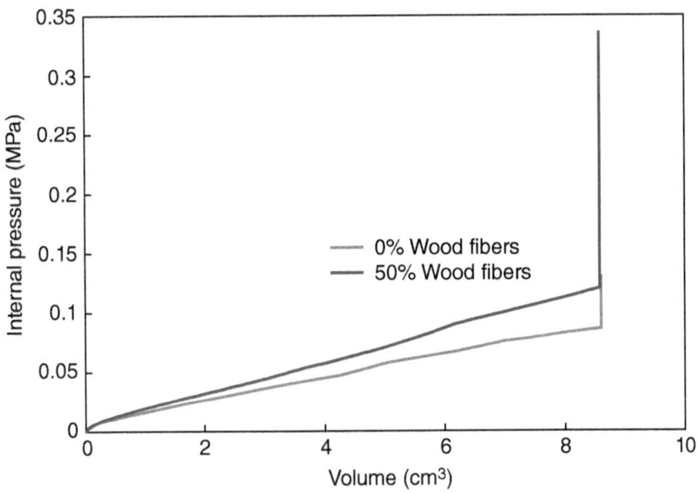

Figure 9.11 Variations of the numerical internal pressure versus the volume for HDPE reinforced with 0% and 50% (w/w) sawdust particles.

biocomposite. Around 50% of the sawdust particles, the maximum pressure induced at the end of the thermoforming cycle is about 250% (0.336 MPa) compared to that used in the virgin HDPE matrix (0.129 MPa). In the cases of 20% and 40% (w/w) sawdust particles, the maximum internal pressures are 0.130 MPa and 0.194 MPa, respectively.

When simulating the thermoforming of a thin part, it is important to predict the thickness and stress distribution in the molded part. In fact, the predictions of the residual stresses and dimensional stability of the

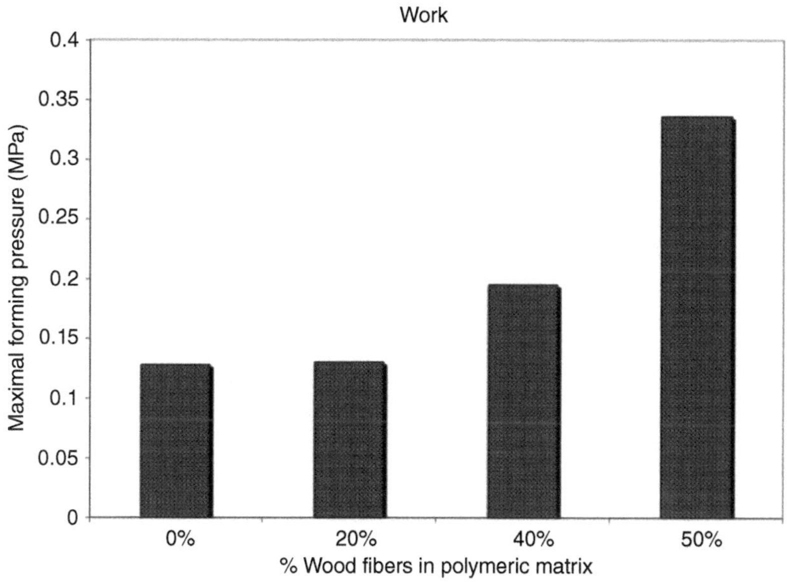

Figure 9.12 Maxima internal pressure versus different proportions of the reinforcement.

final shape of the molded part are closely related to the estimated stresses. In addition, the effect of localized thinning of the deformed membrane is usually accompanied by an increase in the Cauchy stresses (or actual stresses). The distribution of the von Mises stresses and the main extensions on the two perpendicular traces of the thermoformed blade as a function of the sawdust particles content are shown in Figs. 9.13 and 9.14, respectively.

Moreover, Fig. 9.13A and B represents the final von Mises stress distribution, respectively, on the *XZ* and *YZ* half plane of symmetries of the blade predicted using different proportions of the sawdust particles alongside the Lodge constitutive model. For clarity of presentation, only critical values of the von Mises stresses induced at the extrados blade (in the form of a drone blade) are presented for clarity purposes in Table 9.5 for the case of the trace according to the *XZ* plane and Table 9.6 for the case of the trace according to the *YZ* plane. The von Mises stress distribution on the *XZ* half plane exhibits a maximum of 0.117 MPa for 50% reinforcement and the minimum of 0.0253 MPa for virgin HDPE. These values are not identical in the *YZ* half plane. Indeed, as shown in Fig. 9.12B, the von Mises stress distribution on the *YZ* half plane shows the maximum of 0.398 MPa

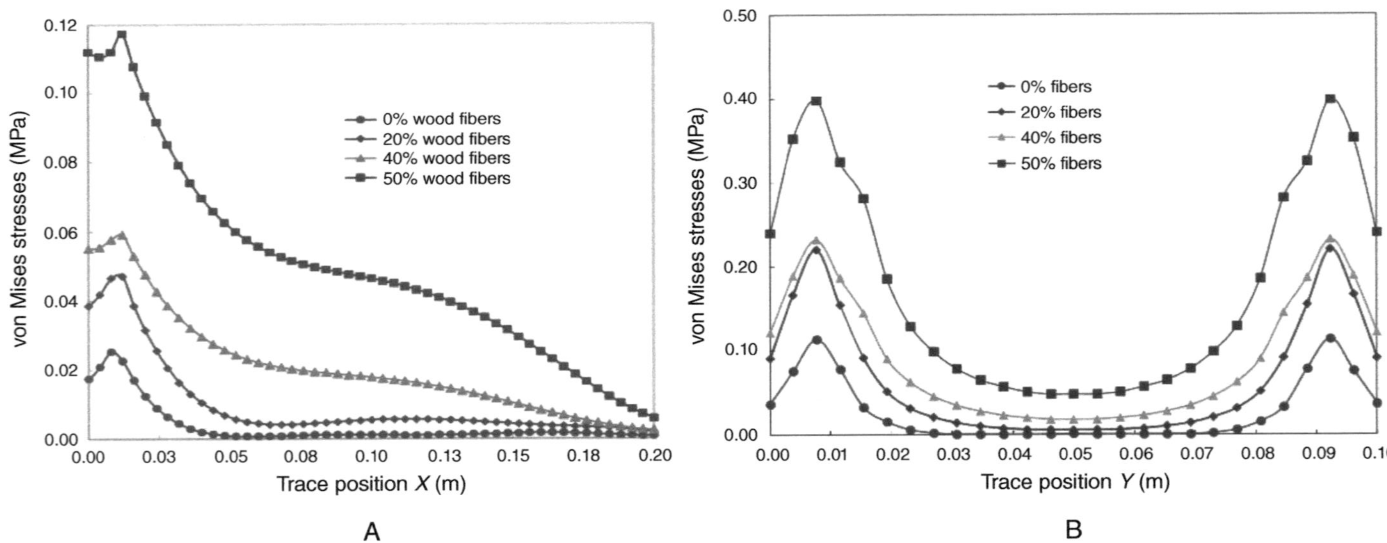

Figure 9.13 Von Mises stress on *XZ* and *YZ* the half plane of symmetry of the blade. (A) Half symmetry plane in *XZ* direction. (B) Half symmetry plane in *YZ* direction.

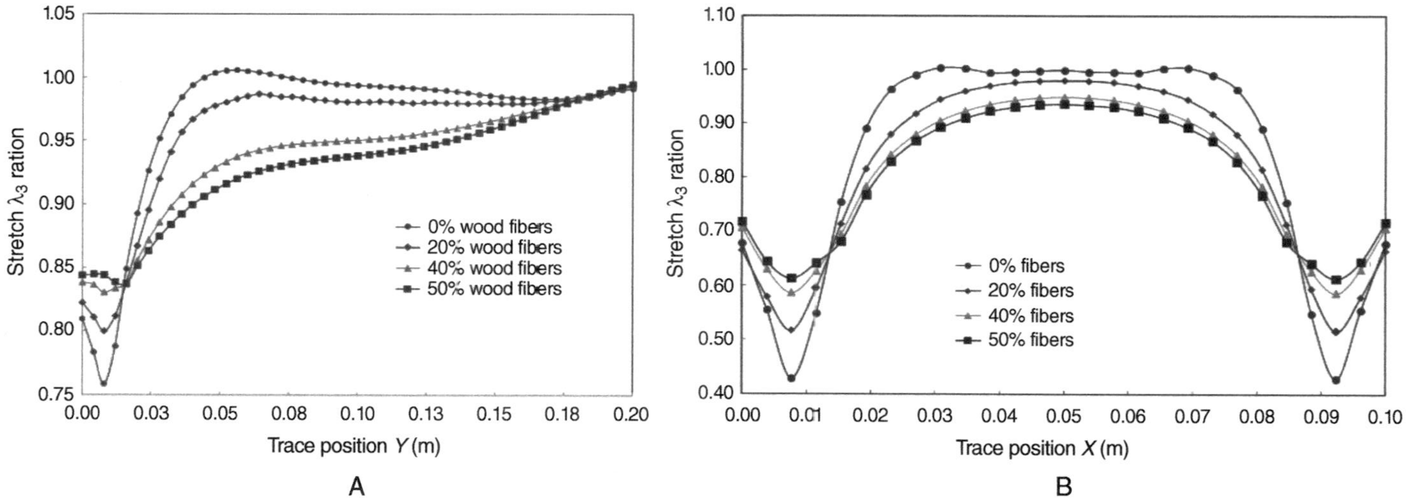

Figure 9.14 Main stretch (λ_3) on the *XZ* and *YZ* half planes of symmetry. (A) Half symmetry plane in *XZ* direction. (B) Half symmetry plane in *XZ* direction.

Table 9.5 Critical values of the von Mises stresses in the *XZ* half plane.

% Sawdust particles	σ_{crit} (MPa)
50%	0.117
40%	0.0591
20%	0.0471
0%	0.0253

Table 9.6 Critical values of the von Mises stresses in the *YZ* half plane.

% Sawdust particles	σ_{crit} (MPa)
50%	0.389
40%	0.232
20%	0.220
0%	0.113

for 50% reinforcement and the minimum of 0.011 MPa for virgin HDPE. Finally, the Von Mises stress distribution and the localized thinning effect indicate that material failure due to large inflation-induced deformations is more likely to occur in the left tip area of the blade as shown in Fig. 9.14A and B, respectively.

These observations show that there is a significant distribution of the von Mises stresses in the biocomposites blade reinforced with 50% (w/w) sawdust particles. This is explained by the high rigidity of the biocomposite material, which requires a powerful airflow for the shaping of the blade. Consequently, this type of biocomposite is poorly suited for both the thermoforming and the blowing of thin and hollow parts. Fig. 9.13A and B, respectively, shows the distribution of the thickness in the final shape of the thermoformed part for each composite according to *XZ* and *YZ* traces.

Fig. 9.15 shows the variations of the energy required by the molding process for HDPE reinforced with 0% and 50% (w/w) sawdust particles with time. It shows that the amount of energy used during the shaping phase of the biocomposite blade depends on the sawdust particles content. It takes the values of 14.6 and 21.1 J for virgin HPDE and HDPE reinforced with 50% (w/w) sawdust particles, respectively. This represents an energy cost of the biocomposite 44.5% as compared to the virgin material. In the other cases, the percentages are 1.6 and 9.6 for 20% and 40% reinforcements, respectively.

From a cost–benefit standpoint, the forming time is also an important issue especially for mass production. The results shown in Table 9.7 show that the forming time increases with the amount of sawdust particles contained

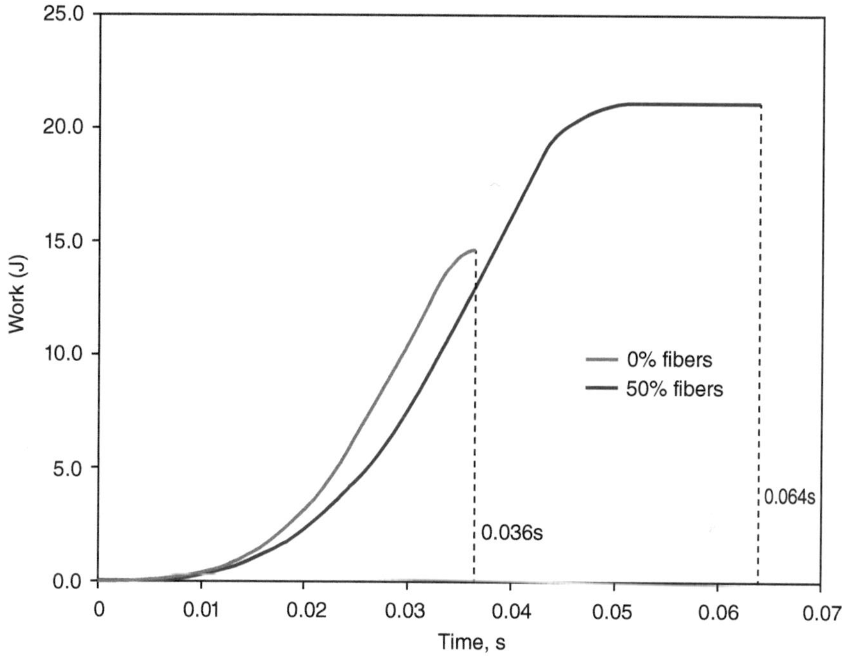

Figure 9.15 Work done by the gas at the end of the forming cycle for the biocomposites of HBPE reinforced with 0% and 50% (w/w) sawdust particles.

Table 9.7 Forming time of the extrados blade part.

% (w/w) Sawdust particles	Time (s)
50%	0.0362
40%	0.0375
20%	0.0457
0%	0.0638

in the biocomposite material. The time required for the forming of the biocomposite material containing 50% (w/w) sawdust particles is 1.75 times longer than the time needed for virgin HPDE.

It therefore appears that based on both the forming energy and the process duration, the thermoforming profitability decreases with an increase in the sawdust particles content. Moreover, it is important to note that these results were obtained without considering the energy dissipated by frictional contact between the mold and the polymer/biocomposite materials.

Fig. 9.16 shows the power and work required to shape each biocomposite sheet. It can be seen that the work increases with the reinforcement

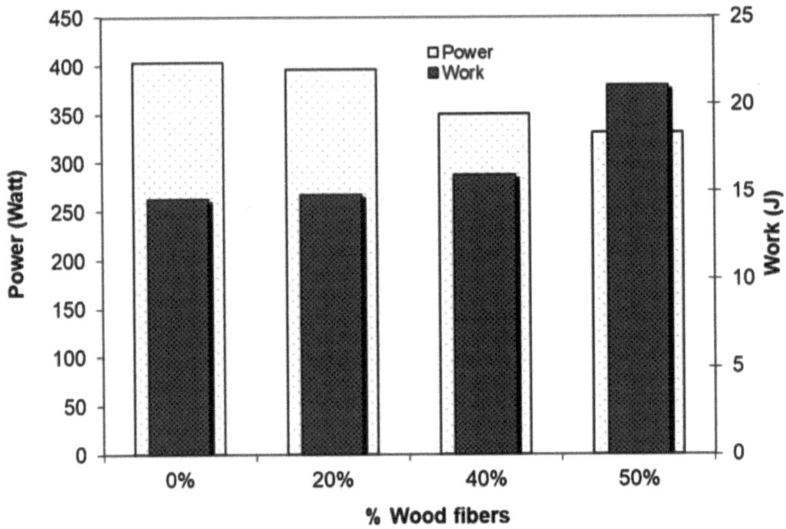

Figure 9.16 Values of the work and power at the end of the forming cycle for different formulations of HDPE reinforced with sawdust particles.

concentration. However, the potency seems to decrease with the wood content probably due to the presence of rigid reinforcements such as sawdust particles which increase the stiffness of the biocomposite material. While requiring a longer forming process time.

It can be noticed that the thickness and stress distributions differ for different sawdust particles contents unlike the work on the thermoforming of isotropic and incompressible materials, where the thicknesses distribution is less dependent on the behavior of the material than on the geometry of the mold, this study on isotropic viscoelastic materials containing sawdust particles shows that the constitutive behavior of these materials influences the thickness and stress distribution in the final product. In addition, it is noted that the energy and the time used for the thermoforming of a composite part increase with the proportion of the reinforcement. The deeper the part is formed, the higher the energy expenditure.

In the light of the results given above, the following remarks can be formulated:

- At ambient temperature, the modulus of elasticity of biocomposite increases with the content of sawdust particles. However, the reinforcement of the matrix has led to a decrease in the mechanical strength of the biocomposites.

- Numerical results show the potential use of biocomposites for the manufacture of drone blades by thermoforming.
- The work and power required to form the UAV blade for virgin polymer and biocomposites with 20% and 40% wood are almost similar.
- The thickness and the von Mises stress obtained from the different biocomposites are not similar and the prediction exhibits some discrepancy, particularly the stress distribution.
- The quantity of matter in the central area of the thermoformed board increases slightly, compared to the virgin polymer, with the mass concentration of the wood in the composite blade.
- The quantity of matter in the leading edge of the thermoformed of UAV blade is more significant than that of the virgin polymer but it increases with the mass concentration of the wood.

The numerical results show the potential use of wood–plastic composites for the manufacture of a UAV blade (upper and lower surface) by thermoforming. Also, the choice to use the finite element method with a pressure load, which is derived from a thermodynamic law, is judicious for the integrated analysis in large deformations of the forming of a thin part.

9.5 Conclusion

A family of biocomposites based on HDPE and 0%, 20%, 40%, and 50% (w/w) sawdust particles was developed, characterized, and numerically analyzed for their potential used for the manufacture of small UAV blade (upper or lower surface), through the thermoforming process. The viscoelastic behavior of the Lodge model was considered. Moreover, the Lagrangian equation together with an assumption of the membrane theory was used in the finite element implementation. This study shows that:

1. The constitutive behavior of HDPE reinforced with sawdust particles has an influence on the final thickness distribution;
2. The time required to for the thermoforming of a biocomposite part increases with an increase of the sawdust particles content;
3. The energy used for the thermoforming of a biocomposite part increases with sawdust particles content;
4. Numerical results show the potential use of biocomposites for the manufacture of drone blades by thermoforming.

References

1. M. Lachance, Prévisions de marchés des plastiques et bioplastiques, Premier Colloque québécois sur les bioplastiques composables, Sherbrooke, Québec, Canada, 2011.
2. D.A. Kaddami, B. Khelifi, et al., Short palm tree fibers – Thermoset matrices composites, Compos. Part A 37 (2006) 1413.
3. S.J. Eichhorn, C.A. Baillie, N. Zafeiropoulos, et al., Current international research into cellulosic fibres and composites, J. Mater. Sci. 36 (2001) 2107.
4. S. Migneault, A. Koubaa, F. Erchiqui, et al., Effects of processing method and fiber size on the structure and properties of wood–plastic composites, Compos. Part A 40 (2009) 80.
5. I.B Amor, H. Rekik, H. Kaddami, M. Raihane, et al., Studies of dielectric relaxation in natural fiber–polymer composites, J. Electrostat. 67 (5) (2009) 717.
6. P. Hogg, Wind Turbine Blade Materials, Wind Turbine Blade Materials, University of Loughborough: SUPERGEN Wind - Wind Energy Technology, 2010 https://pdfcoffee.com/wind-turbine-blade-materials-pdf-free.html.
7. P. Brøndsted, H. Lilholt, A. Lystruo, Composite Materials for Wind Power Turbine Blades. Materials Research, Department Risoe National Laboratory, Roskilde, Denmark, 2005.
8. Derdouri A., Erchiqui F, et al, XIII International Congress on Rheology, Cambridge, UK, pp. 394 (2000).
9. F. Erchiqui, Z. Ozdemir, M. Souli, et al., Neural networks approach for characterisation of viscoelastic polymers, Can. J. Chem. Eng. 89 (5) (2011) 1303.
10. J. Meissner, Polymer melt elongation—methods, results, and recent developments, Polym. Eng. Sci. 27 (8) (1987) 537.
11. A. Benjeddou, E. Jankovich, W.H. Yang, Determination of the parameters of Ogden's law using biaxial data and Levenberg-Marquardt-Fletcher algorithm, J. Elastomers Plastics 25 (1993) 224.
12. M. Tazi, M. S. Sukiman, F. Erchiqui, A Imad, T Kanit, Effect of Wood Fillers on the Viscoelastic and Thermophysical Properties of HDPE-Wood Composite, International Journal of Polymer Science 2016 (2016).
13. A.K. Bledzki, O. Faruk, Wood fibre reinforced polypropylene composites: effect of fibre geometry and coupling agent on physico-mechanical properties, Appl. Compos. Mater. 10 (6) (2003) 365.
14. F. Erchiqui, A.S. Fotso Talla, H. Kaddami, A numerical investigation of the use of novel melt processed PET-hemp fiber composites for thermoforming applications, Polym. Eng. Sci. Vol. 56 (9) (2016) 1021.
15. H.A. Fini Soroush, F. Erchiqui, M. Farzaneh, Investigating the elastic deformation of wood–plastic composites at cold temperature using the bubble inflation technique, J. Thermoplast. Compos. Mater. 28 (3) (2015) 431.
16. A.S. Lodge, Elastic Liquids, Academic Press, London, 1964.
17. F. Erchiqui, A. Gakwaya, M. Rachik, Dynamic Finite Element Analysis of Nonlinear Isotropic Hyperelastic and Viscoelastic Materials for Thermoforming Applications, Polymer Engineering and Sciences 45 (1) (2005) 125.
18. T. Keener, R. Stuart, T. Brown, Maleated Coupling Agents for Natural Fiber Composites, Compos. Part A: Appl. Sci. Manuf. 35 (2004) 357.
19. N. Sombatsompop, C. Yotinwattanakumtorn, C. Thongpin, Influence of type and concentration of maleic anhydride grafted polypropylene and impact modifiers on mechanical properties of PP/wood sawdust composites, J. Appl. Polym. Sci. 97 (2005) 475.
20. C. Burgstaller, Processing of thermal sensitive materials—a case study for wood plastic composites, Monats. Chem.—Chem. Monthly 138 (4) (2007) 343.
21. M. Tazi, F. Erchiqui, H. Kaddami, Influence of SOFTWOOD-fillers content on the biodegradability and morphological properties of WOOD–polyethylene composites, Polym. Compos. 39 (1) (2018) 29, doi:10.1002/app.40495.

22. A. Bendahou, H. Kaddami, H. Sautereau, M. Raihane, F. Erchiqui, A. Dufresne, Short Palm Tree Fibers Polyolefin Composites: Effect of Filler Content and Coupling Agent on Physical Properties, Macromol. Mater. Eng. 293 (2008) 140.
23. M.S. Sukimana, E. Erchiqui. et al., Design and numerical modeling of the thermoforming process of a WPC based formwork structure, Mater. Today Commun. 220 (2020), doi.org/10.1016/j.mtcomm.2019.100805
24. F. Erchiqui, A. Gakwaya, A. Koubaa, H. Vincent, H. Kaddami, Engineering investigations on the potentiality of the thermoformability of HDPE charged by wood flours in the thermoforming part, Polym. Eng. Sci. 49 (8) (2009) 1594.
25. J.D. Ferry, Viscoelastic Properties of Polymers, Wiley, New York, 1980.
26. O. Redlich, J.N.S. Kwong, V—an equation of state, Chem. Rev. 44 (1949) 233.
27. F. Erchiqui, A new hybrid approach using the explicit dynamic finite element method and thermodynamic law for the analysis of the thermoforming and blow molding processes for polymer materials, Polym. Eng. Sci. 46 (11) (2006) 233.

CHAPTER TEN

Blast loading effects on aircraft fuselage

Navid Mehreganian[a], Arash Soleiman Fallah[b]
[a] Department of Civil and Environmental Engineering, Imperial College London, London, United Kingdom
[b] Department of Mechanical, Electronic, and Chemical Engineering, OsloMet, Oslo, Norway

10.1 Introduction

Nonaccidental aircraft explosions have occurred on numerous occasions since 1933 on B247D when a nitroglycerine bomb detonated in the baggage compartment of the United Air Lines, while a more recent incident was the detonation of a bomb concealed in a laptop on Airbus A321-111, Daallo Airlines [1]. Close-in blasts of explosive charges can lead to catastrophic consequences of damage to the fuselage and potential crash of the passenger or commercial aircraft with disrupted transportation and little or no possibility of commuters' survival. Such events have raised public awareness of the significance of the protective systems against extensive shock and blast loads.

An element of the aircraft most susceptible to the detrimental effects of blasts is the fuselage. This is due to the fact that it is both more accessible and weaker. Thin-walled structural forms such as plates and shells have wide applications in aerospace engineering in manufacturing parts of aircraft vehicles. Such structural forms are generally made of ductile, isotropic materials such as aluminum alloys owing to the benefits of, low cost, lightweight, ductility, and the ease of manufacturing and inspection techniques [2]. In fact, although in the recent decade composites have been used in the manufacture of commercial jet airliners for the fuselage, wings, and other structural components in lieu of the aluminum alloys, the latter remain a widely used material due to the benefits of lightweight, ductility, and cost-effectiveness.

Depending on the proximity of the blast source, the loading can be classified as global (i.e., far-field explosions) or localized (i.e., close-in blasts). While a few definitions have been used for the sake of classification of loading depending on certain attributes [3–5], the main concern has been the structural response [5,6], rather than the classifications and definitions.

Multiphysics Simulations in Automotive and Aerospace Applications.
DOI: 10.1016/B978-0-12-817899-7.00004-6

The performance of beams under global blast loading and associated failure modes were first investigated by Menkes and Opat [7] and later extended to plates by Nurick and Martin [8,9] and subsequently compared with those of the localized blasts [5,10]. Hitherto, the performance of beams and plates subjected to localized and globalized blasts have been extensively examined experimentally [11–17]. In recent decades, advanced numerical techniques have enabled researchers to investigate the phenomenon of blasts on various structural forms [18–23]. To this end, a number of numerical studies exist on the dynamic process of blast phenomenon which retains the influence of fluid–structure interaction (FSI) [1,24,25]. During the bast phenomenon at a cruise altitude, pressurization and gravity play a non-negligible role together with the inertia of the rigid-body component of airplane motion.

10.1.1 Fuselage design

Cylindrical shells used in the design of aircraft fuselage are the simplest type of structural forms used for this purpose which possess a single nonzero curvature. Fuselage hells may be fabricated in a variety of shapes and subjected to a wide range of dynamic loads during the operation of the aircraft. They may possess, as they usually do, stiffeners to further enhance them. In some cases, they are subject to extensive shock and/or blast loads. The design of the fuselage entails three main methods, namely truss, monocoque, and semimonocoque.

- Truss design is a conventional method of incorporating steel tubes and frames in a framework, such as Pratt truss or Warren truss framework, commonly used for the first generation of aircraft.
- Monocoque design relies entirely on the skin to provide the load-carrying capacities against bending, compression, and torsional moments and comprises formers, frames, and bulkheads.
- The semi-monocoque design is a modification of monocoque design to provide sufficient strength while maintaining a lightweight framework. The increased stiffness of this design is appreciably contributed to the integration of longerons and stringers. A schematic of such a design is illustrated in Fig. 10.1.

This chapter is, in outline form, organized as follows. The next section describes the basic concepts of blast loads and the general parameters that influence generated overpressure and velocity thus FSI. The influence of FSI on the aircraft fuselage induced by a single internal blast is examined in Section 10.2 with the aid of finite element analysis, while the target is

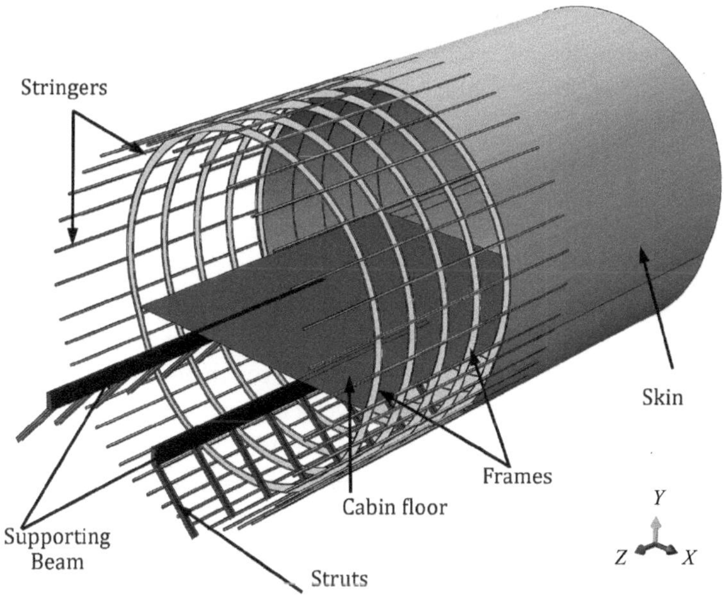

Figure 10.1 Schematic of semimonocoque design.

assumed as rigid. The same problem is extended to multiple blast loading case scenarios in Section 10.3. Section 10.4 examines a theoretical analysis for the elastic behavior of the flexible target subject to a blast load, which is extended to the dynamic behavior of a rigid-perfectly plastic shell subject to pulse pressures of such kind. The sections form the basis for the analysis and design of aircraft fuselage to resist internal explosions of bounded size.

10.1.2 Basic concepts

A blast load is typically generated as a consequence of a fast reaction and may be categorized as a physical, nuclear, or chemical explosion.

- Physical explosion

Physical explosions occur, for example, in the event of a sudden change of state due to external factors, for instance, in a reactor tube subject to a violent transient overpressure in sodium-cooled fast-breeder reactors, sudden bursting of a sealed container due to the high internal pressure or when a hot liquid flows through a frozen pipe.

- Nuclear explosions

Nuclear explosions occur due to the change of state in the nucleus of an atom via nuclear fission or fusion. Nuclear explosions in atomic bombs

(A-bombs) are accompanied by a large release of energy from the uncontrolled chain reaction in a plutonium or uranium atoms. Thermonuclear bombs (H-bombs) release a much higher amount of energy due to the fusion of hydrogen atoms.

- Chemical explosion

Chemical explosions can emanate from either deflagration or detonation. During the deflagration process, for example, propagation of flame in a gas explosion, the explosive product consumes oxygen as it combusts and propagates at a subsonic speed. This depends on the stoichiometric properties of the ambience. Due to the slow rate of chemical reaction, deflagration processes are characterized with much smaller pressure, finite rise time, and lower flame propagation velocity than detonation [26,27].

Detonation occurs when in high explosives, such as TNT (Trinitrotoluene), the material decomposes violently and rapidly, generating substantial heat and a supersonic exothermic shock front that propagates through a medium (e.g., air or water). The presence of air is not required as the oxygen for the chemical reaction is present within the explosive compound.

The usefulness of the explosive compound for industrial or military applications depends on its inertness and stability. For example, high explosives such as composition-C4 (plastic explosive [PE4]) are widely used due to higher stability to either friction or heating at room temperature, while fulminated mercury is somewhat unstable and sensitive to friction. Ammonium nitrate also, potentially, decomposes explosively when large stockpiles oxidize.

The duration of detonation is generally much lower than that of deflagration and the rise time to maximum pressure for detonation is virtually zero in contradistinction to the finite rise time for deflagration processes. There exist, however, events when the deflagration–detonation transition occurs.

The magnitude of the overpressure and the propagation of the blast wave depends on the proximity of the source and the medium through which wave propagation occurs. In air blasts, the flow of gaseous products are assumed inviscid, thus the viscous forces are not considered for explosive modeling [28]. In underwater explosions, the high propagation velocity of shock in water (due to instantaneous compression) levels off rapidly. In this study, only the air blast explosion is considered.

The evolution of high explosives overpressure in time is defined by Friedlander equation as plotted in Fig. 10.2. The main parameters of the blast are [29,30]:

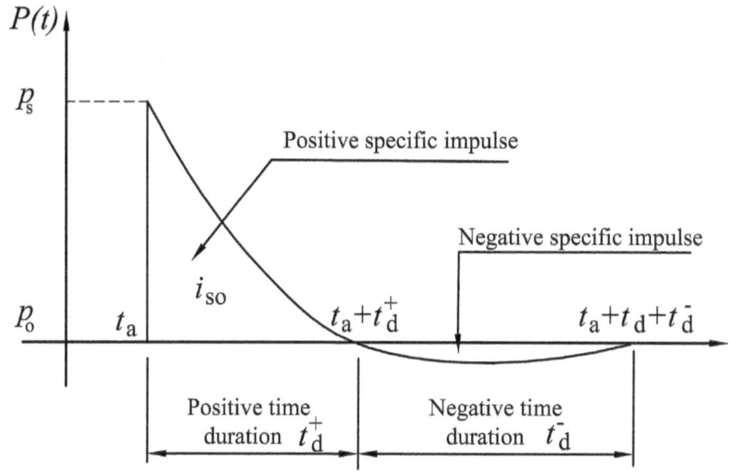

Figure 10.2 Friedlander blast pressure profile P_0 and p_{so} represent the atmospheric and absolute pressure, respectively [31].

- The arrival time t_a, determined by the delay time from the beginning of detonation.
- The air–explosive product interaction, whereupon the detonation of the explosive, the air particles undergo an instantaneous rise in the pressure above the ambient pressure. The gauge pressure $p_s = P_{so} - P_0$ is defined as the peak incident overpressure.
- The positive phase blast duration t_d^+ . As the blast wave expands, the pressure at a point away from the source decays exponentially, back to the ambient pressure P_0.
- Beyond t_d^+, the pressure decays further to a negative pressure, which is due to momentum conservation between the (air) medium and explosive shock front. The overexpansion of air causes the absolute pressure at the tail of the blast to fall below the atmospheric pressure. This reduction generates a reversed flow back toward the center of detonation [31].

The idealized pressure wave is sensitive to the morphology of a medium through which it propagates. The pressure profile describing the positive phase is written as:

$$p(t) = p_s\left(1 - \frac{t}{t_d}\right)e^{-Yt/t_d}, \tag{10.1}$$

where t_d is the decay period (pulse duration). The reflected pressure has the same profile as the incident pressure and may be expressed by replacing p_{s0}

with p_r in Eq. (10.1). The positive phase impulse per unit area of the structural element is obtained by an integration of Eq. (10.1), that is,

$$I = \int_0^{t_d} p(t)dt = \frac{p_s t_d}{Y^2}\left(Y - 1 + e^{-Y}\right). \tag{10.2}$$

Various profiles are proposed for the negative phase, including the extended Friedlander, cubic, and modified cubic with reflection coefficient [32]. The negative phase depends on the stand-off, charge weight, and the explosive topology. The influence of the negative phase increases with the increase of the stand-off distance, that is, distal charges in some cases may lead to the reversed snap buckling phenomenon, a counter intuitive structural behavior that causes the final configuration of the structure to be in opposite direction to the incident wave [20,32]. However, the negative phase becomes insignificant for proximal charges thus its effect is not examined here.

Typical blast loading functions may be truncated as a single term, multiplicative decomposition of their temporal and spatial parts. While the former was discussed earlier, the latter also plays a significant role in the modes of failure and the structural response. The variations in the charge geometry and stand-off determine the type of the plastic hinges formed when the material is assumed as elastic-perfectly plastic or rigid-perfectly plastic [33].

Upon detonation, state variables such as the detonation velocity, the overpressure, and the air density behind the wavefront characterize the detonation wavefront of the incident pressure through the Rankine–Hugoniot [28,34] expressions, as:

$$\rho_s = \rho_0 \frac{(\gamma + 1)p_s + 2\gamma p_0}{(\gamma - 1)p_s + 2\gamma p_0}, \tag{10.3}$$

$$U_s = a_0 \sqrt{\frac{(\gamma + 1)p_s + 2\gamma p_0}{2\gamma p_0}}, \tag{10.4}$$

$$u_s = a_0 p_s \sqrt{\frac{2}{\gamma p_0\left((\gamma + 1)p_s + 2\gamma p_0\right)}}. \tag{10.5}$$

When the pressure impinges upon the target, it is reinforced and reflected. The Rankine–Hugoniot relations for the reflected wave are similarly expressed as:

$$\rho_r = \rho_0 \frac{\gamma\left(p_s + p_0\right)}{(\gamma - 1)p_s + \gamma p_0}, \tag{10.6}$$

$$U_r = u_s \frac{(\gamma - 1)p_s + \gamma p_0}{p_s}, \tag{10.7}$$

$$p_r = p_s \frac{(3\gamma - 1)p_s + 4\gamma p_0}{(\gamma - 1)p_s + 2\gamma p_0}. \tag{10.8}$$

The potential heat and detonation pressure of high explosives may be estimated using Chapman–Jouguet conditions [35,36]. For example, the detonation pressure of TNT and composition-B are 210 kbar and 294 kbar, respectively, the detonation temperatures are 3511 K and 3918 K, while the detonation velocities are 6930 m/s and 7920 m/s, respectively [37,38]. A most prevalent expression to describe the state variables (energy, volume, temperature, and pressure) of the detonation process is the Jones–Wilkins–Lee equation of state as follows:

$$P = A\left(1 - \frac{\omega\widehat{\rho_e}}{R_1\rho_e}\right)e^{-\frac{R_1\rho_e}{\widehat{\rho_e}}} + B\left(1 - \frac{\omega\widehat{\rho_e}}{R_2\rho_e}\right)e^{-\frac{R_2\rho_e}{\widehat{\rho_e}}} + \frac{\omega\widehat{\rho_e}^2}{\rho_0}E_{m0} \tag{10.9}$$

where $\widehat{\rho_e}$ is the explosive product density, ρ_e is the density of the explosive at the beginning of the process, A, B, R_1,R_2, ω are the material constants an E_{m0} is the initial energy density of the explosive.

The expansion of the explosive depends on its geometry, a cylindrical explosive expands primarily in axial direction rather than the radial one, which is the case for a spherical explosive. The inverted cone generates higher impulse than the cylindrical one, but yields the lower permanent deformations on the target, as the impulse is spread over a larger area resulting in a lower impulse density (impulse per unit area) [39]. For air blast loading , the behavior of the medium is commonly characterized by the ideal gas equation of state as:

$$P = \left(C_p - C_v\right)\rho_0 T, \tag{10.10}$$

where C_p and C_v are the specific heats at constant pressure and volume, respectively.

10.1.3 Material behavior

Structural forms made of ductile isotropic materials such as aluminum alloys AL6013, or AL7075, have been extensively utilized in the fabrication of aircraft fuselage. Such materials exhibit high strength-to-weight ratio which enables them to store reasonable amount of energy per unit of volume within the bounds of their elastic region. Beyond the elastic limit, the energy imparted to the structure is mostly dissipated as heat while the structure is

Table 10.1 Material properties of AL70-75, where *c* represents the specific heat capacity.

Material properties				
ρ_p	ν	E (GPa)	χ	c (kJ/kg K)
2700	0.3	70	0.9	910
Strain hardening			**Strain rate hardening**	
A_1 (MPA)	B_1 (MPA)	n	C_1	$\dot{\varepsilon}_0\ \left(\frac{1}{\text{sec}}\right)$
520	477	0.52	0.001	5×10^{-4}
Damage parameters				
D_1	D_2	D_3	D_4	D_5
0.096	0.049	3.465	0.016	1.099
Temperature softening				
T_{melt} (K)	T_t (K)	m		
893	293	1		

undergoing plastic deformation. The material properties of the latter are summarized in Table 10.1.

Common material models such as Ramberg–Osgood and power-law predict exponential growth of the plastic flow in relation to the plastic strain. In dynamically loaded structures with high loading rates, such as those encountering a blast, the plastic flow stress may also be sensitive to the rate of deformation (strain). Various empirical, phenomenological models exist which describe the correspondence between the strain rate and plastic stress flow, such as Cowper–Symonds and Johnson–Cook constitutive models. The latter model estimates the dynamic flow as a function of strain hardening, strain rate, and change of temperature: Johnson–Cook is a multiplicative model, where the work hardening increases concerning viscoplasticity (strain rate sensitivity of the yield) but softening in relation to temperature.

$$\sigma_0 = \left[A_1 + B_1\varepsilon_{\text{pl}}^n\right]\left[1 + C_1 \ln \dot{\varepsilon}_{\text{eq}}\right]\left[1 - \hat{T}^m\right], \tag{10.11a}$$

$$\hat{T} = \left[\frac{T - T_t}{T_{\text{melt}} - T_t}\right], \tag{10.11b}$$

$$\dot{\varepsilon}_{\text{eq}} = \frac{\dot{\varepsilon}}{\dot{\varepsilon}_0}, \tag{10.11c}$$

where T_t and T_{melt} are the room temperature and melting point temperature of the material in Kelvin, respectively. The ratio of strain rate to the reference strain rate is defined as $\dot{\varepsilon}_{\text{eq}}^* = \dot{\varepsilon}/\varepsilon_0$, where the reference strain rate, while the coefficient C_1 determines the sensitivity of the material to strain rate. The hardening coefficients are given as A_1,B_1 while ε_{pl} and n represent the plastic strain and hardening exponent. The value A_1 represents the stress at first yield.

Beyond the plastic limit, ductile damage in the material occurs. At a microscopic level, the terminology "ductile damage" denotes the microfracture as a result of nucleation, growth, and coalescence of the voids within the material particles [40–42], while at a macroscopic level, it has the connotation of large deformation of the material at the onset of necking, which is followed by stress degradation. Rupture (or ductile fracture) of the body results in discontinuity of the total displacement field, whereas dislocation reflects the discontinuity of the elastic/plastic displacement fields [43].

Ductile fracture is characterized by the accumulation of plastic strain at the current stress state described by:

$$\mathcal{D} = \int \frac{\dot{\varepsilon}_{\mathrm{pl}}}{\varepsilon_f}\, dt, \tag{10.12}$$

where $\mathcal{D}$ is the damage parameter which equates to a maximum of 1 when the material element completely erodes, $\dot{\varepsilon}_{\mathrm{pl}}$ denotes the plastic strain rate, while ε_f is the fracture strain, which is a function of the normalized third deviatoric invariant (ζ), temperature (T), and strain rate at any proportional stress state, as well as the stress triaxiality (ς), given by Eq. (10.13). Stress triaxiality is the quotient of mean flow stress (σ_m)—also referred to as the hydrostatic pressure stress—to effective stress (σ_{eff}).

$$\varepsilon_f = \phi(\varsigma,\ \zeta) g\left(\dot{\varepsilon}_{\mathrm{pl}}\right) h(T). \tag{10.13}$$

At the onset of necking, the state of stress is represented by

$$\sigma = (1 - D)\bar{\sigma}. \tag{10.14}$$

In Eq. (10.14), σ and $\bar{\sigma}$ represent the state of stress prior to and after damage, respectively. At the point of maximum damage parameter, the elemental stresses vanish, causing material failure. Studies reveal that stress triaxiality, the state of stress, and material ductility are interrelated. Higher triaxiality increases the speed of void growth which rises [44] rapidly, resulting in reduced ductility and hence lower fracture strains (Fig. 10.3) [45,46].

The functions φ and g can be identified from the Johnson–Cook ductile fracture criterion [42] in Eq. (10.15).

$$\varepsilon_f = \left(D_1 + D_2 e^{-D_3 \varsigma}\right)\left(1 + D_4 \ln \dot{\varepsilon}^*_{\mathrm{eq}}\right)\left(1 + D_5\ \hat{T}\ \right), \tag{10.15}$$

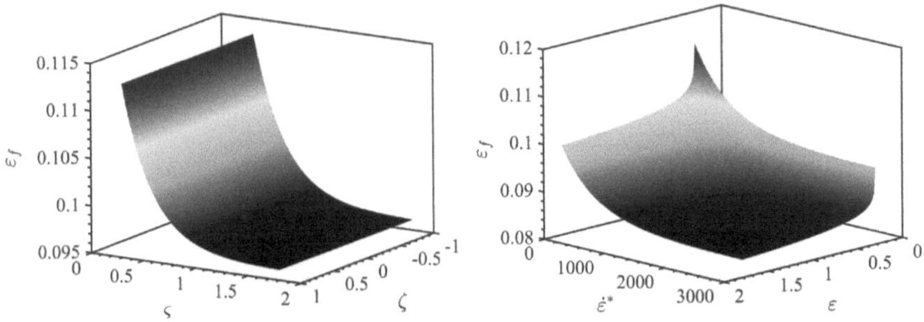

Figure 10.3 Fracture surface for AL7075, (A) the influence of ς and ζ on the fracture strain, (B) the influence of strain rate on the fracture surface.

where $D_1 \ldots D_5$ are material constants determined experimentally and $\hat{T}$ represents the homologous temperature, defined in Eq. (10.11b). Evidently, while the elevated strain rate increases the plastic yield, it lowers the fracture strain (Fig. 10.3).

Basaran [47] proposed a constitutive model for plasticity and fracture where the ductility function $\varphi(\varsigma, \zeta)$ defined by the Johnson–Cook model was developed as a quadratic truncation of the stress triaxiality.

10.2 Fluid–structure interaction during a single blast

Early works of Taylor [48] examined the one-dimensional fluid flow with pressure profile expressed in Eq. (10.1), interacting with an unsupported plate of density ρ_p and thickness h_p. When the blast wave reaches the target plate, the plate gains instantaneous acceleration and departs from the contact interface of the fluid. This is accompanied by a reduction in the transmitted pressure, and hence the impulse, provided by the FSI. The FSI effects were first recognized by Taylor [48] for a blast with exponential decay profile and negligible fluid compressibility. Taylor showed that the transmitted impulse depends on a single dimensionless parameter which is characterized by a time scale of the blast wave and FSI. The light plates have more significant effects on the reduction of the impulse due to the rapid gain in velocity, relieving the plate off the contact interface.

The impulse carried by the reflected pressure is attained by substituting p_r in lieu of $p(t)$ in Eq. (10.2). By the time $t = t_d$ when the loading is complete, the target has acquired maximum velocity and hence the total impulse is transferred to the target. Ignoring the linear part of the pressure profile in Eq. (10.1), the transferred impulse on a flat plate is expressed in dimensionless form [48] as:

$$\frac{I_p}{I_i} = 2\beta_0^{\frac{\beta_0}{1-\beta_0}}, \tag{10.16}$$

where I_i and I_p are the incident and reflected (transmitted) impulse per unit area of the plate. In what follows the subscripts r, s, o, and p denote the states behind the reflected shock, the incident shock, atmospheric (undisturbed), and the plate, respectively. The quantity β_0 represents relative time constant of the FSI and the incident wave, which may be interpreted as the relative volume ratio of the compressed gas to the plate, that is,

$$\beta_0 = \frac{\rho_0 h_f}{\rho_p h_p} = \frac{t_d}{t_0^*}, \tag{10.17}$$

$$t_0^* = \frac{\rho_p h_p}{\rho_0 a_0}. \tag{10.18}$$

The degree of pressure reflection, aside from the physics of the medium, depends on the target mass and rigidity. For example, in the circumstances where $\lim\limits_{\beta_0 \to 0} (I_p/I_i) = 2$ the plate is assumed as rigid and the incident wave is reflected completely off the target plate, the case which occurs in the heavy target limits. In other words, as the limiting case of a plate with infinite mass is approached, the target hardly moves and all of the incident pressure is reflected off it. Consequently, the impulse imparted to the plate reaches its maximum. In the circumstances of the light plates we have [49]:

$$\lim_{\beta_0 \to \infty} (I_p/I_i) = \lim_{\beta_0 \to \infty} \frac{2}{\beta_0} = 0 \tag{10.19}$$

Upon integration of the equation of motion, the pressure acting on the plate p_p, the displacement w_p, and velocity of the plate u_p are derived [50] as:

$$p_p = \frac{2p_{so}}{\beta_0 - 1}\left(\beta_0 e^{-\frac{\beta_0 t}{t_i}} - e^{-\frac{t}{t_i}}\right), \tag{10.20}$$

$$w_p = \frac{2\beta_0 p_s t_i}{\rho_0 a_0 (\beta_0 - 1)}\left(1 - \beta_0 + \frac{1}{\beta_0} e^{-\frac{\beta_0 t}{t_i}} - e^{-\frac{t}{t_i}}\right), \tag{10.21}$$

$$u_p = \frac{2\beta_0 p_{so} t_i}{\rho_0 a_0 (\beta_0 - 1)}\left(e^{-\frac{t}{t_i}} - e^{-\frac{\beta_0 t}{t_i}}\right), \tag{10.22}$$

where t_i is an arbitrary time scale which may be chosen equivalent to t_d/Y. These variables are plotted in Fig. 10.4. Monotonic reduction in the pressure is discernible with both reduction in the volume of compressed gas (and/or

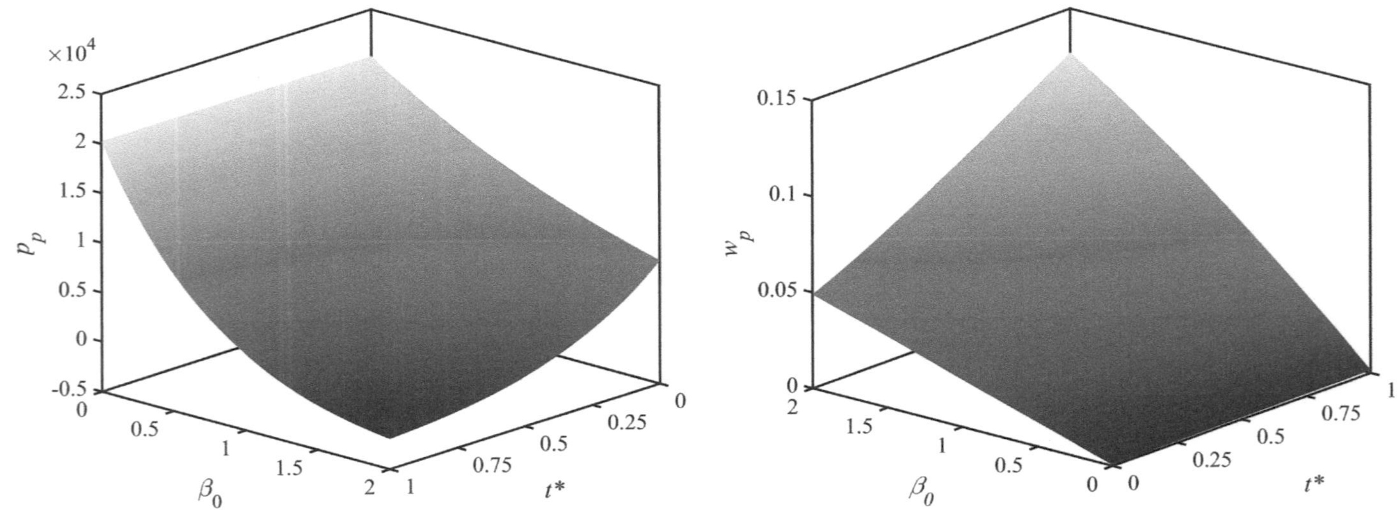

Figure 10.4 Surface interaction of the state variables influenced by the relative time of the blast $t^* = t/t_i$ and the dimensionless number β_0.

reduction in the plate thickness) as well as the increase of relative time of the load. Conversely, the plate displacement markedly increases as the plate thickness reduces ($\beta > 1$).

10.2.1 Finite element analysis

To investigate the influence of FSI on the aircraft fuselage, a number of finite element models were setup and analyses were performed using ABAQUS ® Explicit hydrocode. Hydocodes such as ABAQUS® Explicit and LS_DYNA are pragmatic tools to handle complex problems of the FSI where the Lagrange processor and Euler processor work in tandem.

The motion of a body may be described using a pure Lagrangian or an Eulerian reference frame, in the former case, the material points move with the mesh, such as the axial crushing of a beam or plates or deformation of the structure due to seismic loads. In the latter, the material points move through a mesh of Eulerian body which may be fixed in space.

Common finite element techniques utilize the Eulerian reference frame to predict the FSI problems, such as motion of fluid in a tank or, as investigated herein, the internal air blast loading in the aircraft fuselage. Among the techniques is the Multi-Material Arbitrary Eulerian Lagrangian (MMAEL) method. This method employs continuum mechanics algorithms to model the fluid advection through the Eulerian mesh in space. In other words, the MMAEL equations are defined with regards to the material time derivative relative to the reference configuration [22,51,52]. This technique may either be performed on rigid (referred to as the uncoupled Eulerian Lagrangian method) or flexible (referred to as coupled Eulerian Lagrangian method) target. Both methods obtain solutions for the arbitrary motion of fluid mesh points with respect to the reference Eulerian points that are fixed in space, as follows.

In the uncoupled models, the multimaterial consists only of the medium (air) and the explosive (e.g., PE4), while any surface boundary of the Eulerian medium maybe fixed to monitor the pressure at the gage points of the associated surface. This procedure is referred to as "uncoupled" since the captured pressure is implemented in a separate, pure-Lagrangian model.

The coupled Eulerian Lagrangian method entails a similar procedure, but with the flexible target. The MMAEL algorithms are enforced at the contact interface to solve for the deformation of the target.

The geometry of the fuselage cross section is illustrated in Fig. 10.5, which mimics the half symmetric model of B737 interior. Spherical PE4 charges

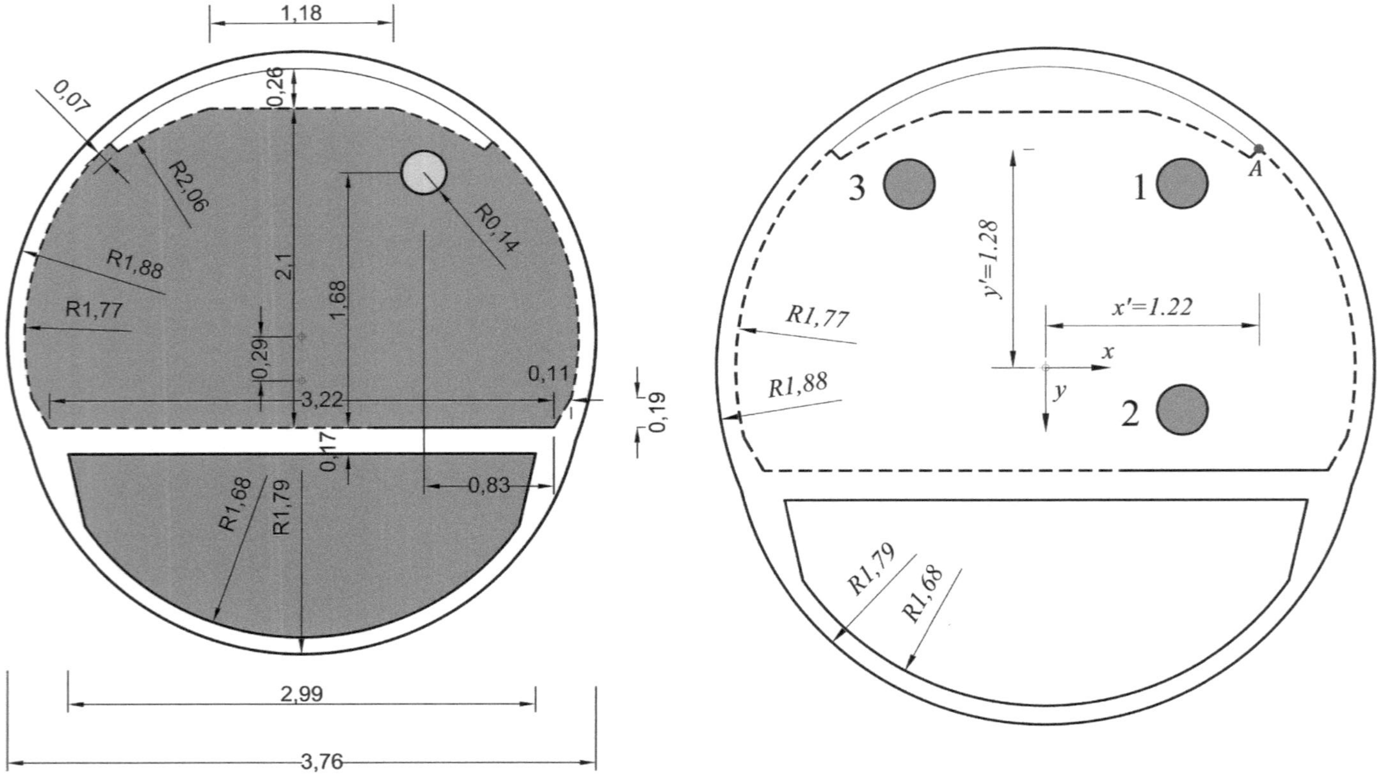

(A) (B)

Figure 10.5 (A) Geometry of the model and (B) arrangements of the PE4 explosives, as (1) or (3) luggage compartment, (2) under passenger seats. For multiple blasts loading, a combination of (1, 3) or (1, 2) is utilized, as discussed in Section 10.4.

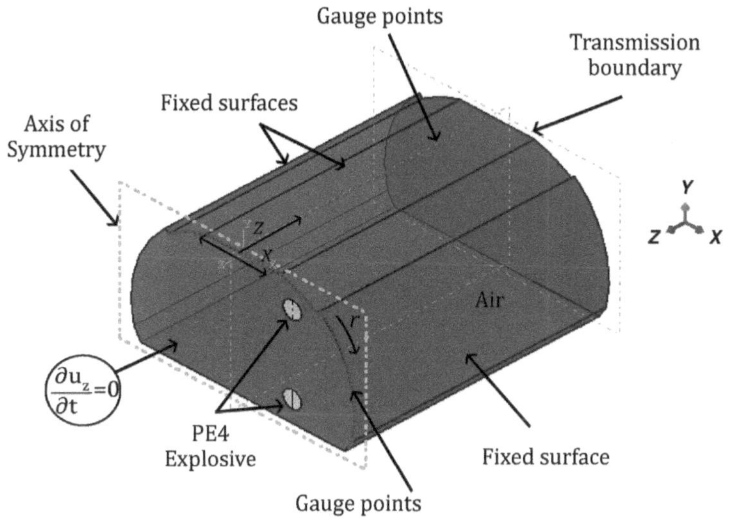

Figure 10.6 Schematic of the finite element model.

of 20 kg and 5 kg were placed at 20 cm distance from the internal skin of the fuselage, once in the luggage area and another time in the sitting area, as arranged in Fig. 10.5B. As the length of a typical aircraft is an order of magnitude higher than its radius and blast waves decay exponentially with distance, the magnitude of the blast pressure becomes infinitesimal along the fuselage axis. Heuristically, the length to radius ratio of 2.35 (length of 4 m) was found sufficient to monitor the blast pressure across the cabin.

A half symmetric schematic of the finite element model is shown in Fig. 10.6. For simplicity, only the cabin section of the plane is modeled. Flow out boundary conditions are assigned to the far edge of the model while the curved surfaces at either side and the top surface are fixed to monitor the pressure. The pressure is captured from the gauge points of such surfaces, along generalized coordinates x, z and arc length $\hat{l}$. The arc length of the gauge points from the tip A is derived as:

$$\hat{l} = R\cos^{-1}\left(1 - \frac{(\hat{x}^2 + \hat{y}^2)}{2R^2}\right), \tag{10.23}$$

where $\hat{x} = x - x'$ and $\hat{y} = y - y'$.

The explosive and air material properties are summarized in Table 10.2. With the increase of B737 cruise altitude to 12,192 m, the outside air

Table 10.2 Explosive and air parameters for Eulerian models—the parameters marked with asterisks are temperature dependent, which data may be found from barometric formula of Engineering Toolbox [53].

Material	ρ_e **(kg/m³)**	**Detonation. wave speed v (m/s)**	A	B	R_1	R_2	ω	E_m	**Pre detonation. bulk modulus**
PE-4	1601	8193	6.0977×10 [11]	1.295×10 [10]	4.5	1.4	0.25	6.057×10 [6]	0
Air	ρ_0 (kg/m³)	$C_p - C_v{}^*$ gas constant (J/kg K)		P (Pa)	Internal energy (kJ/kg)	μ_k dynamic viscosity* (μPa s)			
	1.007	287		79500	156.82	17.26			

temperature decreases to −52°C, which reduces the density, viscosity, and pressure of air correspondingly. To overcome the pressure difference, the pressurization system in the cabin maintains the absolute pressure at 79.5 kPa (˜11.2 psi), which is the pressure at the 2000 m altitude, while the pressure outside the aircraft reduces to 20.19 kPa.

The propagation of the blast wave due to detonation of the charge placed in a luggage compartment is illustrated in Fig. 10.7. Following detonation, three main stages may be associated with the subsequent FSI:

- Phase "i": expansion of the explosive, from the moment of detonation which lasts until the detonation products reach the target interface.
- Phase "ii": interaction of the blast wave with the target, including the incident and reflected pressures.
- Phase "iii": expansion of the blast wave from the time of separation from the target to the time of equilibrium with the deformation of the plate.

Upon contact with the detonation front, the material point within the unreacted explosive instantaneously transfers to gaseous material which is accompanied with sudden release of energy from the chemical reaction [38], resulting in advection of the gaseous fluid through the mesh points. Advection of the fluid mesh points in the Eulerian medium can be of first or second order. By default, ABAQUS algorithm incorporates second-order advection to transfer the state variables (pressure, velocity, and explosive product density). This advection method incorporates a process referred to as "flux limiting" to ensure the advection is monotonic and to reduce the oscillations occurring after the shock wave.

The distribution of pressure along the arc length is illustrated in Fig. 10.8, which may be visualized as a zone of maximal pressure occurring along the points of the arc projected from proximal edge of the charge onto the arc, followed by exponential decay to the right and left of the projected points. The pressure profile is maintained as the blast wave propagates along the arc until the pressure build-up occurs as the blast wave recirculates from the boundary edges. Such boundary conditions are due to the change of curvature of target surface, such as the chamfered edges at the bottom of the arc herein (Fig. 10.8B), or imposed structural boundary conditions such as clamps [54,55]. The impulse density generated from pressure recirculation is infinitesimal compared to its maximum value in the most proximal gauge points of the target. However, provided that the quotient of the charge characteristic length to target length be more than a third, the pressure recirculation contributes to impulse absorption by the boundaries, hence

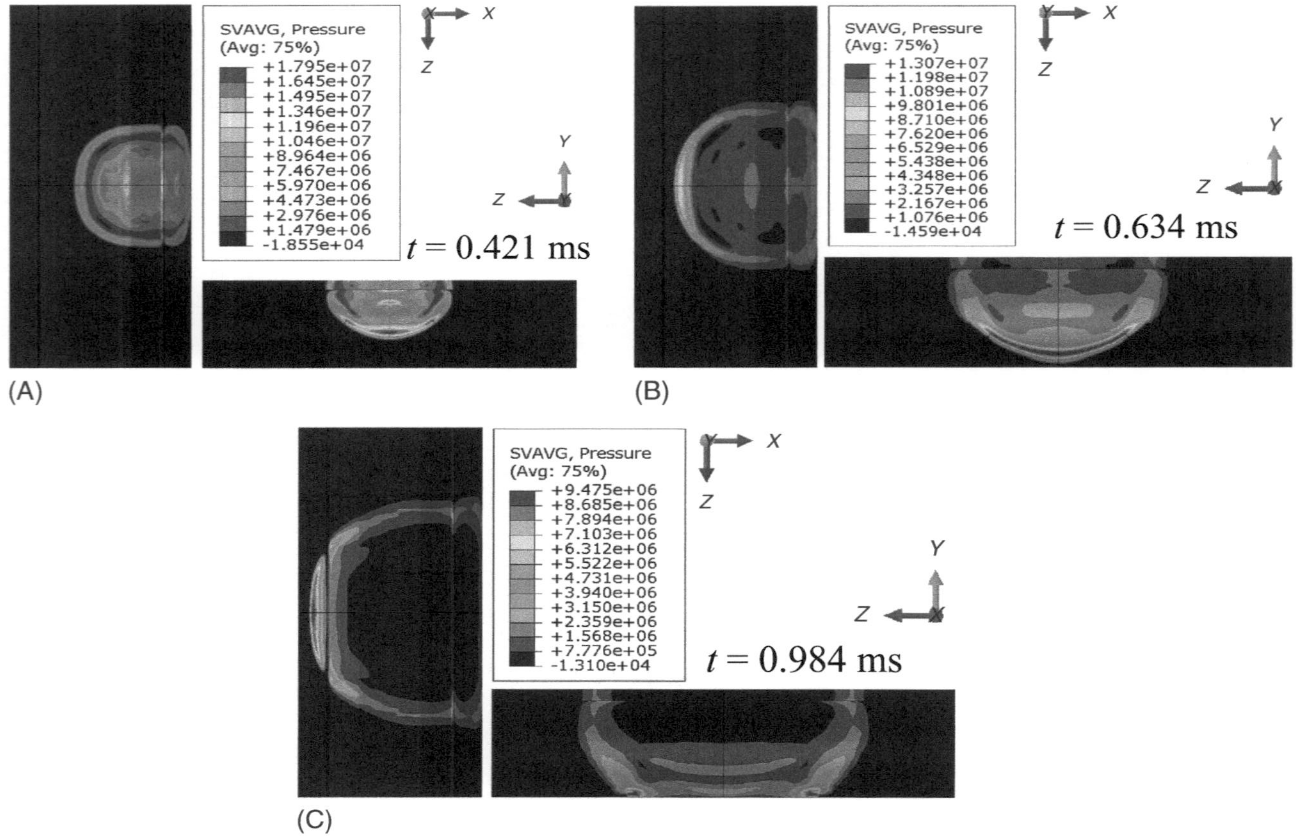

Figure 10.7 Blast wave propagation and interaction with the rigid target (A) at the time of contact with the fuselage roof, (B) at the time point of reaching the cabin floor, (C) propagating across the cabin floor.

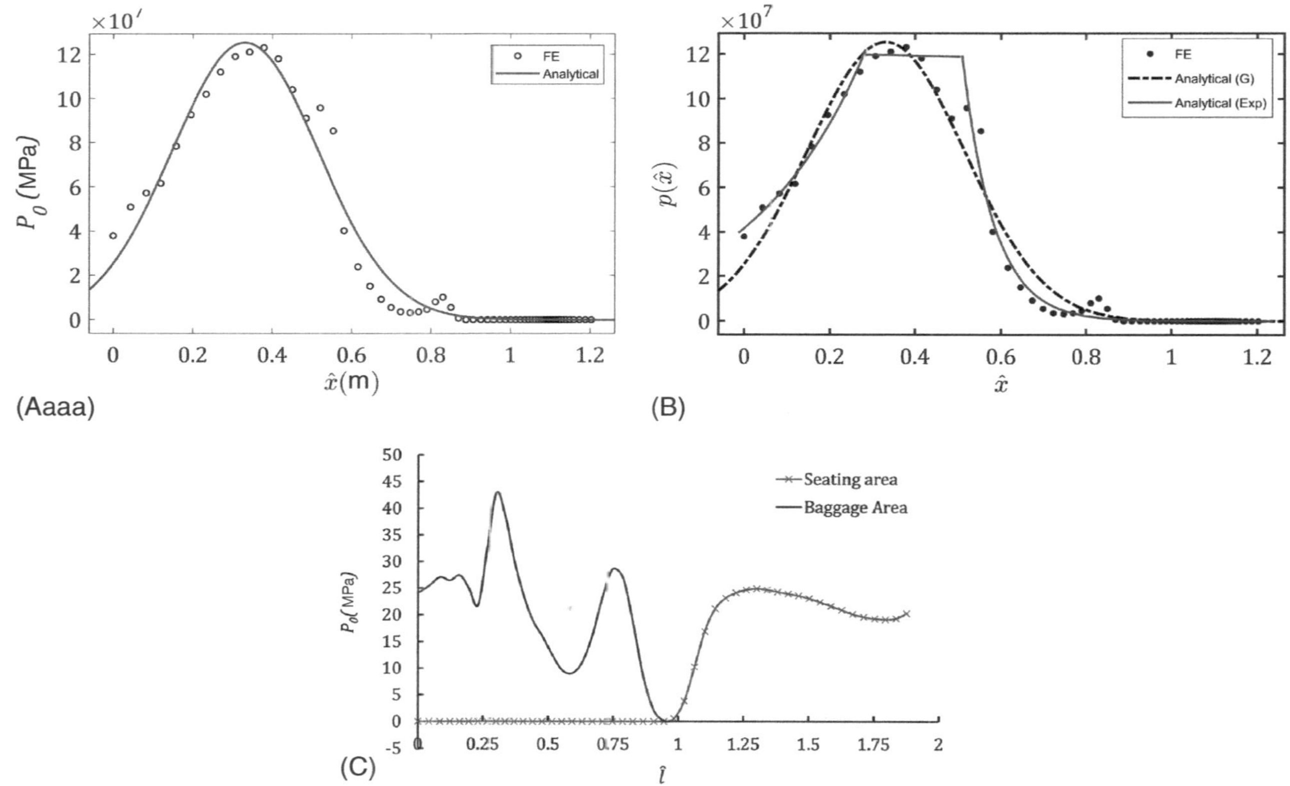

Figure 10.8 Proposed empirical model for (A) the distribution of the blast pressure on curved surface, (B) comparison of the Gaussian curve fit (G) against the exponential profile (Exp.) of Eq. (10.24), (C) distribution of the pressure across the arc length at $t = 297$ μs.

reducing the total impulse imparted onto the plate. As a result, the pressure build up for the charge placed in the luggage compartment becomes more significant than the one placed in the seating area. Further, the former imparts higher pressure on the fuselage skin to the right compared to the one in the seating area, due to the lower stand-off.

A commonly accepted spatial part for the localized blasts for targets of zero curvature (beams and plates) is attributed to a uniform pressure distribution maintained up to the radius R_0 from the point of projection, followed by an exponential decay beyond this value, expressed as:

$$p(\hat{x}) = \begin{cases} P_0 e^{-b(\hat{x}-R_0)}, & 0 \leq \hat{x} \leq R_0, \\ 0\ R_0 \leq \hat{x}, & \end{cases} \tag{10.24}$$

where $\hat{x}$ is the arc length horizontal distance from the corner A in Fig. 10.5B. Based on the numerical results and using curve-fitting tools, the distribution of the close-in blast pressure along an arc length $\hat{l}$ is derived as bell curved, Gaussian distribution:

$$p(\hat{x}) = P_0 e^{-\frac{(\hat{x}-\mu)^2}{c^2}}. \tag{10.25}$$

For the pressure distribution drawn in Fig. 10.8, the parameters are $a = 128\text{MPa}$, $\mu = 0.331$, and $c = 0.2615$. It should be emphasized that the form of Eq. (10.24) which describes the load shape imparted on the plate may alternatively be used to describe the load shape on a curved surface, with a notion that every point on the flat target maps a corresponding point on curved surface with the same stand-off, while assuming the pressure imparted on the two targets of identical stand-off is the same.

10.3 Fluid–structure interaction in multiple blast loading

The foregoing analysis of a single blast load is extended to the circumstances concerning multiple blasts. Two cases of horizontal (arrangement 1 and 3 in Fig. 10.5B) and vertical (arrangement 1 and 2 in Fig. 10.5B) arrangements have been examined.

When two identical charges placed sideways detonate at the proximity of a flat target plane, they generate a zone of high pressure at the midpoint between them which covers an elliptical area. The combined blast pressure then expands and travels in orthogonal directions to the virtual axis of the

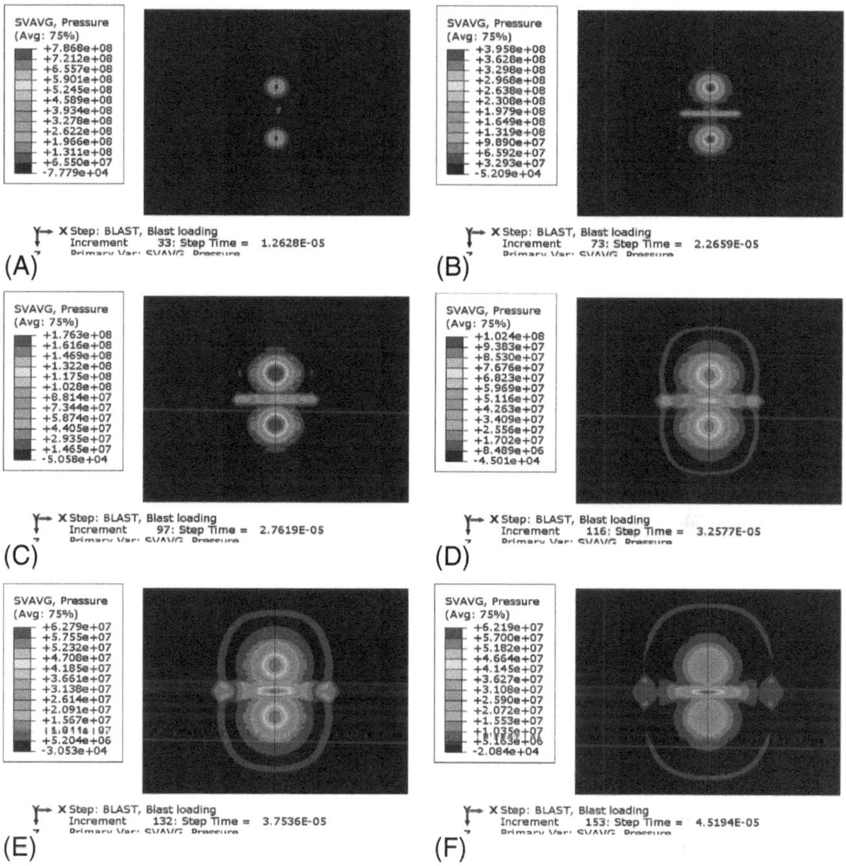

Figure 10.9 Schematic of the blast pressure propagation across a flat target generated from the sideways arrangements of the explosives, (A–C) at the time point of reaching the target detonation, (D–F) initial superposition of the blast waves, (G–L) propagation of the superposed shock front perpendicular to the virtual axis of the charge arrangements.

charges (Fig. 10.9). In the case of horizontal arrangements, the blast wave propagation across the two target interfaces are of interest: the first, being the blast waves from the explosives which superimpose constructively on the curved skin above the midpoint between the charges, creating a region of extensive pressure on the top edge point of projection. This pressure travels across the fuselage with a shock-front. The second represents a shock front traveling downward across the curved surfaces—to the right and left of the fuselage skin, as depicted in Fig. 10.10.

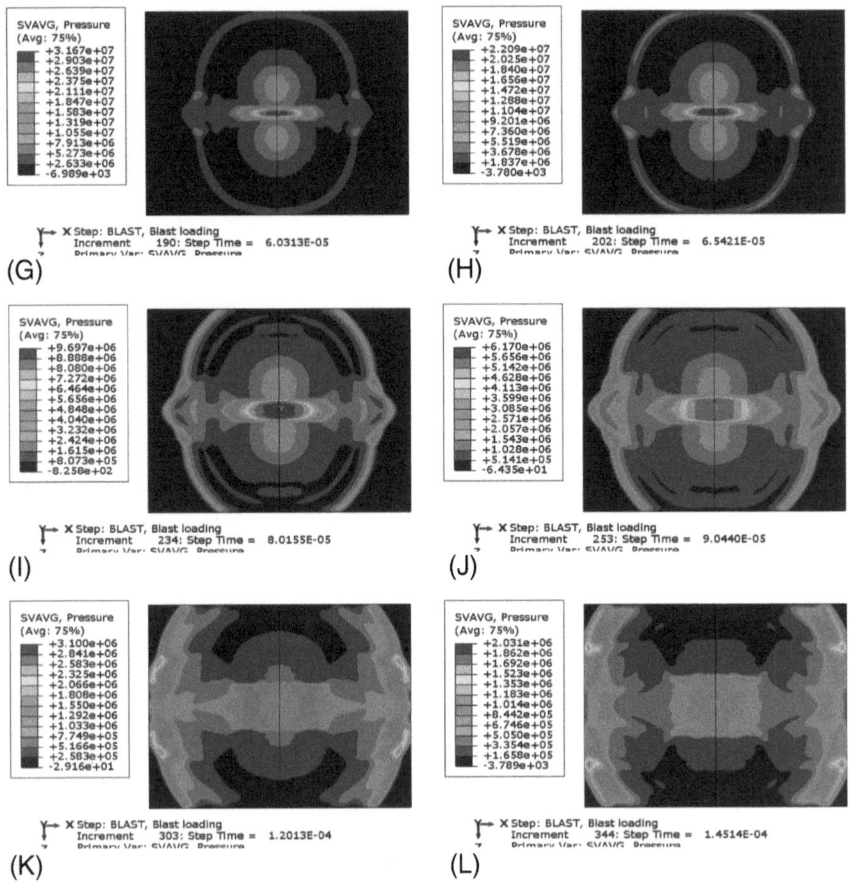

Figure 10.9 *(continued)*

In contrast to the pressure distribution registered at the gauge points of the arc length, the multiple charges elevate the radius of uniform pressure across the top surface of the fuselage (Fig. 10.11). A bulge in the pressure profile occurs near the boundary edge between the two arcs above the cabin.

In the vertical arrangements ((1) and (2)), similarly, as a result of blast waves' constructive interference, two waves are formed, the first is a region of high pressure occurred at midway between the charges and on the curved surfaces of the fuselage skin at the side of the charge, which travels obliquely toward either side of the interfacing edges of the cabin floor and the fuselage skin (Fig. 10.12). The second wave constructs a region of lower pressure along the plane of the charges which dissipates quickly.

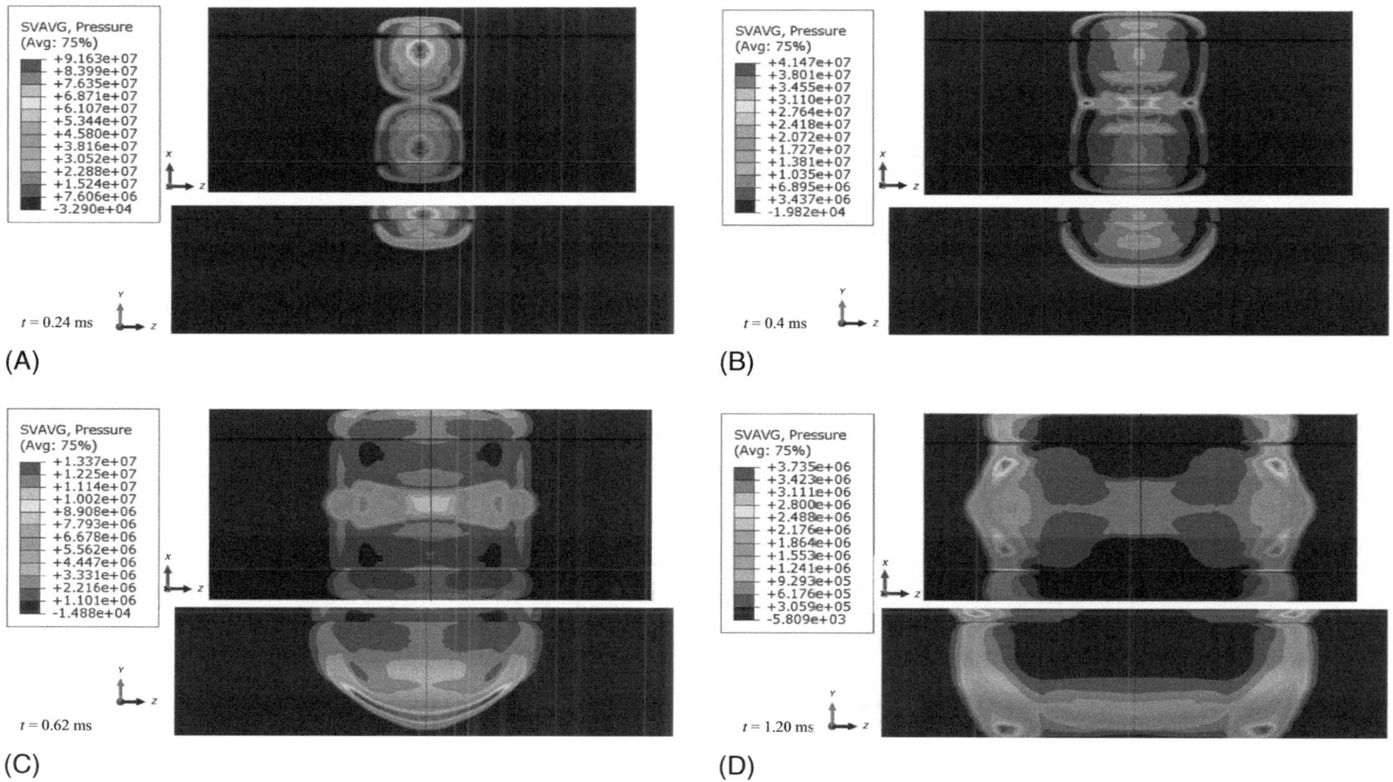

Figure 10.10 Schematic of the pressure distribution across the generalized coordinates $\bar{l}$ and z. (A) At the onset of the FSI. (B) At the onset of superposition. (C–D) Propagation of the generated shock front at 0.62 m and 1.2 ms, respectively.

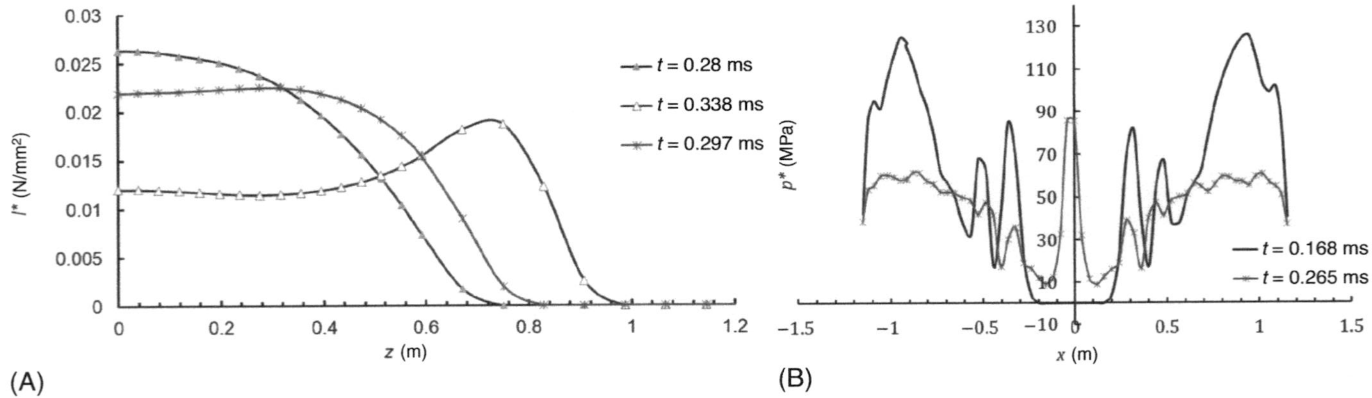

Figure 10.11 Impulse density (i^*) and reflected pressure (P_r) across the gauge points (A) i^* across z, (B) P_r across x.

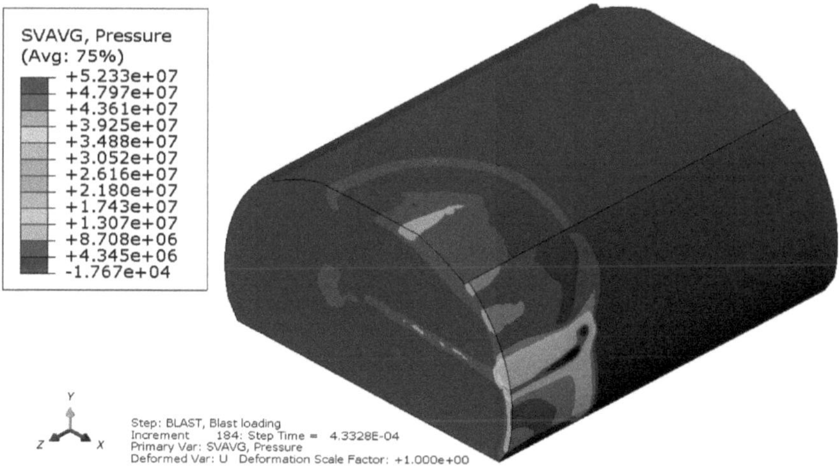

Figure 10.12 Schematic of the blast wave advection in vertical arrangements of multiple charges.

The wave from the lower charge never reaches the top edge/ceiling of the fuselage and the distribution of pressure on the gage points of the edge mimics the half of which was formed from two charges arranged in sideway configurations.

10.4 Elastic behavior of the aircraft fuselage due to internal blasts

Consider a long, reinforced, circular cylindrical shell of length $2L$ and thickness of h which is clamped at both ends as shown in Fig. 10.13. The shell is assumed thin, such that the bending moment predominates the motion and the influence of membrane action, transverse shear, and rotator inertia may be ignored. The shell is subjected to an internal pulse pressure of $p^*(r, t)$. As discussed earlier, the pressure is truncated into the single term of its multiplicative decomposition of spatial (load shape) and temporal (pulse shape) parts, that is, $p^*(r, t) = P(r)p_2(t)$. It is assumed that the load is distributed uniformly across the shell internal surface, that is, $P(r) = P_0$, while associated with the pulse shape is an exponential part of the Eq. (10.1) which lasts for the duration of t_d. It is further assumed that the inward direction is negative.

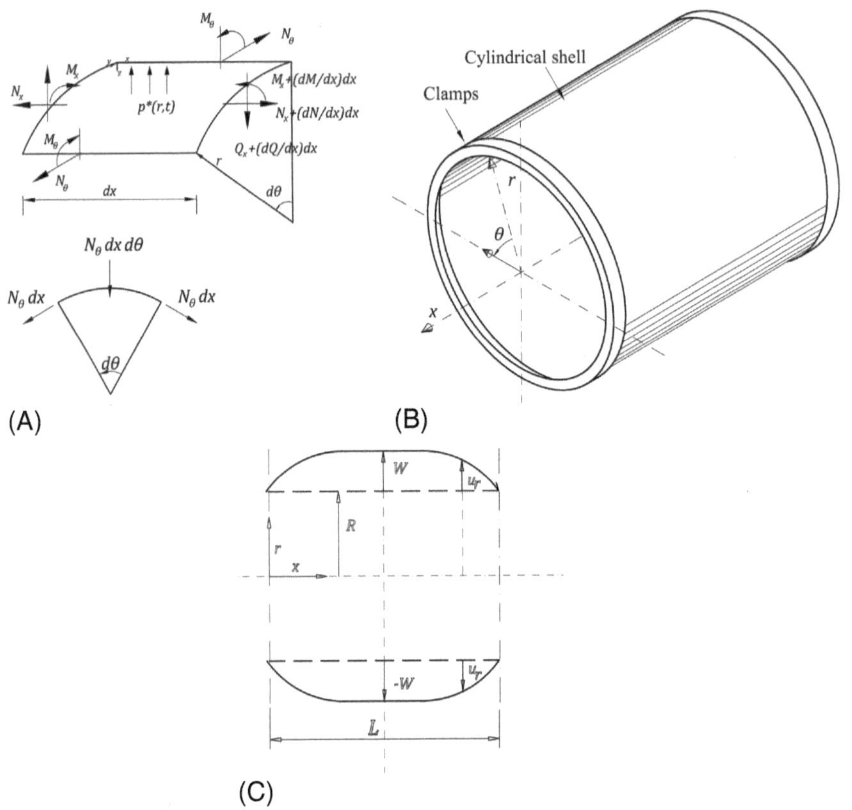

Figure 10.13 (A) Free body diaELSST050-10-gram of the shell, (B) Generalized coordinates in the shell, (C) Displacement profile of the cylindrical shell.

By ignoring the Föppl–Von-Kármán strains, the components of strain tensor in cylindrical shell are expressed as:

$$\varepsilon_{rr} = \frac{\partial u_r}{\partial r}, \tag{10.26a}$$

$$\varepsilon_{\theta\theta} = \frac{1}{r}\left(\frac{\partial u_\theta}{\partial_\theta} + u_r\right), \tag{10.26b}$$

$$\varepsilon_{xx} = \frac{\partial u_x}{\partial x}, \tag{10.26c}$$

$$\varepsilon_{r\theta} = \frac{1}{2}\left(\frac{1}{r}\frac{\partial u_r}{\partial \theta} + \frac{\partial u_\theta}{\partial r} - \frac{u_\theta}{r}\right), \tag{10.26d}$$

$$\varepsilon_{rx} = \frac{1}{2}\left(\frac{1}{r}\frac{\partial u_r}{\partial x} + \frac{\partial u_x}{\partial r}\right). \tag{10.26e}$$

Using Hook's law, the corresponding components of stress tensor are furnished as:

$$\sigma_{rr} = \mu_l(2\varepsilon_{rr} + \beta_1(\varepsilon_{rr} + \varepsilon_{\theta\theta} + \varepsilon_{xx})), \tag{10.27a}$$

$$\sigma_{\theta\theta} = \mu_l(2\varepsilon_{\theta\theta} + \beta_1(\varepsilon_{rr} + \varepsilon_{\theta\theta} + \varepsilon_{xx})), \tag{10.27b}$$

$$\sigma_{xx} = \mu_l(2\varepsilon_{xx} + \beta_1(\varepsilon_{rr} + \varepsilon_{\theta\theta} + \varepsilon_{xx})). \tag{10.27c}$$

where $\mu_l = E/2(1+\nu)$, and $\lambda_l = E\nu/((1+\nu)(1-2\nu))$ are the Lame's parameters, while $\beta_1 = \lambda_l/\mu_l$. Notably, the shell structure examined hereunder is axisymmetric so the equilibrium equations are produced for axisymmetric response. Thus, all derivatives with respect to θ, including shear stress components, vanish. In the same spirit, warping is ignored in the study. The generalized stresses are expressed as:

$$M_x = \int_R^{R+h} \sigma_x r dr, \tag{10.28a}$$

$$N_x = \int_R^{R+h} \sigma_x dr, \tag{10.28b}$$

$$M_\theta = \int_R^{R+h} \sigma_\theta r dr, \tag{10.28c}$$

$$N_\theta = \int_R^{R+h} \sigma_\theta dr, \tag{10.28d}$$

$$Q_x = \int_R^{R+h} \sigma_{rr} dr. \tag{10.28e}$$

The bending moment, shear, and membrane forces in Eqs. (10.28a–10.28e) are referred to as the generalized stresses and generalized strains because they stress resultants through the thickness of an element. It is assumed that these are the only nonzero components (e.g., $M_{x\theta} = \int_R^{R+h} \sigma_{x\theta} r dr = 0$).

From the equilibrium of the elemental forces in the cylindrical shell of radius R, the governing equations for axisymmetric loading are derived as follows[56].

$$\frac{\partial N_x}{\partial x} R = 0, \tag{10.29}$$

$$\frac{\partial M_x}{\partial x} - Q_x = 0, \tag{10.30}$$

$$R\frac{\partial Q_x}{\partial x} + N_\theta - Rp^*(t) + \frac{\mu R \partial^2 u_r}{\partial t^2} = 0. \tag{10.31}$$

Eq. (10.29) dictates that the membrane force $N_x = \bar{N}_0 Eh$ be constant along the shell central axis.

From the stress–strain constitutive relations (Eqs. (10.27a) and (10.28a–10.28e)), the generalized strains N_x and N_θ are derived as:

$$\begin{bmatrix} N_x \\ N_\theta \end{bmatrix} = \frac{Eh}{1-\nu^2}\begin{bmatrix} 1 & \nu \\ \nu & 1 \end{bmatrix}\begin{bmatrix} \varepsilon_{xx} \\ \varepsilon_{\theta\theta} \end{bmatrix}. \tag{10.32}$$

Substituting Eqs. (10.26b–10.26c) in the first element of the strain vector yields:

$$\frac{\partial u_x}{\partial x} = \frac{-u_r\nu + \left(1-\nu^2\right)\bar{N}_0 R}{R}. \tag{10.33}$$

From which it transpires that:

$$N_\theta = \frac{\left(\nu r\bar{N}_0 + u_r\right)Eh}{R}. \tag{10.34}$$

Thus, substituting Eq. (10.30) and (10.33) in Eq. (10.31) furnishes the latter into:

$$R\frac{\partial M_x^2}{\partial x^2} + \frac{\left(\nu R\bar{N}_0 + u_r\right)Eh}{R} - Rp^* - \frac{\mu R\partial^2 u_r}{\partial t^2} = 0, \tag{10.35}$$

where $\beta^4 = Eh/(4DR^2)$ and $D = Eh^3/\{12(1-\nu^2)\}$ is the flexural rigidity.

By integration of Eq. (10.28a) through thickness the generalized stress is expressible as:

$$M_x = D\frac{\partial u_r^2}{\partial x^2}. \tag{10.36}$$

Which furnishes Eq. (10.35) into:

$$\frac{Eh}{4R\beta^2}\frac{\partial u_r^4}{\partial x^4} + \frac{\left(\nu r\bar{N}_0 + u_r\right)Eh}{R} - Rp^*(t) - \frac{\mu R\partial^2 u_r}{\partial t^2} = 0. \tag{10.37}$$

In the sequel, the solutions to the equation of motion expressed in Eq. (10.37) are sought for the cases of static and dynamic loading scenarios.

10.4.1 Static analysis

In the case of static loading, the inertia term of Eq. (10.37) vanishes, reducing this governing partial differential equation into a fourth-order ordinary differential equation (ODE). A general solution to this ODE is expressible as:

$$u_r = (c_1 \cos(\beta x) + c_2 \sin(\beta x))e^{-\beta x} - \bar{N}_0 \nu R + (c_3 \cos(\beta x) + c_4 \sin(\beta x))e^{\beta x} + \frac{p_{st} R^2}{Eh} \tag{10.38}$$

where p_{st} represents the static load. For shells subject to internal pressure, it would seem reasonable to assume the coefficients c_3 and c_4 to be zero in accordance with Saint Venant's principle, as the term $e^{\beta x}$ would increase the stresses in the shell unrealistically (unboundedly for a long cylinder). Hence, the following Dirichlet boundary conditions of long reinforced shell must be satisfied:

$$M_x(0) = -M_0, \tag{10.39a}$$

$$Q_x(0) = Q_0, \tag{10.39b}$$

$$u_r|_{x=0} = 0, \tag{10.39c}$$

$$\left.\frac{\partial u_r}{\partial x}\right|_{x=0} = 0. \tag{10.39d}$$

Using Eqs. (10.39a, 10.26b), it transpires that:

$$c_1 = \frac{Q_0 - M_0 \beta}{2D_1 \beta^3}, \tag{10.40a}$$

$$c_2 = \frac{M_0}{2D_1 \beta^2}. \tag{10.40b}$$

We assume that the membrane force $\bar{N}_0 = 0$. From Eqs. (10.39c, 10.26d) it follows that

$$M_0 = \frac{Q_0}{2\beta} = -\frac{p_{st}}{2\beta^2}. \tag{10.41}$$

Hence $c_1 = c_2$. The negative sign is due to the assumption of positive sign for the outward lateral displacement.

10.4.2 Dynamic analysis

i. First phase of motion

Under the circumstances of dynamic loading, the governing equations of motion remain valid as they are. It is assumed that the profile of the shell deformation in dynamic loading remains identical to that of the static one. Notably, ignoring the negative phase of the blast reduces the mathematical treatment to two phases of motion, the first associated with a forced vibration, during which the plate acquires maximum inertia within a duration of t_d, followed by a second phase of motion wherein the vibration is merely due to the initial kinetic energy of the structure. Consequently, the form of Eq. (10.38) entails temporal parts representing modes of deformation as:

$$u_r = \cos(\beta x)e^{-\beta x}A_1(t) + \sin(\beta x)e^{-\beta x}A_2(t) + \frac{p_s R^2}{Eh}A_3(t). \tag{10.42}$$

An exponential pulse shape is assumed for the blast load such that $p^*(t) = P_0 e^{-Y\frac{t}{t_d}}$. Substituting Eq. (10.42) in (10.37) yields:

$$\mathcal{L}(x,t) = e^{-\beta x}\cos(\beta x)\frac{d^2A_1(t)}{dt^2} + e^{-\beta x}\sin(\beta x)\frac{d^2A_2(t)}{dt^2} + \frac{p_{st}R^2}{Eh}\frac{d^2A_3(t)}{dt^2} + p_{st}/\mu A_3(t) - P_0/\mu e^{-\frac{Yt}{t_d}} = 0. \tag{10.43}$$

Due to the presence of spatial and temporal terms, the partial differential equation in Eq. (10.43) may be reduced to a set of ODEs using the Galerkin–Ritz variational technique:

$$S = \int_0^L \mathcal{L}(x,t)\delta u_j dx = 0. \tag{10.44}$$

where δu_j represent the weight functions to minimize $\mathcal{L}$, and may be assumed as the coefficients of jth mode of deformations, expressed as:

$$\delta u_1 = e^{-\beta x}\cos(\beta x),\ \ \delta u_2 = \mu R e^{-\beta x}\sin(\beta x),\ \ \delta u_3 = 1. \tag{10.45}$$

Performing the integration by applying the three weight functions above results in three independent expressions. Heuristically, from solving the set of ODEs, it turns out that the modes of deformation are uncoupled as the inertia terms of the first and second modes both equate to zero, while the vibration of the third mode is represented by a linearized ODE as:

$$\frac{d^2A_3(t)}{dt^2} + \omega^2 A_3(t) - \omega^2\eta e^{-\frac{Yt}{t_d}} = 0, \tag{10.46a}$$

$$\omega^2 = \frac{Eh}{R^2\mu}, \tag{10.46b}$$

$$\eta = \frac{P_0}{p_{st}}. \tag{10.46c}$$

The generalized solutions of each mode is given as

$$A_1(t) = a_1 t + b_1, \tag{10.47a}$$

$$A_2(t) = a_2 t + b_2, \tag{10.47b}$$

$$A_3(t) = a_3 \sin(\omega t) + b_3 \cos(\omega t) + \frac{\omega^2 t_d^2 \eta e^{-\frac{Yt}{t_d}}}{\left(Y^2 + \omega^2 t_d^2\right)}. \tag{10.47c}$$

From the Dirichlet (Eqs. (10.39a–10.39d)) and kinematic conditions of the displacement and velocity fields at the onset of loading, that is, $u_r|_{t=0} = \left.\frac{\partial u_r}{\partial t}\right|_{t=0} = 0$ we find:

$$b_2 = c_2, \tag{10.48a}$$

$$b_1 = c_1, \tag{10.48b}$$

$$a_1 = a_2 = a_3 = 0, \tag{10.48c}$$

$$b_3 = -\frac{t_d^2 \omega^2 \eta}{\left(Y^2 + \omega^2 t_d^2\right)} - \frac{\mu \omega^2 c_1}{p_{st}}. \tag{10.48d}$$

Since the moment and shear cannot exceed the corresponding values of M_0 and Q_0 determined from static analysis. The final form of the displacement field is expressed as:

$$u_r = \frac{M_0}{2D_1 \beta^2}\left((\cos(\beta x) \mid \sin(\beta x))e^{-\beta x} - \cos(\omega t)\right) + \frac{P_0 t_d^2}{\mu\left(Y^2 + \omega^2 t_d^2\right)}\left(e^{-\frac{Yt}{t_d}} - \cos(\omega t)\right). \tag{10.49}$$

i. Second phase of motion

The loading is complete at the time $t = t_d$, however, the system continues motion due to the initial momentum associated with the forced vibration in the first phase. The solution to the ODE (10.43) is expressed as:

$$A_3(t) = c_5 \cos(\omega t) + c_6 \sin(\omega t). \tag{10.50}$$

By setting $f(x) = c_1 \cos(\beta x)e^{-\beta x} + c_2 \sin(\beta x)e^{-\beta x}$ an employing the kinematic continuity of displacement and velocity fields at $t = t_d$, the coefficients $c_5 - c_6$ are determined. Hence the displacement is furnished as:

$$u_r = f(x) + \frac{\eta p_{st} e^{-Y} t_d}{\mu \omega\left(\omega^2 t_d^2 + Y^2\right)}(\omega t_d \cos(\omega(t - t_d)) - Y \sin(\omega(t - t_d))) + \frac{p_{st} b_3}{\mu \omega^2}\cos(\omega t). \tag{10.51}$$

10.5 Plastic behavior of the aircraft fuselage due to internal blasts

Materials such as aluminum alloys have a considerable reserved capacity beyond the initial yield conditions. Beyond the yield limit, the motion of the structure is characterized with a complex response as plastic flow occurs in some regions while the rest of the structure remains elastic. When the plastic work is completely dissipated through heat in the structure, it continues to vibrate elastically which is gradually damped out.

As the plastic deformations are interspersed with elastic ones, predictions of elastic–plastic response of the structure theoretically is inherently difficult. To this end, simplified methods, such as the method of limit plastic analysis, are employed, where the elastic effects are ignored for simplicity and the material is assumed as rigid-perfectly plastic. Such a method provides an appropriate assessment of the dynamic response of blast-loaded structures provided the loading is treated as a short duration impulse and the ratio of kinetic energy to elastic strain energy is considerably high. To estimate the elastic–plastic response, the solutions of purely elastic response with those of the perfectly plastic one can be coupled.

The mathematical procedure for a cylindrical shell made of rigid-perfectly plastic material and subjected to a uniform pulse pressure is sought in a similar fashion to the elastic analysis. Thus, the influence of transverse shear deformations, membrane action, and rotary inertia effects are ignored in the study.

The rate of advection of the blast load in the air varies in accordance with the type of explosive, such as gas explosion or high explosive detonation. Accordingly, pressure pulses assume sinusoidal, triangular, exponential, linear, or rectangular profiles. While the pulse shape would have significant effect on the overall response of the dynamically loaded systems, for a rigid-perfectly plastic response, its profound effect can be eliminated by utilizing Youngdahl's correlation parameters [57,58], namely: effective impulse $I_e = \int_0^{t_f} p^* dt$, mean time of the pulse t_m (centroid of the pulse profile), and effective pressure $p_e = I_e/2t_m$. For impulsive blasts, the pulse shape has no effect on the system. A general expression of the temporal part of the blast which describes the linear, rectangular, and exponential pulse is expressed by Li and Meng [59] as:

$$p_2(t) = \begin{cases} \left(1 - X\frac{t}{t_d}\right)e^{-Y\frac{t}{t_d}}, & 0 \le t \le t_d, \\ 0, & t_d \le t, \end{cases} \tag{10.52}$$

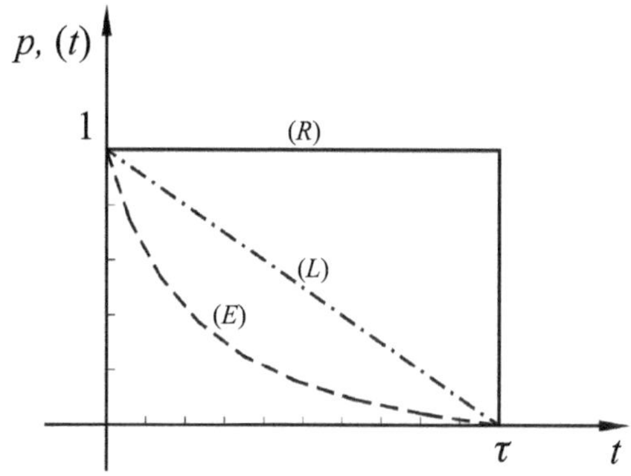

Figure 10.14 Pulse shapes of the rectangular, linear, and exponential profiles.

where Y is a constant to be determined by experimental or numerical models. It must be noted that for $X \neq 1$, t_d represents the pulse duration as the second rule forces the pressure to vanish. The various pulse shapes from this expressions are plotted in Fig. 10.14, where τ is a dimensionless time parameter.

The principal axiom of rigid-perfectly plastic analysis is the development of (bending or shear) plastic hinges. From Drucker's stability postulate, plastic flow follows along a path on the convex yield surface. At each point on the curve, the state of stress (or curvature) and the normal to the yield surface are co-directional. It has been established that, based on the Tresca yield criterion, the yield curve of a cylindrical shell subject to axisymmetric loading, satisfies the condition [60]:

$$\frac{M_x}{M_0} - \frac{4N_\theta}{N_0} + 4\left(\frac{N_\theta}{N_0}\right)^2 = 0, \tag{10.53}$$

where M_0 and N_0 are the plastic capacity of the cross section when subject to pure bending and the membrane force to produce the plastic flow throughout the entire thickness, respectively.

$$M_0 = \frac{\sigma_0 h^2}{4}, \tag{10.54a}$$

$$N_0 = \sigma_0 h. \tag{10.54b}$$

Although the yield curve of Eq. (10.53) is nonlinear, the associated theoretical solution of the shell may be approximated by a lower bound hexagonal curve which inscribes the exact yield curve ABCDEFGH

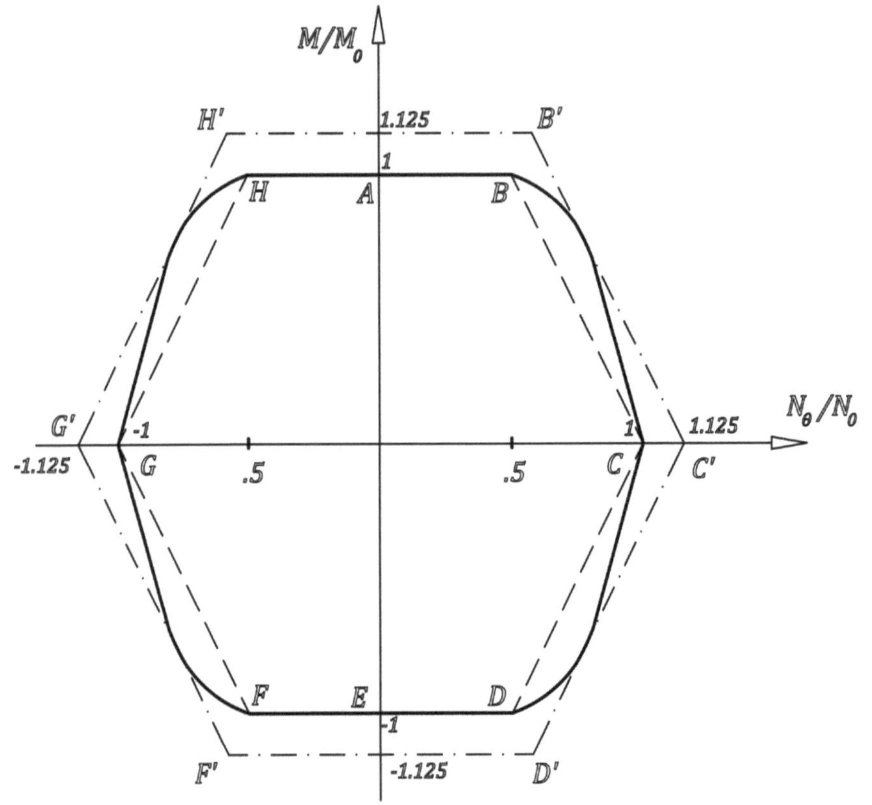

Figure 10.15 Exact yield surface (continuous line) versus the circumscribing and inscribing hexagonal yield surfaces (broken lines).

(Fig. 10.15), as well as an upper bound curve which circumscribes the exact yield curve by a scale of 1.125.

For further simplifications, however, it is pragmatic to approximate the solution by assuming an inscribing rectangular yield curve which is 0.75 times larger than the exact curve, as drawn in Fig. 10.16.

10.5.1 Static plastic collapse

The cylindrical shell examined here is assumed long and reinforced with rigid rings at intervals of $2L$ implying clamped boundary conditions. The assumed internal pressure acts uniformly across the shell internal surface. Provided the internal pressure does not vary axially, the mathematical analysis is reduced to one bay, thus $N_x = 0$ (Fig. 10.17).

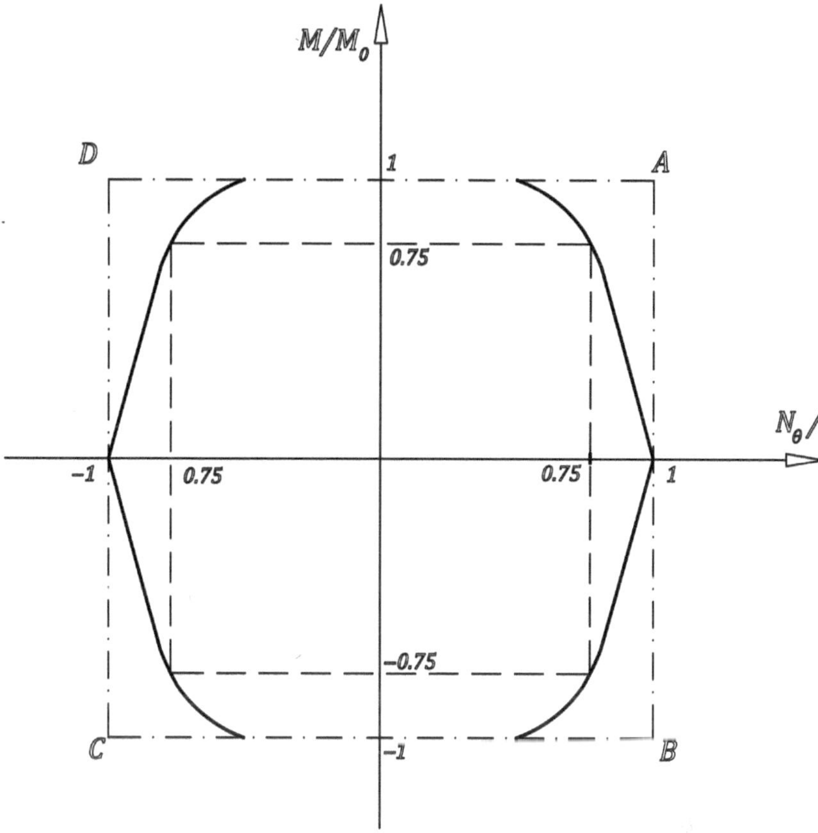

Figure 10.16 Exact yield surface versus (continuous line). The inscribing and circumscribing rectangular yield surfaces (broken lines).

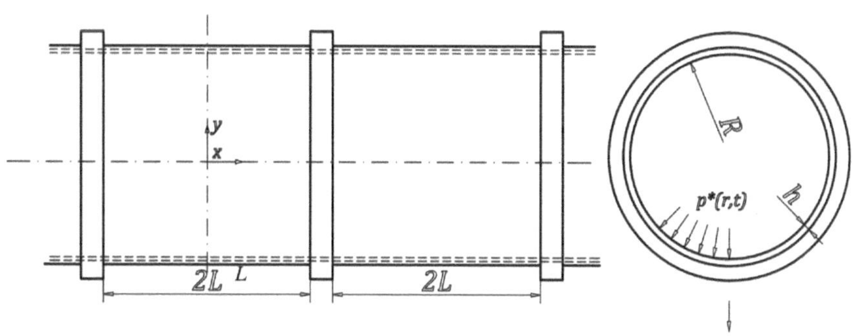

Figure 10.17 Schematic of the cylindrical shell with rigid rings.

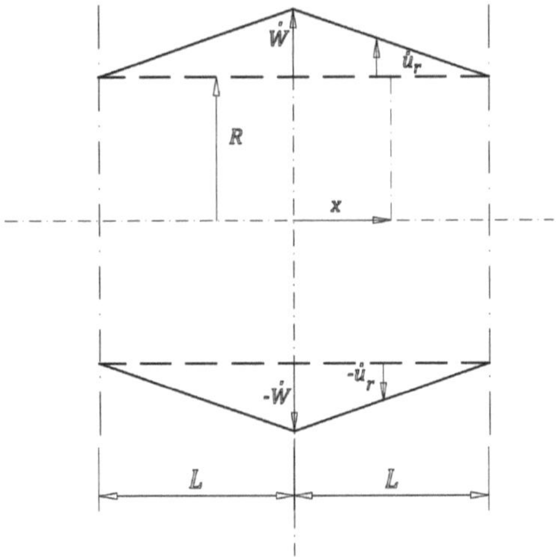

Figure 10.18 Deformation profile of the cylindrical shell.

The assumed displacement profile of the shell is drawn in Fig. 10.18, with two bending plastic hinges at each support and another circumferential hinge in the center, which the associated plastic flow initiates from corner B. The stress profile AB that governs the response dictates that $N_\theta = N_0$ and $-M_0 \le M_x \le M_0$.

Now, combining Eqs. (10.29) and (10.31) and noting that $N_x = \ddot{u} = 0$ followed by two integrations over x yields:

$$M_x = \left(p^* - \frac{N_0}{R}\right)\frac{x^2}{2} - M_0. \tag{10.55}$$

The integration constants are derived by imposing the conditions $Q_x(0) = 0$ and $M_x(0) = -M_0$. With the notion that $M_x = M_0$ at $x = L$ and using Eqs. (10.54a–10.54b), the static plastic collapse pressure of the shell is expressed as:

$$p_c = \frac{N_0}{R}\left(1 + \frac{Rh}{L^2}\right). \tag{10.56}$$

10.5.2 Dynamic analysis

The mathematical treatment in the case of rigid-perfectly plastic material is analogous to the case of elastic response, that is, consideration of the solution to the equilibrium equations in the two phases of motion, and applying the kinematic continuity of the state variables at the transition point of load completion. The constitutive behavior is obviously different. The dynamic equilibrium equations are expressed as:

$$\frac{\partial M_x}{\partial x} - Q_x = 0, \tag{10.57}$$

$$R\frac{\partial Q_x}{\partial x} + N_\theta - Rp^*(r,t) - \frac{\mu R \partial^2 w}{\partial t^2} = 0. \tag{10.58}$$

1. First phase of motion

The velocity profile is assumed as follows:

$$\dot{u} = -\dot{W}\left(1 - \frac{x}{L}\right), \tag{10.59}$$

with inward direction taken as positive. Eq. (10.59) renders Eq. (10.58) as:

$$R\frac{\partial^2 M_x}{\partial x^2} + N_0 - RP_0\left(1 - X\frac{t}{t_d}\right)e^{-\frac{Yt}{t_d}} + \frac{\mu R \partial^2 W}{\partial t^2}\left(1 - \frac{x}{L}\right) = 0, \tag{10.60}$$

which leads to

$$M_x = \left(RP_0\left(1 - X\frac{t}{t_d}\right)e^{-\frac{Yt}{t_d}} - N_0\right)\frac{x^2}{2R} - \mu\ddot{W}\left(\frac{x^2}{2} - \frac{x^3}{6L}\right) - M_0. \tag{10.61}$$

Given the conditions $M_x = -M_0$ and $\frac{\partial M_x}{\partial x} = Q_x = 0$ at $x = 0$ apply for the formation of plastic hinge to satisfy the symmetry requirements. Assuming the shell is fully clamped, $M_x = M_0$ at $x = L$ which refurnishes Eq. (10.61) into:

$$\frac{\mu\ddot{W}L^2}{3} = \left(P_0 R\left(1 - X\frac{t}{t_d}\right)e^{-\frac{Yt}{t_d}} - N_0\right)\frac{L^2}{2R} - 2M_0. \tag{10.62}$$

Upon two time integrations the temporal part of the displacement field is derived:

$$W = \frac{-3P_0((Xt - t_d)Y + 2Xt_d)t_d e^{-\frac{Yt}{t_d}}}{2\mu Y^3} - \frac{3t^2}{4\mu}p_c + c_1 t + c_2, \tag{10.63}$$

where the integration constants are derived from the kinematic continuity of the displacement and velocity fields at the start of loading as:

$$c_1 = \frac{3t_d(Y - X)P_0}{2\mu Y^2}, \tag{10.64a}$$

$$c_2 = \frac{3t_d^2(2X - Y)P_0}{2\mu Y^3}. \tag{10.64b}$$

The solution for exponential pulse shape can be readily recovered by evaluating $X=0$. Similarly, for a linear pulse shape, $c_1=c_2=0$ and $X=1$, while the maximum transverse displacement from the solution of Eq. (10.62) furnishes into:

$$W = \frac{t^2((3\eta - 3)t_d - \eta t)p_c}{4\mu t_d}, \tag{10.65}$$

where $\eta = P_0/p_c \geq 1$ is the overloading factor. This phase terminates at $t=t_d$ of Eq. (10.65).

1. Second phase of motion

The loading is complete at $t=t_d$. However, the motion of the shell continues until all the reserved kinetic energy is dissipated. Eqs. (10.57) and (10.58) remain valid but with $P_0=0$ thus:

$$W = -\frac{3p_c t^2}{2\mu} + c_3 t + c_4, \tag{10.66}$$

$$c_3 = \frac{3(((X-1)Y + X)e^{-Y} + Y - X)P_0 t_d}{2\mu Y^2}, \tag{10.67}$$

$$c_4 = -\frac{3(((X-1)Y^2 + (2X-1)Y + 2X)e^{-Y} - 2X + Y)P_0 t_d^2}{2Y^3\mu}. \tag{10.68}$$

For a linear pulse, $c_3 = 3P_0 t_d/4\mu$ and $c_4 = -P_0 t_d^2/4\mu$. The displacement field is derived as:

$$W = -\frac{3p_c}{4\mu}\left(t^2 - \eta t_d\left(t - \frac{t_d}{3}\right)\right). \tag{10.69}$$

The shell reaches its permanent defamation when the momentum vanishes ($\dot{W} = 0$), which occurs at the time:

$$t_f = \frac{(((X-1)Y + X)e^{-Y} - X + Y)\eta t_d}{Y^2}. \tag{10.70}$$

The time of completion of the motion is $\eta t_d/2$, which is half the duration of the rectangular pulse. It is of interest that the foregoing expressions for the transverse displacements due to linear pulse may have alternatively been derived by setting $X=1$, followed by an approximation of the exponential function as a McLaurin series with an order of 3, that is, $e^{-\frac{Yt}{t_d}} = 1 - \frac{Yt}{t_d} + (\frac{Yt}{t_d})^2 + o((\frac{Yt}{t_d})^2)$ and then ignoring the remaining terms of Y or by taking a limit $\lim_{Y\to 0} W$.

The influence of the pulse shape on the deformation of the shell is plotted in Fig. 10.19. Using Youngdahl's correlation parameters, the response due to various loading profiles fits into a single curve as shown in Fig. 10.19B.

Jones [60] derived an expression for the plastic response of short clamped cylindrical shell subjected to a rectangular pulse as:

$$W_f = -p_0 t_d^2 \left(\eta\left(8+\lambda^2\right) - \left(6+\lambda^2\right)\right) / \left\{2\mu\left(6+\lambda^2\right)\right\}, \quad (10.71)$$

where $\lambda^2 = 2L^2/Rh$. In such a case, the motion is broken down into three phases where during the first phase the two incipient bending hinges gain the initial inertia, then travel inward toward the midpoint of the shell in phase 2, followed by the final phase (phase 3) during which all the reserved kinetic energy is dissipated before the shell comes to rest. A comparison of the analytical model with numerical one shows good agreement between the two. The divergence of the analytical estimate from the numerical one at higher load magnitudes is due to quadratic dependence of the displacement field to the overpressure. This is shown in Fig. 10.20.

10.6 Conclusions and remarks

This chapter examined the influence of FSI in an explosion scenario on the B737 fuselage made of AL7075 aluminum alloy. Following the work of Taylor [48], a number of dimensionless parameters were outlined and the parameters which influence the FSI were discussed. With the aid of advanced numerical techniques, the propagation of the blast wave(s) generated from a single and couple charges in the fuselage cabin was (were) monitored.

In the case of a single blast, the charge placed in the luggage compartment had a significant influence on the fuselage skin with a maximum pressure traveling obliquely downward along the curved surface of the cabin, while the one placed at the seating area traveled obliquely upward toward it. Based on the curve-fitting tools in MATLAB, an empirical expression was derived to estimate the pressure distribution along the arc length of the shell.

In the case of coupled charges, the horizontal arrangement led to the superposition of the waves with a shock front which traveled along the

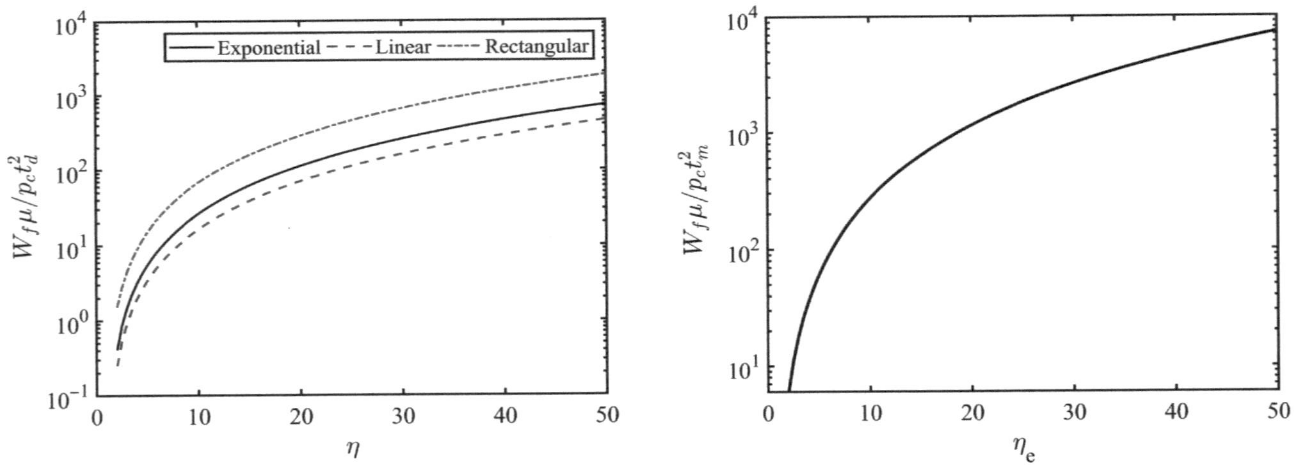

Figure 10.19 (A) Influence of pulse shape on the response of the cylindrical shell. (B) Elimination of pulse shape effects using Youngdahl's correlation parameters, where $\eta_e = p_e/p_c$.

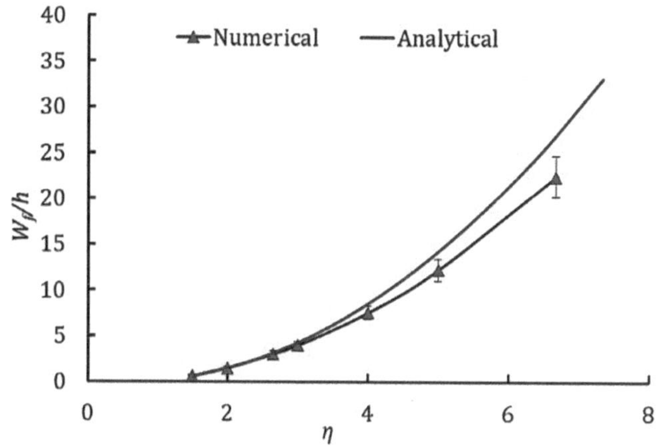

Figure 10.20 Comparison of the analytical and numerical models.

ceiling. On the other hand, the shock traveled in the same direction, but on the curved surface of the fuselage skin, when the two charges were placed in vertical arrangement.

Mathematical treatment of a thin cylindrical shell resembling the aircraft fuselage was performed by assuming a purely elastic response, while ignoring the influence of finite displacements (or geometry changes). The expressions of the displacement were obtained in both static loading and dynamic loading cases. In the latter case, the transient deformation was obtained for exponentially decaying pulse.

Maintaining the assumptions and limitations of the study, the analyses were extended to plastic behavior of a long reinforced cylindrical shell. The influence of pulse shape on the response was examined and a commentary on the elimination of pulse shape effects was given.

Nomenclature

Latin uppercase

A_1 Johnson–Cook first yield (true) stress $[ML^{-1}T^{-2}]$
B_1 Johnson–Cook hardening parameter; $[ML^{-1}T^{-2}]$
C_1 Johnson–Cook strain rate coefficient [1]
C_p Ideal gas specific heat capacity at constant pressure $[L^2T^{-2}K^{-1}]$
C_v Ideal gas specific heat capacity at constant volume $[L^2T^{-2}K^{-1}]$
$D_1 - D_5$ Johnson–Cook damage parameters [1]
D Flexural rigidity $[ML^2T^{-2}]$
$\mathcal{D}$ Damage number [1]
E Young's modulus $[ML^{-1}T^{-2}]$

I_i Impulse density of the incident wave $[ML^{-1}T^{-1}]$
I_p Impulse transmitted to the plate $[MLT^{-1}]$
L Half-length of cylindrical shell bay $[L]$
M_x Generalized bending moment along x direction per unit length $[MLT^{-2}]$
M_θ Generalized circumferential bending moment per unit length $[MLT^{-2}]$
M_0 Maximum bending moment per unit length $[MLT^{-1}]$
N_x Generalized membrane force along x direction per unit length $[MT^{-2}]$
N_θ Generalized circumferential membrane force per unit length $[MT^{-2}]$
P_0 Maximum dynamic overpressure $[ML^{-1}T^{-2}]$
Q_x Generalized shear force per unit length $[MT^{-2}]$
Q_0 Maximum shear force per unit length $[MT^{-2}]$
T Temperature $[K]$
U_s Shock propagation speed $[LT^{-1}]$
W Maximum transverse displacement of the cylindrical shell $[L]$
X, Linear pulse decay coefficient [1]
Y Exponential pulse decay coefficient [1]
a_0 Sound speed $[LT^{-1}]$

Latin lowercase

$a_i - c_i$ Integration constants [Various]
c Specific het of aluminum alloy $[L^2T^{-2}K^{-1}]$
h_f Characteristic dimension (depth) of the fluid ahead of the shock $[L]$
h_p Characteristic thickness of the plate $[L]$
h_p Characteristic thickness of the cylindrical shell $[L]$
u_p Plate velocity due to FSI $[LT^{-1}]$
u_r Transverse displacement of the shell at any point $[L]$
w_p Plate displacement due to FSI $[L\}$
$\hat{l}$ Arc length $[L]$
p_0 Atmospheric pressure $[ML^{-1}T^{-2}]$
p_s Incident pressure $[ML^{-1}T^{-2}]$
p_r Reflected pressure $[ML^{-1}T^{-2}]$
p_c, Static overpressure $[ML^{-1}T^{-2}]$
r p_{st} Characteristic radial distance from center $[L]$
t_i Arbitrary time scale of the decay period $[T]$
t_d Duration of the positive phase of the blast load $[T]$
$\hat{x}$ Generalized coordinate $[L]$
$\hat{y}$ Generalized coordinate $[L]$

Greek upper and lower case

χ Taylor–Quinney coefficient [1]
β Static deformation profile coefficient $[L^{-1}]$
β_0 Quotient of the compressed gas to plate volume [1]
ς Stress triaxiality [1]
ε Strain tensor [1]
$\dot{\varepsilon}$ Strain rate tensor $[T^{-1}]$
$\dot{\varepsilon}_{eq}$ Normalized plastic strain rate
ε_f Strain at tensile fracture [1]

ε_{rr}	Radial strain [L]
$\varepsilon_{\theta\theta}$	Circumferential strain [L]
$\varepsilon_{r\theta}$,	Shear strain components [1]
λ ε_{rx},	Shell slenderness coefficient [1]
μ_k	Dynamic viscosity of air $[ML^{-1}T^{-1}]$
ν	Poisson's ratio [1]
σ_0	Plastic flow stress $[ML^{-1}T^{-2}]$
ρ_p	Plate density $[ML^{-3}]$
ρ	Shell density $[ML^{-3}]$
ρ_r	Fluid density behind the reflected shock $[ML^{-3}]$
ρ_e	Explosive density $[ML^{-3}]$
ρ_0	Fluid (air) initial density $[ML^{-3}]$
ρ_s	Density of the compressed gas behind the incident shock $[ML^{-3}]$
γ	$(=C_p/C_v)$ Isentropic expansion factor [1]
ζ	Normalized third deviatoric invariant [1]

References

1. F. Masi, P. Maria Mariano, P. Vannucci, Blast actions in aircrafts: an integrated methodology for designing protection devices, Eng. Struct. 175 (2018) 895–911.
2. T. Dursun, C. Soutis, Recent developments in advanced aircraft aluminium alloys, Mater. Des. 56 (2014) 862–871.
3. N. Mehreganian, A.S.S. Fallah, G.K.K. Boiger, et al., Response of Armour steel square plates to localised air blast load—a dimensional analysis, Int. J. Multiphys. 11 (2017) 1–20.
4. K. Micallef, A. Fallah, D.D.J.D. Pope, et al., On dimensionless loading parameters for close-in blasts, Int. J. Multiphys. 9 (2015) 193.
5. N. Jacob, G.N. Nurick, G.S. Langdon, The effect of stand-off distance on the failure of fully clamped circular mild steel plates subjected to blast loads, Eng. Struct. 29 (2007) 2723–2736.
6. A. Tyas, Experimental Measurement of Pressure Loading from Near-Field Blast Events: Techniques, Findings and Future Challenges, Proceedings 2 (2018) 471, https://doi.org/10.3390/ICEM18-05364.
7. S.B. Menkes, H.J. Opat, Broken beams—tearing and shear failures in explosively loaded clamped beams, Exp. Mech. 13 (1973) 480–486.
8. G.N. Nurick, J.B. Martin, Deformation of thin plates subjected to impulsive loading—a review. Part I: theoretical considerations, Int. J. Impact Eng. 8 (1989) 159–170.
9. G.N. Nurick, J.B. Martin, Deformation of thin plates subjected to impulsive loading—a review—part II—experimental results, Int. J. Impact Eng. 8 (1989) 171–186.
10. N. Jacob, S.C.K. Yuen, G.N. Nurick, et al., Scaling aspects of quadrangular plates subjected to localised blast loads—experiments and predictions, Int. J. Impact Eng. 30 (2004) 1179–1208.
11. G.N.N. Nurick, M.D.D. Olson, J.R.R. Fagnan, et al., Deformation and tearing of blast-loaded stiffened square plates, Int. J. Impact Eng. 16 (1995) 273–291.
12. S. Chung Kim Yuen, G.N. Nurick, G.S. Langdon, et al., Deformation of thin plates subjected to impulsive load: part III – an update 25 years on, Int. J. Impact Eng. 107 (2017) 1339–1351.
13. G.S. Langdon, G.N. Nurick, S.L. Lemanski, et al., Failure characterisation of blast-loaded fibre-metal laminate panels based on aluminium and glass-fibre reinforced polypropylene, Compos. Sci. Technol. 67 (2007) 1385–1405.

14. G.S. Langdon, S.C.K. Yuen, G.N. Nurick, Experimental and numerical studies on the response of quadrangular stiffened plates. Part II: localised blast loading, Int. J. Impact Eng. 31 (2005) 85–111.
15. R.G. Teeling-Smith, G.N. Nurick, The deformation and tearing of thin circular plates subjected to impulsive loads, Int. J. Impact Eng. 11 (1991) 77–91.
16. V.H. Balden, G.N. Nurick, Numerical simulation of the post-failure motion of steel plates subjected to blast loading, Int. J. Impact Eng. 32 (2006) 14–34.
17. G.S. Langdon, A. Curry, A. Siddiqui, Improving the impulse transfer and response characteristics of explosion loaded compound angle V-plates, Thin Walled Struct. 148 (2020) 106609.
18. T. Børvik, A.G. Hanssen, M. Langseth, et al., Response of structures to planar blast loads – a finite element engineering approach, Comput. Struct. 87 (2009) 507–520.
19. B. Zakrisson, H.-A. Haggblad, P. Jonsen, Modelling and simulation of explosions in soil interacting with deformable structures, Cent. Eur. J. Eng. 2 (2012) 532–550.
20. V. Aune, G. Valsamos, F. Casadei, et al., Numerical study on the structural response of blast-loaded thin aluminium and steel plates, Int. J. Impact Eng. 99 (2017) 131–144.
21. D. Bonorchis, G.N. Nurick, The analysis and simulation of welded stiffener plates subjected to localised blast loading, Int. J. Impact Eng. 37 (2010) 260–273.
22. N. Mehreganian, L.A. Louca, G.S. Langdon, et al., The response of mild steel and armour steel plates to localised air-blast loading-comparison of numerical modelling techniques, Int. J. Impact Eng. 115 (2018) 81–93.
23. N. Mehreganian, A.S. Fallah, L.A. Louca, Plastic dynamic response of simply supported thick square plates subject to localised blast loading, Int. J. Impact Eng. 126 (2019) 85–100.
24. T. Kotzakolios, D.E. Vlachos, V. Kostopoulos, Blast response of metal composite laminate fuselage structures using finite element modelling, Compos. Struct. 93 (2011) 665–681.
25. G.S. Settles, B.T. Keane, B.W. Anderson, et al., Shock waves in aviation security and safety, Shock Waves 12 (2003) 267–275.
26. W.E. Baker, P.A. Cox, P.S. Westine, et al., Explosion Hazards and Evaluation, Elsevier Scientific Publication Company, Amsterdam-Oxford-New York, 1983.
27. Micallef K., Eng B., Melit A.H., et al. The dynamic response of blast-loaded monolithic and composite plated structures. A thesis submitted for the degree of Doctor of Philosophy (Ph.D.).
28. R. Rajendran, J.M. Lee, Blast loaded plates, Mar. Struct. 22 (2009) 99–127.
29. T. Ngo, P. Mendis, A. Gupta, et al., Blast loading and blast effects on structures – an overview, Electron. J. Struct. Eng. 7 (2007) 76–91.
30. M. Larcher, F. Casadei, Explosions in Complex Geometries A Comparison of Several Approaches International Journal of Protective Structures. 1(2) (2010) 169195, doi:10.1260/2041-4196.1.2.169.
31. N. Mehreganian, Dynamic Response of Steel Monolithic Plated Structures Subjected to Localised Blast Loads, Imperial College, London, 2018.
32. S.E. Rigby, A. Tyas, T. Bennett, et al., The negative phase of the blast load, Int. J. Prot. Struct. 5 (2014) 1–19.
33. N. Mehreganian, A. Fallah, L.A. Louca, Inelastic dynamic response of square membranes subjected to localised blast loading, Int. J. Mech. Sci. 148 (2018) 578–595.
34. P.D. Smith, J. Hetherington, Blast and Ballistic Loading of Structures, 2011th ed., Taylor & Francis, New York, 1994.
35. M. Nettleton, Gaseous Detonations: Their Nature, Effects and Control, first ed., Chapman and Hall, London, 1987.
36. I. Glassman, R.A. Yetter, *Combustion*. Fourth, Elsevier, London, 2008.
37. M.H. Keshavarz, H.R. Nazari, A simple method to assess detonation temperature without using any experimental data and computer code, J. Hazard. Mater. 133 (2006) 129–134.

38. M.H. Keshavarz, H.R. Pouretedal, An empirical method for predicting detonation pressure of CHNOFCl explosives, Thermochim. Acta 414 (2004) 203–208.
39. G.N. Nurick, S. Mahoi, G.S. Langdon, The response of plates subjected to loading arising from the detonation of different shapes of plastic explosive, Int. J. Impact Eng. 89 (2016) 102–113.
40. A.C. Mackenzie, J.W. Hancock, D.K. Brown, On the influence of state of stress on ductile failure initiation in high strength steels, Eng. Fract. Mech. 9 (1977) 167–188.
41. J.W. Hancock, A.C. Mackenzie, On the mechanisms of ductile failure in high-strength steels subjected to multi-axial stress-states, J. Mech. Phys. Solids 24 (1976) 147–160.
42. G.R. Johnson, W.H. Cook, Fracture characteristics of three metals subjected to various strains, strain rates, temperatures and pressures, Eng. Fract. Mech. 21 (1985) 31–48.
43. C. Fressengeas, V. Taupin, A field theory of strain/curvature incompatibility for coupled fracture and plasticity, Int. J. Solids Struct. 82 (2016) 16–38.
44. H. Levanger, Simulating Ductile Fracture in Steel using the Finite Element Method: Comparison of Two Models for Describing Local Instability due to Ductile Fracture (MSc Thesis). University of Oslo, Oslo, 2012, p. 135.
45. T. Børvik, O.S. Hopperstad, T. Berstad, On the influence of stress triaxiality and strain rate on the behaviour of a structural steel. Part II. Numerical study, Eur. J. Mech. A/Solids 22 (2003) 15–32.
46. M.A. Iqbal, K. Senthil, P. Sharma, et al., An investigation of the constitutive behavior of Armox 500T steel and armor piercing incendiary projectile material, Int. J. Impact Eng. 96 (2016) 146–164.
47. M. Basaran, Stress State Dependent Damage Modeling with a Focus on the Lode Angel Influence, RWTH Aachen university, Aachen, Germany, 2011 doi:urn:nbn:de:hbz:82-opus-38332.
48. G.I. Taylor, et al., The pressure and impulse of submarine explosion waves on plates, in: G.K. Batchelor, C.F. Sharman, J.W. Maccoll, et al. (Eds.), The Scientific Papers of Sir Geoffrey Ingram Taylor: Vol 3 Aerodynamics and the Mechanics of Projectiles and Explosions, Cambridge University Press, Cambridge, UK, 2011, p. 590.
49. N. Kambouchev, R. Radovitzky, L. Noels, Fluid–structure interaction effects in the dynamic response of free-standing plates to uniform shock loading, J. Appl. Mech. 74 (2007) 1042.
50. N. Kambouchev, L. Noels, R. Radovitzky, Nonlinear compressibility effects in fluid-structure interaction and their implications on the air-blast loading of structures. J. Appl. Phys.; 100 1–11. Epub ahead of print 2006. doi:10.1063/1.2349483.
51. R. van Loon, P.D. Anderson, F.N. van de Vosse, et al., Comparison of various fluid-structure interaction methods for deformable bodies, Comput. Struct. 85 (2007) 833–843.
52. M. Souli, A. Ouahsine, L. Lewin, ALE formulation for fluid-structure interaction problems, Comput. Methods Appl. Mech. Eng. 190 (2000) 659–675.
53. Engineering Toolbox. Air - Density, Specific Weight and Thermal Expansion Coefficient at Varying Temperature and Constant Pressures [online].
54. D. Bonorchis, G.N. Nurick, The influence of boundary conditions on the loading of rectangular plates subjected to localised blast loading – importance in numerical simulations, Int. J. Impact Eng. 36 (2009) 40–52.
55. N. Mehreganian, A.S.S. Fallah, L.A. Louca, Dynamic performance of simply supported rigid-plastic square plates subject to localized blast loading, J. Eng. Mech. 140 (2018) 159–171.
56. S. Timoshenko, S. Woinosky-Kreiger, Theory of Plates and Shells, second ed., McGraw Hill Book Company, Inc., New York, 1959.
57. C.K. Youngdahl, Correlation parameters for eliminating the effects of pulse shape on dynamic plate deformation, Trans. ASME J. Appl. Mech. 37 (1970) 744–752.

58. C.K. Youngdahl, Influence of pulse shape on the final plastic deformation of a circular plate, Int. J. Solids Struct. 7 (1971) 1127–1142.
59. Q.M. Li, H. Meng, Pulse loading shape effects on pressure-impulse diagram of an elastic-plastic, single-degree-of-freedom structural model, Int. J. Mech. Sci. 44 (2002) 1985–1998.
60. N. Jones, *Structural Impact*, Cambridge: Cambridge University Press, 1990. doi:10.1017/CBO9780511624285.

Index

Page numbers followed by "*f*" and "*t*" indicate, figures and tables respectively.

N

P

Q

R

S

Kolibri 360 GmbH
Ferdinand-Jühlke-Straße 7
99095 Erfurt, Deutschland
produktsicherheit@kolibri360.de